# Knowledge Modelling and Big Data Analytics in Healthcare

# Knowledge Modelling and Big Data Analytics in Healthcare

## Advances and Applications

Edited by

Mayuri Mehta,
Kalpdrum Passi,
Indranath Chatterjee, and
Rajan Patel

CRC Press is an imprint of the
Taylor & Francis Group, an **informa** business

First edition published 2022
by CRC Press
6000 Broken Sound Parkway NW, Suite 300, Boca Raton, FL 33487–2742

and by CRC Press
2 Park Square, Milton Park, Abingdon, Oxon, OX14 4RN



CRC Press is an imprint of Taylor & Francis Group, LLC



*Library of Congress Cataloging-in-Publication Data*
Names: Mehta, Mayuri, editor.
Title: Knowledge modelling and big data analytics in healthcare : advances and applications / edited by Mayuri Mehta, Kalpdrum Passi, Indranath Chatterjee, and Rajan Patel.
Description: First edition. | Boca Raton : CRC Press, 2022. | Includes bibliographical references and index.
Identifiers: LCCN 2021025540 (print) | LCCN 2021025541 (ebook) | ISBN 9780367696610 (hardback) | ISBN 9780367696634 (paperback) | ISBN 9781003142751 (ebook)
Subjects: LCSH: Medical informatics. | Medical care—Information technology. | Big data.
Classification: LCC R858 .K66 2022 (print) | LCC R858 (ebook) | DDC 610.285—dc23
LC record available at https://lccn.loc.gov/2021025540
LC ebook record available at https://lccn.loc.gov/2021025541

ISBN: 978-0-367-69661-0 (hbk)
ISBN: 978-0-367-69663-4 (pbk)
ISBN: 978-1-003-14275-1 (ebk)

DOI: 10.1201/9781003142751

Typeset in Times
by Apex CoVantage, LLC

# Contents

## SECTION V Issues in Security and Informatics in Healthcare

# Preface

We are living in an era of information and knowledge modelling through big data analytics. Big data analytics has been widely applied to fields such as agriculture, e-commerce, education, finance, manufacturing industry and sports to extract information and knowledge from a voluminous amount of data. An enormous volume of structured as well as unstructured data produced by the healthcare industry also offers ample opportunities to convert them into real and actionable insights, thereby providing new understanding and ways for quicker and enhanced healthcare services. Due to the increasing availability of electronic healthcare data and the rapid progress of analytic techniques, much research is being carried out in this area. Popular methods include machine learning/deep learning for structured data and natural language processing for unstructured data. Guided by relevant clinical questions, knowledge modelling through big data analytic techniques can unlock clinically relevant information hidden in the massive amount of data, which, in turn, can assist clinical decision-making.

This book provides state-of-the-art research in the healthcare sector, along with related subdomains, to ensure completeness and quality. Specifically, it highlights modern study and developments, challenges, opportunities and future research directions in healthcare. The focus of the book is on automated analytical techniques used to extract new knowledge from a huge amount of medical data for healthcare applications. The book connects four contemporary areas of research: artificial intelligence, big data analytics, knowledge modelling and healthcare.

The book is structured into five sections that cover key areas of healthcare, such as big data in healthcare, medical imaging, computational genomics, applications in clinical diagnosis and issues in security and informatics in healthcare.

Section I—Big Data in Healthcare. This section focuses on healthcare systems, opportunities and challenges, clinical decision support systems and intelligent software systems.

Section II—Medical Imaging. This section presents the latest research on the application of artificial intelligence on various applications of image analysis for the diagnosis and prediction of severe diseases such as cancer, schizophrenia and neuroinformatics.

Section III—Computational Genomics. Genetics and genomics are emerging sub-domains of healthcare. Genetics and genomics datasets are complex and large, wherein deep learning algorithms have a plethora of uses. This section reports progress in genetics and genomics, including genetic diagnosis, genotype–phenotype correlations and sequencing of the human genome.

Section IV—Applications in Clinical Diagnosis. The purpose of this section is to report the clinical applications of tracking modulations in epilepsy and the use of the Internet of Things for monitoring diabetes.

Section V—Issues in Security and Informatics in Healthcare. This section presents security and privacy issues in healthcare information systems and the use of data science in healthcare.

This book is primarily intended for data scientists, machine learning enthusiasts, industry professionals from the healthcare sector, researchers, doctors, students and academicians. By reading this book, they will gain essential insights on modern big data analytic techniques needed to advance innovation in both the healthcare industry and patients.

# Editors

**Mayuri Mehta, PhD,** is a passionate learner, teacher, and researcher. She is a professor in the Department of Computer Engineering, Sarvajanik College of Engineering and Technology, Surat, India. She earned her PhD in computer engineering from Sardar Vallabhbhai National Institute of Technology, India. Her areas of teaching and research include data science, machine learning and deep learning, health informatics, computer algorithms, and Python programming. Dr Mehta has worked on several academic assignments in collaboration with professors from universities across the globe. Her 20 years of professional experience includes several academic and research achievements along with administrative and organizational capabilities. She has also co-edited a book, *Tracking and Preventing Diseases using Artificial Intelligence.* With the noble intention of applying her technical knowledge for societal impact, she is working on several research projects in the healthcare domain in association with doctors engaged in private practice and doctors working with medical colleges, which reflect her research outlook. Dr Mehta is an active member of professional bodies such as the IEEE, Computer Society of India (CSI) and the Indian Society for Technical Education (ISTE).

**Kalpdrum Passi, PhD,** earned his PhD in parallel numerical algorithms from the Indian Institute of Technology, Delhi, India in 1993. He is an associate professor in the Department of Mathematics & Computer Science at Laurentian University, Ontario, Canada. He has published many papers on parallel numerical algorithms in international journals and conferences. Dr Passi has collaborated in research work with faculty in Canada and the United States, and the work was tested on the CRAY XMP's and CRAY YMP's. He transitioned his research to web technology and, more recently, has been involved in machine learning and data mining applications in bioinformatics, social media, and other data science areas. His research in bioinformatics has been on improving the accuracy of predicting diseases such as different types of cancer using microarray data. Dr Passi has published several papers related to prediction of cancer using microarray data and epigenomic data. He obtained funding from the Natural Sciences and Engineering Research Council of Canada and Laurentian University for his research. He is a member of the Association of Computing Machinery and the IEEE Computer Society.

**Indranath Chatterjee, PhD,** is a professor in the Department of Computer Engineering at Tongmyong University, Busan, South Korea. He earned his PhD in computational neuroscience from the Department of Computer Science, University of Delhi, Delhi, India. His research areas include computational neuroscience, medical imaging, data science, machine learning, and computer vision. Dr Chatterjee is the author of four textbooks on computer science and has published numerous scientific articles in renowned international journals and conferences. He is currently serving as a chief section editor of the *Neuroscience Research Notes* journal and as a member of the advisory board and editorial board of various international journals and open-science organizations worldwide. Dr Chatterjee is presently working with government and nongovernment organizations as principal/co-principal investigator on several projects related to medical imaging and machine learning for a broader societal impact, in collaboration with more than 15 universities globally. He is an active professional member of the Association of Computing Machinery (USA), the Organization of Human Brain Mapping (USA), Federations of European Neuroscience Society (Belgium), the International Association of Neuroscience (India), and the International Neuroinformatics Coordinating Facility (Sweden).

**Rajan Patel, PhD,** is a professor at the Gandhinagar Institute of Technology, Gandhinagar, Gujarat, India. He earned a PhD in computer engineering from R. K. University, Rajkot, India, and an M.Tech. in computer engineering from S. V. National Institute of Technology (NIT), Surat, India. Dr Patel has more than 16 years of teaching experience in the fields of computer science and engineering and research experience mainly in the domain of networking, security, and intelligent applications. He has more than 51 collaborative publications in journals and conferences and presented 17 articles in national/international conferences including IEEE, Science Direct, Springer, and Elsevier. As a co-editor, he has published an edited international book, *Data Science and Intelligent Applications*, in Springer's Lecture Notes on Data Engineering and Communication Technologies series. He also worked for Information Security Education & Awareness Program-sponsored and Ministry of Human Resource and Development-funded project during his postgraduation period at NIT, Surat, India. Dr Patel is a member of several professional bodies, including CSI, the International Society for Technology in Education, and the Universal Association of Computer and Electronics Engineers. He has also received numerous awards, honors, and certificates of excellence. His main area of interest includes artificial intelligence, data science, and intelligent communication and its security.

# Contributors

**Zahra Aminoroaya**, Department of Computer Engineering, Allameh Naieni Higher Education Institute.

**Dr. P. Anbalagan** is an assistant professor in Department of Computer Science and Engineering, Annamalai University with over 12 years of experience. He completed his PhD in computer science and engineering in 2016. He published several papers in international conferences and journals. His research interests include data mining, big data analytics, mobile ad hoc networks and software engineering.

**Anindya Banerjee**, Master's student at Friedrich-Alexander-Universität Erlangen-Nürnberg, Germany. Previously he has worked in Capgemini India private limited and holds a Master's degree (M.Tech) in communication engineering from Kalyani Government Engineering College, India. He has few years of both research and industrial experience.

**Manu Banga**, researcher at Amity University Uttar Pradesh Noida. His area of research is artificial intelligence and data mining. He earned his PhD in computer science and engineering from ASET Amity University Noida, M.Tech in computer science and engineering from CDAC Noida and B.Tech from Gurukul Kangri Vishwavidyalaya Haridwar. He published 13 research papers in SCI and Scopus journals, 2 patents and 2 books.

**Dulari Bhatt**, is an assistant professor in GTU affiliated Engineering College in Gujarat. She has 10 years of teaching experience to UG and PG students. She has published a book titled *Big Data Analytics* with TechKnowledge Publication. She has also published more than 15 research papers in high quality journals. Machine learning, deep learning and computer vision are her areas of expertise.

**Ameyaa Biwalkar** is currently working as an assistant professor in the Computer Engineering Department at Mukesh Patel School of Technology Management and Engineering, NMIMS University. She has co-authored various papers in reputed conferences and journals. Her research interests include data mining, natural language processing, and machine learning. She has completed her Master's in information technology and is pursuing her PhD.

**Mariella Bonomo** is a PhD student in information and communication technologies (ICT) at University of Palermo. She earned her MS degree in computer science in 2019. Her main research interests are in big data, bioinformatics, precision medicine and data integration. In particular, she works on the analysis of complex networks in different application contexts.

**Indranath Chatterjee** is a professor in the Department of Computer Engineering, Tongmyong University, South Korea. He earned his PhD from University of Delhi, India. His areas of interest include schizophrenia, computational neuroscience, medical imaging, machine learning, natural language processing, and deep learning. He is an author of 5 textbooks and more than 40 scientific papers.

**Mrs. Madhuri Chopade** is a research scholar in Indus University and an assistant professor at Gandhinagar Institute of Technology, Gandhinagar since February 2010. She has 13 years of academic experience and her areas of interest include data science machine learning, deep learning, and image processing. She has published five papers in international and two papers in national journals.

**Dr. Praveen Dhyani** is an honorary professor of computer science at Banasthali Vidyapith. Served in various academic and administrative capacities in BITS Pilani, BIT Mesra, Ranchi and Banasthali Vidyapith.

**Dr. Cecil Donald**, is an assistant professor in the Department of Computer Science, CHRIST (Deemed to be University), Bengaluru, Karnataka. He has a Master's degree in software engineering and a PhD in computer science. He holds an Australian Patent and a copyright to his credit. His research interests include cloud computing, IoT and big data.

**Saeed Doostali** earned a PhD in software engineering from University of Kashan. He was the head of the computer department at Mahallat Institute of Higher Education from 2014 to 2019. He is currently the manager of the *Soft Computing Journal*, Kashan, Iran. His research interests include soft computing and its applications in different fields of science, especially computers and medicine.

**Jos Dumortier** is an honorary professor at the Faculty of Law of KU Leuven. He is the founder and for 24 years was the first director of the Interdisciplinary Centre for Law and ICT (KU Leuven – ICRI, today CITIP). He is the senior editor of the *International Encyclopaedia of Cyber Law.* Since 2014, he is a full-time partner of TIMELEX, the Brussels-based law firm specialising in ICT and data protection law.

**Himadri Sekhar Dutta** Department of Electronics and Communication Engineering, Kalyani Government Engineering College Kalyani, India.

**Vivek Gaur**, assistant professor at BIT Mesra Jaipur Campus with over 17 years of teaching experience. He earned his PhD from Banasthali University and has a Master's degree from Birla Institute of Technology and Science (BITS), Pilani. Currently, he is an assistant professor in the Computer Science Department at BIT, Mesra Jaipur Campus. His research interests include distributed systems, cloud computing and data analytics technologies.

**Akansha Gautam** is a guest lecturer in Miranda House, University of Delhi. She has authored several papers in the field of big data and cloud computing. She earned a Master's degree in computer science from University of Delhi, New Delhi.

**Dr. Gustavo Gonzalez-Granadillo** is a research engineer at Atos Research and Innovation Spain, where he participates as a project director and proposal coordinator on various EU initiatives related to cybersecurity. His research interests include risk assessment methods, SIEM, critical infrastructures, artificial intelligence, and attack impact models.

**Anastasios Gounaris** is an associate professor in the Department of Informatics of the Aristotle University of Thessaloniki, Greece. He earned his PhD from the University of Manchester (UK) in 2005. He is involved in several data science projects and his research interests are in the areas of big data management and analytics.

**Chakresh Kumar Jain** earned his PhD in the field of bioinformatics from Jiwaji University, Gwalior, India, focusing on computational designing of non-coding RNAs using machine learning methods. He is an assistant professor, Department of Biotechnology, Jaypee Institute of Information Technology, Noida, India. He is CSIR-UGC-NET [LS] qualified and a member of International Association of Engineers (IAENG)

**Namrata Jawanjal**, is a PhD student at Pukyong National University, Busan, Republic of Korea. She has completed her master's degree in medical biotechnology. Her areas of interest are genetics, expression of genes, RNA mis-splicing, single nucleotide polymorphism, and designing a genetic marker kit.

**Majid Khoda Karami** is an English Teacher. He is currently working at IELTS TEHRAN. He earned an MA in general linguistics. He graduated from the University of Birjand.

**Georgia Kougka** is a postdoc in Aristotle University of Thessaloniki and her research interests lie in the field of data-centric flow modelling, optimization and execution, but also in business process optimization. Other research interests are the data pre-processing for predictive maintenance purposes in Industry 4.0 scenarios. Her published research results provide an advanced insight in data analytics optimization.

**Dr. A. Dalvin Vinoth Kumar**, assistant professor, Department of Computer Science, Kristu Jayanti College Bengaluru, has 4 years of teaching and 7 years of research experience. He earned his PhD in IoT from Bharathidasan University. His areas of interest include MANET, IoT, routing protocols, computer vision and IoT data analytics. He filed three Indian Patents and one computer software copyright granted.

**Dr. Prabhat K. Mahanti** is a professor of computer science at the University of New Brunswick, Canada. He has 43 years of teaching and research experience and has published more than 200 papers in international journals and guided several graduate students. He worked as a reviewer for international conferences and editorial board member of several international journals of repute.

**Khushboo Mittal** has earned her M. Sc. in computer science from the Department of Computer Science, University of Delhi, India. Her research areas of interest include schizophrenia, machine learning, and image processing.

**Diana Navarro-Llobet**, is the head of Research and Innovation at FPHAG and chair of the board of the Spanish network REGIC. She is an evaluator expert for the European Commission and other agencies. She earned a PhD in chemistry and has worked as a scientist before turning to research and innovation management.

**Apostolos N. Papadopoulos** is an associate professor of computer science at the School of Informatics of Aristotle University of Thessaloniki (AUTH). He earned his 5-year Diploma Degree in computer engineering and informatics from the University of Patras, and his PhD in Informatics from the School of Informatics (AUTH). His research interests include data management, data mining and big data analytics.

**Kalpdrum Passi** is an associate professor, Department of Mathematics & Computer Science, at Laurentian University, Ontario, Canada. He has been involved in machine learning and data mining applications in bioinformatics, social media and other data science areas. His research in bioinformatics has focused on improving the accuracy of predicting cancer using microarray data.

**Prof. Prakash Patel**, is an assistant professor in information technology at GIT. Prof. Patel earned a master's degree in computer science & engineering from Nirma University. His areas of interest are software defined networking, DBMS, system programming, and big data analytics. He has published four international and two national papers. He attended various conferences, STTP, and workshops. He published 10 articles in technical magazines.

**Rajan Patel** is a professor at Gandhinagar Institute of Technology with over 17 years of experience. He earned a PhD from RK University, Rajkot, and M.Tech. from NIT Surat. He worked under the ISEAP sponsored project. He has more than 53 publications in conferences and journals. He is a co-author of edited books and he has two granted patents.

**Vishwambhar Pathak** is an assistant professor in the Department of CSE. He is working on a project funded by the government of India on the topic of deep learning with application to biophysiological signals. Dr. Pathak earned a doctoral degree in computer science along with Master's degree in computer applications and MSc in physics.

**Armando La Placa** is a PhD student in information and communication technologies (ICT) at University of Palermo. His research topics include the study of bioinformatics problems and the extraction of information from heterogeneous data. He is interested in developing solutions using big data technologies such as NoSQL databases and distributed computational frameworks (e.g., Apache Spark).

**Simona E. Rombo** is associate professor of computer science at University of Palermo. Her research interests span from bioinformatics to big data management and artificial intelligence. She leads several projects in these fields. She is CEO and co-founder of a startup working on decision support for precision medicine.

**Dr. R. Saminathan** is an associate professor in the Department of Computer Science and Engineering, Annamalai University with over 21 years of experience. He earned his PhD in computer science and engineering in 2012. He published several papers in international conferences and journals. His research interests include computer networks, network security, mobile ad hoc networks and big data.

**Dr. S. Saravanan** is an assistant professor in the Department of Computer Science and Engineering, Annamalai University with over 15 years of experience. He earned his PhD in computer science and engineering in 2015. He published several papers in international conferences and journals. His research interests include computer networks, network security, mobile ad hoc networks and big data.

**Arjun Sarkar** FH Aachen University of Applied Sciences, Germany.

**Zhengxin Shi** is a graduate student in computational sciences. He has completed his Master's program and is currently working as a software developer in an IT company in China.

**Mohammad Shiralizadeh Dezfoli** earned his BSc. and MSc. in Software Engineering from Shahid Ashrafi Esfahani University, Esfahan, Iran, in 2016 and 2018, respectively. His thesis was on educational data mining. He also studied and published several papers in the realms of robotics, IoT, embedded systems, and data mining.

**Swapnil Singh** is currently pursuing a B Tech in computer engineering and MBA in technology management from Mukesh Patel School of Technology Management and Engineering, NMIMS University, Mumbai, India. He is also working on a B Tech Honors in artificial intelligence and machine learning from IBM. His research interests include machine learning, deep learning, medical imaging and natural language programming.

**Dr. Behzad Soleimani Neysiani** earned his BSc and MSc in software engineering from the Islamic Azad University of Najafabad in 2009 and 2012, and his PhD in the same field from the University of Kashan in 2019. He published more than 20 articles on software and artificial intelligence-related fields such as software testing and modelling, distributed systems, text, and data mining.

**Nasim Soltani** earned her B.Sc. in computer engineering from Payam Noor University, and her M.Sc. from Allame Naeini Higher Education Institute, in 2013 and 2016 respectively. She has published several papers in cloud computing field in many conferences and journals. Her main research interests include distributed systems, data mining and artificial intelligence.

**Dr. Margaret Mary T**, assistant professor at Kristu Jayanti College (Autonomous), Bangalore, India, She has authored various papers and book chapters in reputed national and international journals in Scopus, and was granted two patent from India and Australia. Additionally, she was deputy custodian in BU A.Y 2018–2019 for MCA and MSc, and was a BOE Member for BU and North University in A.Y 2019–2020 (MCA and MSc).

**Mr. Rahul A. Vaghela** is pursuing his PhD from PARUL University and working as assistant professor & Head of IT Department at Gandhinagar Institute of Technology, Moti Bhoyan. He has more than 12 years of academic and 3 years of industrial experience. He published articles in eight research publications. He was awarded with Pedagogical Innovation Award from Gujarat Technological University and Letter of Appreciation from IIT Bombay for Open Source Technology Club (OSTC) coordinator.

**Athena Vakali**, is a professor at the School of Informatics, Aristotle University of Thessaloniki, Greece, and is also leading the Laboratory on Data and Web science. She has co-authored 5 books, 15 book chapters and more than 160 papers in high quality journals and conferences. She has coordinated and participated in more than 25 national, EU and international research and innovation projects.

**Vidhi Vazirani** is currently pursuing a B.Tech in computer engineering along with an MBA in technology management from Mukesh Patel School of Technology Management and Engineering NMIMS University, Mumbai, India. She is also working on a B.Tech Honors in artificial intelligence and machine learning from IBM. Her research interests include machine learning, deep learning and natural language programming.

**Eleni Veroni** is a research associate and member of the Systems Security Laboratory at the University of Piraeus, Greece. She earned an M.Sc. degree in Digital Systems Security from the same university. She has participated in several EU-funded research and innovation projects. Her research interests lie in the areas of data security and privacy.

**Prof. Christos Xenakis** is a faculty member in the Department of Digital Systems of the University of Piraeus, Greece, since 2007, where he currently is a professor, a member of the Systems Security Laboratory and the director of the Postgraduate Degree Programme on Digital Systems Security. He has authored more than 90 papers in peer-reviewed journals and international conferences.

# *Section I*

## *Big Data in Healthcare*

# 1 Intelligent Healthcare Systems

## *A Design Overview of Existing Systems and Technologies*

*S. Saravanan, R. Saminathan, and P. Anbalagan*

## CONTENTS



## 1.1 INTRODUCTION

As the world population increases, chronic diseases and critical health issues also increase day by day. To seek out health problems, a smart health monitoring system should be developed throughout the world. In today's world, people are busy with their work schedules; they do not have time to visit hospitals or healthcare providers. This affects people's health status and creates greater problems. In order to improve the health status, a smart healthcare monitoring system should be installed for periodic health checkups (Al Brashdi et al. 2018). A smart system sends an alert to the concerned healthcare provider or hospital or to the concerned patient about the patient's health status. The smart healthcare system (Figure 1.1) monitors the patient's temperature, blood pressure, heart rate, respiration rate, oxygen rate

DOI: 10.1201/9781003142751-2

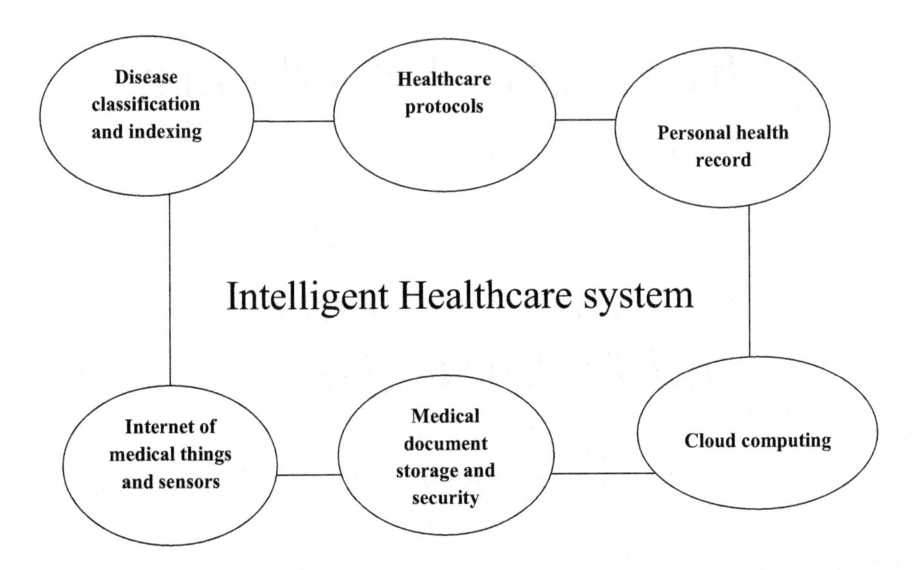


**FIGURE 1.1** Intelligent Healthcare.

through SpO2 and the like. If there is a change in these parameters, the alert short message service (SMS) will be transferred to the patient's caretaker or doctors or healthcare provider. For an efficient, smart healthcare system, the Internet of Things (IoT) is implemented. Generally, IoT is an efficient and promising technology in this modern world. The IoT consists of a collection of physical devices, such as sensors, computing devices, and storage devices, that work based on the data collected from the sensors, and the data are stored in the form of a database (Dineshkumar et al. 2016; Hossain et al. 2016; Medvediev et al. 2018). The stored database in the IoT can be accessed and controlled anywhere anytime by Internet technologies. Table 1.1 depicts the pros and cons of the present healthcare system.

The IoT framework consists of three architectural layers, namely, (1) the IoT device layer, which is on the client side; (2) the IoT gateway layer, which is on the server side; and (3) the IoT platform layer, which works as a pathway between the clients and the operator. The building blocks of the IoT are sensors, actuators, data acquisition devices,

**TABLE 1.1**
**Pros and Cons of Present-Day Healthcare System**

| S. No | Pros | Cons |
|---|---|---|
| 1. | Downsizing | Hospital downsizing |
| 2. | Cost reduction | Staff workload |
| 3. | Reduce the cost of labor | No foresight in the management |
| 4. | Reduced budget | No adequate care for patient |

preprocessing devices and cloud platform. The sensors in the system can be used either as an embedded device or as self-supporting devices that are used to collect the telemetric data. There is another device called an actuator. Actuators are the devices that are used to convert physical movement of data into another form. When actuators are connected with sensors, the data collected are converted into physical data. Based on the collected data, the IoT device analyzes the data and orders the actuators to perform a certain task or activity. The data acquired from the sensors are data and resource constrained, and the data consume a large amount of power, which leads to device failure. Even though the sensors and the actuators are acting together (Ananth et al. 2019), the data acquisition process is the most important stage in the IoT, where the collection of data, filtration and storage-based platforms like cloud computing are used. Gateways in the IoT are used to convert the sensor data into some format that can be transferrable to some other format. Gateways are used to control, transfer, filter and select data to minimize the storage level in the cloud-based platform. Gateways act as local preprocessors of sensor data that are ready for further processing. The gateway acts as a security. The gateway is accountable for monitoring the data stream on both the client and server sides with the proper encryption and authentication. In the IoT, the data transfer speed can be enhanced by using edge computing (Shahidul Islam et al. 2019). Edge computing is used to speed up the data analysis in the IoT platform. In edge computing, the base station is close to the server, and it is easy to collect and process the data. This, in turn, reduces the power consumption of the network infrastructure (Wanjari et al. 2016).

In the IoT, the sensors and gateway act as the neurons and backbone of the system. The cloud acts as the brain of the system. To store and access the massive amount of data, a cloud-based platform or data centers are used. Present-day remote monitoring systems apply deep learning and artificial intelligence to categorize and classify health data. The IoT has a variety of applications, such as agriculture monitoring, health monitoring, traffic monitoring, smart grid and energy-saving management, water supply management, fleet management, maintenance management and the like (Cheol Jeong et al. 2018).

In the healthcare industry, the IoT plays an important and crucial role. The automated features of the IoT in the medical industry are called the Internet of Medical Things. There are two types of applications in the health industry: (1) E-health and (2) M-health. Generally, the Internet of Medical Things is defined as a group of medical devices like blood pressure sensors, pressure sensors and the like that connect to a network of computers through gateways. E-health is implemented with the help of electronic devices and communication. M-health, or mobile health, is defined as the sharing of medicine prescriptions through mobile applications. The basic device used in IoT is radio-frequency identification (RFID). The RFID automatically identifies and tracks an object with a tag. In the RFID method, the chip is enabled with an RFID tag to support the patients. The chip is implanted in the person's body to monitor the health parameters.

In Software Development Life Cycle technology, two different phases are used: (1) the planning phase and (2) the implementation phase. It is a type of waterfall methodology. In this method, each stage is parted into small groups. Testing and implementation are carried out through each stage. Five different phases are implemented: (1) Planning: In planning phase, the requirements are gathered to reach the goal. (2) Analysis: There are two steps in this stage: the preliminary stage and system analysis. In the preliminary stage, the problem, objectives and goals should be defined. In system analysis, the information is gathered, interpreted and used for a diagnosis. (3) Design: In the design stage, the perception of each stage is visible to the users. (4) Implementation: In this stage, the program is developed. (5) Maintenance: The performance of the system is monitored continuously.

## 1.2 CONTRIBUTIONS IN THIS CHAPTER

- This chapter discusses state-of-the-art healthcare systems. It also explores the recent technologies used in healthcare systems.
- Furthermore, cloud platforms, which have deep support towards data population and organization, are also discussed.
- The industrial healthcare platforms used in the present-day market are also discussed, along with geriatric healthcare services.

## 1.3 HEALTHCARE MONITORING IN IoT USING BIG DATA ANALYTICS

In general, big data refers to the collection of a large volume of structured and unstructured data such as unstructured medical notes, medical imaging data, genetic data, DNA sequences and the patient's behavioral data received from the sensors. Big data is used when large volumes of data cannot be processed by using traditional methods. The large volume of data is deployed in the cloud, which can be accessed anytime using the Internet (Viceconti et al. 2015). The definition of big data consists of three Vs: (1) Volume: The data can be collected from various sources, leading to storage insufficiency. By using a cloud platform like deep lakes and Hadoop, the burden of storage can be reduced. (2) Velocity: Due to the large accumulation of data, the speed of data is reduced. By using RFID tags and sensors, the speed of data transfer is increased in real time. (3) Variety: Data can be in different forms such as structured and unstructured formats. The main objective of the system is to collect data from the sensor and analyze the same using the Hadoop platform. The system gathers the health information of the patient using the sensors, the data are processed by microcontrollers and communication is done wirelessly to the other devices in the system. The sensor-captured data are analyzed and stored in the cloud by a healthcare proxy. The IoT agent—Intel Galileo Gen 2—is used as a healthcare proxy (Kumar et al. 2018). A graphical user interface is used to monitor the patient's health. The proposed architecture consists of a blood pressure sensor, a heart rate sensor and a humidity sensor, which are connected with the health proxy. Health parameters such as temperature and blood pressure are stored in the cloud.

Data collected from the different healthcare institutions are accumulated as massive amounts of data that are warehoused in the cloud technology (Zhou et al. 2020). Since the data are stored in the cloud, the past examination of the health status of a specific patient is analyzed. The smartphone is directly connected to the IoT agent through general-purpose input/output pins; therefore, the doctor's mobile phone is used to collect the health data of the patient (Dhayne et al. 2019). If there is an emergency situation, proper medical instructions should be provided. There is connectivity from the sensors to the cloud and from cloud to the end users. Data analytics are executed by Hadoop Distributed File System and MapReduce process. Hadoop is an open-source software used to solve problems with a massive amount of data and computation. It uses the MapReduce programming framework for processing big data with a better fault tolerance. The sensors are to be touched by the patient to test the electrophysiological gestures of the patient. The readings from the sensors are observed using the Arduino processor. From the sensors, the IoT agent is connected to store the data. The GSM module is provided with voice, SMS, fax and data application for doctors. When sensor values like heartbeat rate are noted, they are recorded by the IoT agent, and they are transferred to the ThingSpeak cloud platform (Sethuraman et al. 2020).

The sensor data can be accessed from the cloud with the help of the Hadoop platform using the query command Hive. The MapReduce process is used to process the data from the cloud. A mobile phone is connected to the IoT agent with surface-mount technology package and an unlocked mini subscriber identity module. Based on the threshold value, the alert system is created. Once the emergency is aroused, an alert message is passed on immediately (Kim et al. 2020). When the threshold value exceeds, it sends an alert message to the doctors or healthcare providers. In real time, big data analysis of health information is carried out. Due to big data analytics, the response time of the system is less. By periodical monitoring of health parameters, the alert messages are transferred by means of mobile phones using GPRS or GPS connection. It offers the doctors a huge amount of health data and provides the right medical instructions to the right person at right time. The advantages of big data include data optimization, low cost, data management, the highest level of security and biomedical research, among others (Zhang et al. 2015).

## 1.4 INDUSTRIAL IoT-ENABLED FRAMEWORK FOR HEALTH USING THE CLOUD

Due to the increased number of geriatric patients, it is more important to monitor health parameters such as blood pressure, heartbeat rate, oxygen rate and the like in a periodical manner. The Health Industrial IoT (IIoT) is the amalgamation of sensors and devices, communication devices and mobile applications. Health IIoT is used to track, monitor and accumulate data. In this framework, the health parameters are collected by communicative devices like mobile phones and sensors and the data is shared securely to the cloud; then they can be accessed by the professionals. Healthcare IIoT is a combination of large number of interconnected devices and cloud computing devices (Kumar et al. 2020). Cloud technology is used to collect patients' data in an endless manner. The IIoT is the mixture of big data analytics, Internet-connected sensor devices, machine-to-machine communication, the cloud and the real-time examination of data collected from the sensors. Through the IIoT, a large amount of data can be accessed anywhere anytime.

Data collected from the interconnected devices and applications are gathered and analyzed. The gathered data are analyzed from medical records like electrocardiograms (ECGs), medical imaging equipment and physiological sensors. The collected data are further processed and classified to help doctors make decisions. The data are transferred in an endless manner through gateways to the data centers. In a cloud-based remote health monitoring system, heartbeat signals are monitored and recorded through sensor devices and are stored in remote cloud storage. After monitoring, the signal features are extracted and enhanced; furthermore, it is stored in the cloud. The information stored is validated by the cloud to check whether the entered information is correct. Once the features are extracted and validated, the data get redirected to healthcare experts or professionals (Takahashi et al. 2017).

The health monitoring system (Yang et al. 2018) consists of signal enhancement, extraction of features, analysis of the ECG, reconstruction of signal and watermarking, which are done on the end user. In this system, the ECG signal is produced by the medical equipment, which is recorded by mobile phones and stored in the cloud. Physiological rarity can be a source for muscular activities that result in small skews that causes large oscillations in the collected data. A non-physiological rarity can be caused by electrical intrusion and electrode faults. Electrode faults are caused by misplacing electrodes, having loose connections, not using enough electrode gel, setting a wrong filter, damaged wires and so on. The major error in ECG data accumulation is caused by misplacing electrodes and displacing cables, which result in cardiac abnormalities like an ectopic rhythm. In the stage of enhancement, the ECG signal is recorded and sent to a low-pass filter. The low-pass filter filters the high-frequency components in the signal. The output of the low-pass filter is passed to the 25-point moving average filter to smooth the signal. By passing the ECG signal through these two filters, the signal is free from high-frequency components and looks clear. This preprocessing step is much needed to find the electrical signals in the ECG. An average wavelet transform is used to detect the peak R value. An average wavelet transform is a combination of a Hilbert transform and a wavelet transform. These transforms combine the time-frequency location and the local slope transformation obtained from the Hilbert and wavelet transforms. The next step in ECG monitoring is to protect the data from forgery. Forgery can be prevented by watermarking the ECG signal.

Watermarking refers to inserting some information without altering the information context. There are two types of watermarking, namely, (1) watermark embedding – in this process, the authenticity of the ECG signal is transferred to the cloud – and (2) watermark extraction –it extracts the features of ECG signal for data verification. Watermarking is done on the basis of discrete wavelet transform–singular value decomposition (DWT-SVD). DWT decomposes the signal into multiple levels of time and frequency. SVD is a technique for matrix factorization that decays the matrix into three matrices. From the ECG signal, several features like heartbeat rate, P wave duration, PR interval, T-shape wave and QRS complex are extracted. To map the feature space into higher dimensional space, a one-class support vector machine classification is used. It is the mechanism that trains only one class. It checks whether the detected ECG signal is normal or not.

## 1.5 PATIENT HEALTH MONITORING SYSTEM BASED ON THE IoT

In this system, wireless sensing node technology is used to detect the unforeseen circumstances of health. Nowadays, people are facing serious health issues that may lead to their sudden demise. To avoid this, a special monitoring device has been developed to monitor patients and communicate with their neighbors. The health monitoring system is made up of wireless sensors like temperature sensors and heartbeat sensors to track the patient's temperature and heartbeat rate. It makes use of sensor technology and informs their sensor neighbors about the patient's health status when they are in critical condition. Both the sensor and the system are connected to an Arduino Uno processor (Chaudhury et al. 2017).

A microcontroller is connected with the LCD display and Wi-Fi module to dispatch the data to the wireless sensing node. If there is any sudden change in the patient's heartbeat or temperature, an alert message is conveyed to the neighbor using the IoT. It monitors the patient with a time interval through the Internet. Generally, the IoT-based healthcare system consists of two topologies:(1) IoT net topology, an amalgamation of physical devices like laptops, smartphones and so on and application-oriented devices, and (2) IoT net architecture, which is defined as the specification of physical devices organization and their practical association. The sensor system consists of self-organized devices that utilize sensors to track physiological factors. The sensor network gathers the data from the network and passes the data to clients using the Internet. The architecture consists of an Arduino Uno processor with an Atmega controller. The Atmega controller is interfaced with different equipment. The microcontroller collects the data from the sensor and processes the data (Satapathy et al. 2020). A Wi-Fi module is connected to the microcontroller, which shares the sensor data through the network. LM 35 is a body temperature sensor that measures the temperature in the Celsius scale. The temperature measured is directly proportional to the output signal. It operates at 4V and draws a current less than 60 microamps. It can operate at an optimal temperature and provides good accuracy. The conversion from the output voltage to the centigrade scale is very easy. The LM 35 is attached with the microcontroller. The measured temperature is uploaded to the Internet through the IoT. If the threshold value is exceeded, an alert will be given to the nearby sensors. The LM 358 heartbeat sensor is used to measure the ECG signal, that is, heartbeat rate. The heartbeat sensor is attached to the microcontroller. The measured values from the sensors are sent to the microcontroller through the Internet. If the sensor value crosses the threshold value, an alert notification is shared. The LCD display is interfaced and will display the message (Bi et al. 2019). The Wi-Fi module transfers the message by means of the Internet. Thus, patient monitoring is effectively done by the Internet.

## 1.6 IoT SMART HEALTH MONITORING SYSTEM

The abnormal conditions of the patients can be possible with the use of remote healthcare systems. The instant solution can be provided by wearable smart devices, which screen the health activities and evaluate the patient on day-to-day basis. The sensor data are captured and analyzed by the devices and provides a response to patients. In this system, a diverse range of healthcare metrics are considered and measured using smart devices. To collect all the mentioned parameters, the device should be coupled with the mobile devices. A smart watch is one among the reliable wearable devices that continuously monitor the body looking for the accurate heart rate and photoplethysmography (PPG) signal. It is a highly effective and low-cost device that prevents unsafe activities in human body. It combines various sensor types, such as bio-impedance sensor, accelerometer, temperature, heart rate, capacitive and motion sensor. Accelerometer values are used to predict the mobility of a patient by using ECG sensor to monitor the patient's heart rate, and the oxygen level is measured using SpO2 (Dhayne et al. 2019; Zhang et al. 2015).

A temperature sensor is used to monitor body temperature, and these data are considered to be one of the vitals. A bio-impedance sensor is used to collect the information about cell pathology and physiology. A capacitive sensor is used in touch screen displays. Low-energy Bluetooth is used in smartwatches to communicate with other low-energy devices like mobile phones, Raspberry Pi 3 and others. It is used because of its reduced power consumption. Near-field communication protocol provides communication between the two Bluetooth devices. A microcontroller obtains signals from the sensors through an analog-to-digital convertor. The wireless connection through the Bluetooth device is established by a universal asynchronous receiver transmitter (UART). The signal from the microcontroller is searched by an antenna through the UART. The standby time of the watch is mandatory option. In order to improve the standby time, the display is removed. The display is connected to the smartphone, through which the battery level and applications can be accessed. The smartphone is an Android device with a touchscreen, cellular network and an operating system. To transfer the data from the smartwatch to cloud, the smartphone is used. It acts as a bridge between the wearable device and the smartphone. An Android application has been created to develop a communication between the smartwatch and the server (Yang et al. 2018; Chaudhury et al. 2017).

The data are collected in real time by the smartphone in JSON format. The JSON format is further converted into the form of charts. The application is attached to ThingSpeak API for initiating the smartwatch and ThingSpeak cloud for communicating the data. Finally, the data are sent to the cloud. ThingSpeak technology is used to access the data through smart devices that are worn on the human body and that continuously monitor human health. The ThingSpeak cloud is a MATLAB® analytics cloud where the data can be accessed and processed under different conditions. In a cloud platform, the heart parameters are analyzed whether the cardio condition is normal or not. The data are analyzed to check the patient's heart health. HTTP protocol is used to forward the alert messages to the neighbors of the patients and to the nearby doctors. The server-to-server communication is carried out by a web-based service provider called IFTTT (If This Then That). Applets are simple conditional statements created by IFTTT. When there is a trigger, the applets pop up. It works when there is an emergency; the alert is initiated on the smartphone. The emergency alert is forwarded to a relative of the patient and the doctor (Bi et al. 2019).

## 1.7 IoT-BASED HEALTHCARE MONITORING SYSTEM

IoT emerged from the field of information and communication technologies. It allows the devices to transmit and receive data through an Internet methodology. The IoT plays an important role in the health industry, which includes monitoring of elderly and disabled people (Kumar et al. 2018). In the field of wireless communication, patient monitoring for a health condition is carried out continuously. The monitored data are transferred to the network through a Wi-Fi module. In the case of any emergency or if a critical condition occurs, a notification is sent to the patient's caretaker or doctor. For an efficient monitoring system, cloud computing is used to protect the data and maintain security measures. Wi-Fi modules provide authorization, security and privacy. They allow access to the verified users.

The main concept behind the system is to monitor the patient continuously throughout the interval. It is a three-tier architecture framework that includes a Wireless Body Sensor Network (WBSN), a base station, and a graphical user interface. The WBSN is composed of wearable sensors like ECG sensors, temperature sensors and heartbeat sensors (Sethuraman et al. 2020; Kim et al. 2020). The WBSN acts as a data-gathering unit and collects the physiological signals from the sensors that are attached to the patient's body. The gathered data are passed onto the base station or the server through the Wi-Fi module. The data forwarded from the network are stored in the form of files that can be used in the future. From the base station, the required data can be accessed anytime anywhere by typing in the authorized IP address at the user device. A graphical user interface is used to save and analyze the data in graphical and text formats.

The graphical user interface is used to share the alert message to the health experts through the Global System for Mobile Communication (GSM). The IoT architecture uses an Arduino Uno processor with an Atmega 328 p microcontroller, an LCD, a buzzer and a Wi-Fi module. The alert sound is given through buzzer. The health status of the patient is displayed by the LCD. HTTP protocol provides communication between the Wi-Fi and the web server. The webpage gets refreshed for every 15 seconds. The remote server uses FTP protocol to send emails and SMS. The temperature sensor used here is LM35. These types of sensors are very accurate and have high precision. The output voltage increases linearly with the Celsius scale. It doesn't undergo oxidation process. It is easy to convert the output voltage into temperature. ECG sensors consist of electrodes that are placed on the patient body to monitor ECG signals. The ECG sensor is connected to the AD8232 single-lead heart rate monitor, and it acts as an operational amplifier or op amp. It reduces the noise in the ECG signal. It gives PR and QT signals as output. A heartbeat sensor is used to measure the oxygen level in the hemoglobin, and it also measures the heartbeat per minute. It ensures the blood volume is similar to the heartbeat. Arduino Uno is the microcontroller board. It consists of six analog inputs, 14 digital input and output pins, a 16-MHz quartz oscillator, a USB connector, a power supply jack, an ICSP header and a reset button. A microcontroller can be connected to the computer using the USB connector. The microcontroller has its own memory unit and processor (Satapathy et al. 2020; Bi et al. 2019). The microcontroller is used to control the information. Raspberry pi is a small computer that is used to run the codes. It has a

Linux operating system. Arduino is used as an interface between the operating system and Raspberry pi. When the Arduino Uno is connected to the Wi-Fi module, the Wi-Fi module acts as a modem and Raspberry pi acts as a server. In order to transmit the data, HTTP protocol is used. The buzzer is made up of a piezoelectric crystal that produces sound. Whenever an abnormality is detected, the buzzer produces the sound to alert caretakers and doctors.

## 1.8 MONITORING OF HUMAN BODY SIGNAL BASED ON LoRa WIRELESS NETWORK SYSTEM

The LoRa wireless network consists of a temperature sensor, an ECG sensor, an oxygen pulse rate sensor and an oxygen saturation sensor, along with MySignals (Chushig-Muzo et al. 2020). These sensors are enabled by MySignal to collect data. The main aim is to transfer the data from MySignals to a computer using a LoRa network. The sensors are interfaced with the MySignals platform. Generally, a LoRa network is defined as a long-range network that works on the upper layer of the network. It is based on cloud access of the network layer. It can connect devices up to 50 miles apart. It requires only minimal power for its operation. It provides end-to-end encryption and high security with authentication. Sensors connected in the LoRa network are known as picocells. LoRa is based on spread spectrum modulation. LoRa has unique features that are different from other short-range technologies (Din et al. 2019). LoRa uses mesh networking technology. LoRa can be used in medical field to monitor the patient by transmitting the data through sensors. The collected data are sent to the cloud for further processing. The cloud has high communication range with low data rate. LoRa technology has wide applications in various fields, such as agriculture monitoring, air pollution monitoring, home security, fleet management, indoor air quality, animal processing, industrial temperature monitoring and others. Physiological sensors like temperature sensors, ECG sensors and heart rate sensors are connected to the MySignals hub.

From MySignals, the Arduino Uno is connected to collect the data with wired and wireless connections like Bluetooth and others. The LoRa protocol connects the Arduino and MySignals platform by a radio shield. The communication and transmission of data are carried out through wire, Bluetooth, Wi-Fi or LoRa. The MySignals platform consists of 12 sensor ports. The platform is interfaced with Wi-Fi serial transceiver module ESP8266 (Misran et al. 2019). The data obtained from the sensors are transferred through the LoRa module and from the Waspmote gateway to the computer. LoRa is coupled with a multiprotocol radio shield, which can connect to multiple devices at the same time. The multiprotocol radio shield can be coupled with any short-range wireless modules like Bluetooth, Wi-Fi, Zigbee, RFID and so on. The LoRa protocol is connected to the Waspmote gateway. Waspmote is an ultra-low-power open-source wireless sensor platform. It allows the sensor nodes with a low-power consumption mode to operate autonomously. RealTerm is a hyper terminal of Waspmote that is used to capture, control and debug the binary and other data formats. RealTerm is used to connect LoRa modules to send and receive the sensor data with the specific baud rate for communication. It can operate at the frequency of 868 MHz or 900 MHz.

The physiological signals are collected from the sensors and were selected to predict the health condition of the patient and check the reliability of the data. The sensor data can be assessed by two methods. The sensor data are collected by serial monitor through Bluetooth, and RealTerm collects the data from the Arudino Uno and LoRa modules. The gateway receives the messages from the LoRa module node for a particular period. LoRa operates at a frequency of 900 MHz ISM band with a data rate of 12.5 kbps. The data collected from the sensors are transmitted at the date rate of 8–10 kbps to the cloud. LoRa has an acceptable baud rate of 115200 bps. For a proper transmission, the baud rate should be similar for both transmission and reception. The duty cycle of LoRa is less than 1%. When one or more devices transmit their data packets at the same time, there may be a loss of data packets or network overload. In a LoRa module, the data packets are sent properly if the device is configured at different times to transfer. The microcontroller unit, the sensor unit and the LoRa unit consume more energy (Tayeh et al. 2020). The transmission power of the Waspmote is 14dBm. The MySignals unit is operated by battery power and consumes low power. Due to increased distance, the transmission and receiving power is high, which, in turn, reduces the transmission rate. Secure communication is an important factor at IoT solutions. The data transmission should be reliable, and it should be free from vulnerabilities. In an IoT protocol, each layer is vulnerable to security issues. To maintain the security in the network, LoRa provides end-to-end AES-128 encryption. The application key encrypts all the data with optimal privacy and protection. AES-128 is known as an application key. The App key produces two keys, namely, the Network Session key (NwKSkey) and the Application Session key (AppSkey). The NwKSkey creates a specific signature to ensure message integrity for each and every device. Every application-specific session key is responsible for the encryption and decryption of the application data (Valach et al. 2020).

## 1.9 A FRAMEWORK FOR HEALTHCARE SYSTEM USING CLOUD COMPUTING

The healthcare industry is one of the major industries growing toward development. In the field of medicine, information technology plays a major role. Even though there is growth in the health industry, some organizations are still based on handwritten prescriptions and medical records. These handwritten notes may be lost or damaged. In order to maintain medical records, digital information is used. The patient's information is stored on the Internet and can be accessed when needed. In the digital world, sharing data is more complex and rarer because it may be subject to many kinds of attacks. It has a capacity to connect device to machine, object to object, patient to doctor, patient to machine, sensor to mobile, alerting (smart devices to human), and RF reader. It ensures the effectiveness of the health industry by connecting devices, machines and humans smartly (Al Brashdi et al. 2018; Dineshkumar et al. 2016; Hossain et al. 2016).

The biggest challenges in an IoT-based health industry are security, tracking and monitoring the patients. The cloud IoT consists of different applications like e-medical prescription systems, personal health records, clinical decision-making systems, electronic health records and pharmacy systems. The doctor can use the cloud IoT for the improvisation of clinical records and diagnosis of patients. Self-assessment can be done by the patient about their health condition. They can also find hospitals and clinical services. Physical activity can be monitored by a personal monitoring system. If a person is wearing a monitoring device, they can monitor their physical activities and sleep activities data. These devices can be placed on the human body. It can be used in the shape of tiny devices that can be embedded into the jewelry, under the skin, or under the clothes. Thus, the monitoring devices are embedded into the body, which creates a WBSN. These networks are capable of collecting samples, processing samples and communicating with each other. It helps find the person's location and differentiate the person's state like resting, walking, sleeping and the like.

The data collected are updated to the device through an electronic health records system. The data are shared with the cloud application and get updated to the patient's medical profile. The updated profile can be shared to doctors and hospital. Patient records consist of X-rays, computed tomography (CT) scans, and magnetic resonance imaging (MRI) scans that can be uploaded in real time and shared through cloud platforms. These updates can be accessed by any health expert, such as a cardiologist or a radiologist, on demand of patient. When someone is traveling abroad, they could get a health checkup from local doctors on the ground with immediate access of their health records. The prescription provided by the doctor is available to the associated pharmacist. The health records can be accessed anywhere anytime through network technologies. The advantage of this system enables low-cost technology and speedy recovery of the patient.

## 1.10 WEARABLE DEVICES

Wearable devices are also known as smart devices that can be placed or worn by the patient for their health monitoring. Wearable devices are used to monitor heart rate, blood pressure levels, the distance walked and so on. Day by day, health issues and pandemic issues are arising that need continuous monitoring. Continuous monitoring can be done by smart devices like a smartwatch, a fitness tracker and the like. The devices collect the data in real time and shared through some kind of communication. Personalized health information like physiological signals can be collected from the sensors attached to the human body and can be accumulated by Android-based mobile data acquisition. The data accumulated can be accessed by the smartphone and transferred to the server for future processing. For communication between the devices, Bluetooth with low energy is used. A Bluetooth Low Energy (BLE) device increases the efficiency of the device. The architecture of wearable devices is made up of a sensor network, BLE, an Android application and a server.

The sensor network consists of tiny sensors that can be embedded or implanted in the human body. These sensors collect data in real time and pass the information to a mobile device through BLE. From the mobile application, the data is passed to the server. From the server, the data can be accessed. BLE is a power-consumption technology when compared to Bluetooth. For continuous monitoring, the energy is maintained by BLE. It operates at a frequency of 2.4 GHz ISM band. BLE remains idle until the connection is established. For a short connection, the data rate is 1 Mb/s. To enable BLE in Android applications, mobile phones are used. For this network, Android 4.3 is used to enable BLE. To use BLE on the receiver side, the following steps are used: A Generic Attribute (GATT) Profile is used to send the data over the BLE link. More than one profile can be implemented in the device. An Attribute (ATT) Profile is placed at the top of the GATT Profile. The attributes are converted into characteristics and services. Characteristics mean single values. Descriptors give different values to the characteristics. Services are defined as a collection of characteristics. The features of Bluetooth can be used by the Bluetooth permission BLUETOOTH. The permission is required to perform the request and accept the connection. The device needs to support both peripheral and central to establish the connection. The transfer of GATT metadata is started once the connection is established. Bluetooth permission requires starting, accepting and transferring the data. The BLE device is finding by scanning the nearby devices. To interface with the BLE device, a GATT server connection is established. The GATT server is connected to the Android app, which is used to discover the services. Once the data transfer finishes, the connection gets closed. The data, updated to the mobile application, are transferred to the server. While the data are on the server, they can be accessed by anyone with valid credentials like a username and password. Once the valid credentials are entered, the data can be accessed. The application screen displays the test results obtained from the BLE transmitter.

## 1.11 PRECISION MEDICINE AND HEALTH STATE MONITORING THROUGH WEARABLE DEVICES

Precision medicine main focuses on the tailored treatment for individual-centric characteristics. Precision medicine involves large datasets for discovering the disease mechanisms. A key principle of using the deployment of precision wearable devices influences the seven physiological signals such as heart rate, respiration rate, temperature, blood pressure, physical activity, sweat and emotion (Tembhare et al. 2019; Botero-Valencia et al. 2018; Polonelli et al. 2019).

1. **Measurement of Heart Rate:** The heart is the main organ of the human body that pumps and regulates blood circulation throughout the body. It purifies the blood and supplies enough nutrient content to the body. The heart pumping rate is measured from the physical activity carried out by the body. For an athlete, heart rate varies from 40–60 bpm, and for adults and children, heart rate varies from 60–100 bpm. In an ECG, the electrical activity of the heart can be tracked by placing electrodes on the chest. The pulse rate and the acoustic signal associated with heartbeat can be tracked by stethoscope. The heart rate can be obtained manually by listening the mechanical pulses of the artery. The contraction and enlargement of vessels make changes in optical transmission or reflection which can be observed by PPG (Tembhare et al. 2019).

   *PPG:* It is used to ration the variations in the volume of organ or tissue. It uses an optical light source and detector to measure the volume of the organ or tissue. The enlargement and reduction of blood vessels in the dermis and the hypodermis depend on the changes in the blood volume. The PPG signal measures blood oxygen saturation level, stroke volume, heart rate, and vascular resistance. PPG has a wavelength between 700–1000 nm. PPG sensors are embedded in wrist-worn devices with wireless communication support. The photo detector emits green, red or infrared LEDs. Wrist-worn devices are incorporated by battery and transmission electronics. PPG sensors are portable and convenient. These sensors can be worn on the ear lobe or wrist to measure the heart rate. The major disadvantage of this sensor is that measurement at ear lobe decreases the temperature results in a decrease of the signal-to-noise ratio.

*Impedance Cardiography (ICG):* This device processes variations in the complete electrical conductivity of the thorax. It is used to evaluate the cardio-dynamic bounds like stroke volume, vascular resistance and heart rate. It operates at the frequency of 20–100 kHz, with sinusoidal perturbation of 1–5 mA. These devices can be placed on the user's chest, waist or neck. The disadvantage of ICG is that it could not produce accurate results.

2. **Measurement of Respiration Rate:** The respiratory system in the human body includes cavities, the nose, the mouth, the bronchioles, the lungs and the muscles responsible for respiration, among others. Respiration rate differs from 30–60 breaths per minute for newborns and 12–18 breaths per minute for adults. Due to alcohol consumption and sleep apnea, respiration rate may be less than eight breaths per minute. While exercising, the respiration rate increases, which, in turn, increases oxygen intake. To measure the lungs' volume, capacity and gas exchange are measured using a pulmonary function test (PFT). In a PFT, the single volume breath is measured during inhalation and exhalation. The efficiency of oxygen exchange in vascular system is measured by lung diffusion capacity (Botero-Valencia et al. 2018; Polonelli et al. 2019).

   *Spirometry:* Wearable devices are used to measure the oxygen flow rate and volume during inhalation and exhalation. An ergo-spirometry mask is the most common wearable spirometer device. An ergo-spirometer is used to measure the speed and amount of air inhaled and exhaled during exercise. A spirometer uses an ultrasonic transducer to measure the speed of airflow, and the mask is used to find the pressure difference. These are smaller and lighter in size. Spirometers are mostly used in a clinic.

   *Respiratory Inductance Plethysmography (RIP):* It detects the enlargement and shrinking of the chest and abdomen. It consists of two bands; each band consists of a wire loop. One is kept on the chest, and the other one is on the abdomen. The loop is zigzagged to change its length. During enlargement and reduction, there is a change in the magnetic field prompted in the loop. For continuous monitoring of respiration rate, RIP is used. These devices are light in weight and suitable for kids and infants (Ayaz 2018).

3. **Measurement of Blood Pressure:** For normal heart function, the pressure of the blood varies from maximum systolic to minimum diastolic. The most common three methods to measure blood pressure are palpitation, listening to Korotkoff sounds using a stethoscope and the oscillometric method. An increase in stress, drugs and irregular physical activity could lead to blood pressure in a high volume. The normal blood pressure value varies from 120 mm HG in the case of systolic and less than 80 mm Hg in the case of diastolic. The systolic pressure varies between 160–220 mm Hg, and diastolic pressure varies less than 100mm Hg while exercising (Polonelli et al. 2019; Cheol Jeong et al. 2018).

   *Oscillometric Blood Pressure:* Manual blood pressure measurement is carried out by using palpitation or a stethoscope to listen to Korotkoff sounds. The pressure fluctuations in the cuff can be measured by the oscillometric method. Since arterial elasticity depends on pressure, the magnitude of oscillations changes with pressure. Wearable oscillometric devices include an air bladder cuff, pump, valves, pressure sensors, a power supply, display and communication devices for wireless transmission of data (Bigelow et al. 2016). These are wrist and fingertip devices. These devices are used for their high accuracy and robustness. The systolic and diastolic pressure can be measured from the maximum amplitude.

4. **Measurement of Temperature:** Generally, body temperature is measured manually by using a thermometer. The normal temperature of the human body is 98.63°F. For healthy adults, the body temperature varies from 97.8°F to 99°F. A glass thermometer is used to measure the temperature where the core temperature is displayed on the LCD. Most thermometers are thermistor-based, which are more sensitive. These devices are placed on the wrist, forehead, chest, ears, feet, fingers and other parts of the body. To measure the temperature at the tympanic membrane, an infrared thermopile sensor is used. Surface temperature is based on the exchange of heat at the environment. Athletic performance, fatigue, emotional condition and sleep states are monitored by surface temperature (Bae et al. 2017).
5. **Measurement of Physical Activity:** Physical activity is the movement of body produced by the skeletal system that requires energy. Physical activity is important for individuals with cardiovascular diseases, diabetes and obesity. Fitness trackers and wearable accelerometers are used to monitor the daily activity of the patient. A wearable accelerometer measures the acceleration with one or more axes. A gyroscope measures angular motion. An accelerometer is small in size, and it can be wearable (Cheol Jeong et al. 2018).

## 1.12 CONCLUSION

In this chapter, an IoT-based healthcare system was analyzed along with the different applications. This chapter also addressed the real-time usage of wearable devices. From observations, it is clear that an IoT-based healthcare system reaches different dimensions of the information technology field. The usage of IoT in medical standards makes the system to be cost-effective, and it provides speedy recovery of the patient. Furthermore, it has also been proved that wearable devices also provide effective and efficient monitoring. With the help of these devices, health parameters are monitored, analyzed and transferred to the cloud or server. From the cloud or server, the data can be accessed anywhere by healthcare experts. This makes the entire healthcare services feasible. In the future, there may be a chance of an autonomous healthcare service.

## REFERENCES

Al Brashdi, Z. B. S., Hussain, S. M., Yosof, K. M., Hussain, S. A., & Singh, A. V. (2018). IoT based Health Monitoring System for Critical Patients and Communication through Think Speak Cloud Platform. In *2018 7th International Conference on Reliability, Infocom Technologies and Optimization (Trends and Future Directions) (ICRITO)*, pp. 652–658, IEEE, Noida, India.

Ananth, S., Sathya, P., & Mohan, P. M. (2019). Smart Health Monitoring System through IoT. In *2019 International Conference on Communication and Signal Processing (ICCSP)*, pp. 0968–0970, IEEE, Chennai, India.

Ayaz, H. (2018). Observing the Brain-on-Task using Functional Optical Brain Monitoring. In *2018 IEEE Signal Processing in Medicine and Biology Symposium (SPMB)*, pp. 1–1, IEEE, Philadelphia, PA, USA.

Bae, J., Corbett, J., Kim, J., Lewis, E., Matthews, B., Moore, C., & Patek, S. (2017). CloudConnect: Evaluating the Use of Precision Medicine in Treatment of Type 1 Diabetes. In *2017 Systems and Information Engineering Design Symposium (SIEDS)*, pp. 138–143, IEEE, Charlottesville, VA, USA.

Bi, Z., Wang, M., Ni, L., Ye, G., Zhou, D., Yan, C., & Chen, J. (2019). A Practical Electronic Health Record-Based Dry Weight Supervision Model for Hemodialysis Patients. *IEEE Journal of Translational Engineering in Health and Medicine*, *7*, 1–9.

Bigelow, M. E. G., Jamieson, B. G., Chui, C. O., Mao, Y., Shin, K. S., Huang, T. J., & Iturriaga, E. (2016). Point-of-Care Technologies for the Advancement of Precision Medicine in Heart, Lung, Blood, and Sleep Disorders. *IEEE Journal of Translational Engineering in Health and Medicine*, *4*, 1–10.

Botero-Valencia, J., Castano-Londono, L., Marquez-Viloria, D., & Rico-Garcia, M. (2018). Data Reduction in a Low-Cost Environmental Monitoring System Based on LoRa for WSN. *IEEE Internet of Things Journal*, *6*(2), 3024–3030.

Chaudhury, S., Paul, D., Mukherjee, R., & Haldar, S. (2017). Internet of Things based Healthcare Monitoring System. In *2017 8th Annual Industrial Automation and Electromechanical Engineering Conference (IEMECON)*, pp. 346–349, IEEE, Bangkok, Thailand.

Cheol Jeong, I., Bychkov, D., & Searson, P. C. (2018). Wearable Devices for Precision Medicine and Health State Monitoring. *IEEE Transactions on Biomedical Engineering*, *66*(5), 1242–1258.

Chushig-Muzo, D., Soguero-Ruiz, C., Engelbrecht, A. P., Bohoyo, P. D. M., & Mora-Jiménez, I. (2020). Data-Driven Visual Characterization of Patient Health-Status Using Electronic Health Records and Self-Organizing Maps. *IEEE Access*, *8*, 137019–137031.

Dhayne, H., Haque, R., Kilany, R., & Taher, Y. (2019). In search of Big Medical Data Integration Solutions-a Comprehensive Survey. *IEEE Access*, *7*, 91265–91290.

Din, I. U., Almogren, A., Guizani, M., & Zuair, M. (2019). A decade of Internet of Things: Analysis in the Light of Healthcare Applications. *IEEE Access*, *7*, 89967–89979.

Dineshkumar, P., SenthilKumar, R., Sujatha, K., Ponmagal, R. S., & Rajavarman, V. N. (2016). Big data analytics of IoT based Healthcare Monitoring System. In *2016 IEEE Uttar Pradesh Section International Conference on Electrical, Computer and Electronics Engineering (UPCON)*, pp. 55–60, IEEE, Varanasi, India.

Hossain, M. S., & Muhammad, G. (2016). Cloud-Assisted Industrial Internet Of Things (IIoT) – Enabled Framework for Health Monitoring. *Computer Networks*, *101*, 192–202.

Kim, J. C., & Chung, K. (2020). Multi-Modal Stacked Denoising Auto Encoder for Handling Missing Data in Healthcare Big Data. *IEEE Access*, *8*, 104933–104943.

Kumar, A., Krishnamurthi, R., Nayyar, A., Sharma, K., Grover, V., & Hossain, E. (2020). A Novel Smart Healthcare Design, Simulation, and Implementation Using Healthcare 4.0 Processes. *IEEE Access*, *8*, 118433–118471.

Kumar, S., & Singh, M. (2018). Big Data Analytics for Healthcare Industry: Impact, Applications, and Tools. *Big Data Mining and Analytics*, *2*(1), 48–57.

Medvediev, I., Illiashenko, O., Uzun, D., & Strielkina, A. (2018). IoT Solutions for Health Monitoring: Analysis and Case Study. In *2018 IEEE 9th International Conference on Dependable Systems, Services and Technologies (DESSERT)*, pp. 163–168, IEEE, Kyiv, Ukraine.

Misran, N., Islam, M. S., Beng, G. K., Amin, N., & Islam, M. T. (2019). IoT Based Health Monitoring System with LoRa Communication Technology. In *2019 International Conference on Electrical Engineering and Informatics (ICEEI)*, pp. 514–517, IEEE, Bandung, Indonesia.

Polonelli, T., Brunelli, D., Girolami, A., Demmi, G. N., & Benini, L. (2019). A Multi-Protocol System for Configurable Data Streaming On IoT Healthcare Devices. In *2019 IEEE 8th International Workshop on Advances in Sensors and Interfaces (IWASI)*, pp. 112–117, IEEE, Otranto, Italy.

Satapathy, S. C., Cruz, M., Namburu, A., Chakkaravarthy, S., & Pittendreigh, M. (2020). Skin Cancer Classification using Convolutional Capsule Network (CapsNet). *Journal of Scientific and Industrial Research (JSIR)*, *79*(11), 994–1001.

Sethuraman, S. C., Vijayakumar, V., & Walczak, S. (2020). Cyber attacks on Healthcare Devices Using Unmanned Aerial Vehicles. *Journal of Medical Systems*, *44*(1), 1–10.

Shahidul Islam, M., Islam, M. T., Almutairi, A. F., Beng, G. K., Misran, N., & Amin, N. (2019). Monitoring of the Human Body Signal through the Internet of Things (IoT) based LoRa Wireless Network System. *Applied Sciences*, *9*(9), 1884.

Takahashi, Y., Nishida, Y., Kitamura, K., & Mizoguchi, H. (2017). Handrail IoT sensor for Precision Healthcare of Elderly People in Smart Homes. In *2017 IEEE International Symposium on Robotics and Intelligent Sensors (IRIS)*, pp. 364–368, IEEE, Ottawa, ON, Canada.

Tayeh, G. B., Azar, J., Makhoul, A., Guyeux, C., & Demerjian, J. (2020). A Wearable LoRa-Based Emergency System for Remote Safety Monitoring. In *2020 International Wireless Communications and Mobile Computing (IWCMC)*, pp. 120–125, IEEE, Limassol, Cyprus.

Tembhare, A., Chakkaravarthy, S. S., Sangeetha, D., Vaidehi, V., & Rathnam, M. V. (2019). Role-based Policy to Maintain Privacy of Patient Health Records in Cloud. *The Journal of Supercomputing*, *75*(9), 5866–5881.

Valach, A., & Macko, D. (2020). Optimization of LoRa Devices Communication for Applications in Healthcare. In *2020 43rd International Conference on Telecommunications and Signal Processing (TSP)*, pp. 511–514, IEEE, Milan, Italy.

Viceconti, M., Hunter, P., & Hose, R. (2015). Big data, Big Knowledge: Big Data for Personalized Healthcare. *IEEE\Journal of Biomedical and Health Informatics, 19*(4), 1209–1215.

Wanjari, N. D., & Patil, S. C. (2016). Wearable devices. In *2016 IEEE International Conference on Advances in Electronics, Communication and Computer Technology (ICAECCT)*, pp. 287–290, IEEE, Pune, India.

Yang, Y., Liu, X., & Deng, R. H. (2018). Lightweight Break-Glass Access Control System for Healthcare Internet-Of-Things. *IEEE Transactions on Industrial Informatics, 14*(8), 3610.

Zhang, Y., Qiu, M., Tsai, C. W., Hassan, M. M., & Alamri, A. (2015). Health-CPS: Healthcare Cyber-Physical System Assisted By Cloud and Big Data. *IEEE Systems Journal, 11*(1), 88–95.

Zhou, S., He, J., Yang, H., Chen, D., & Zhang, R. (2020). Big Data-Driven Abnormal Behavior Detection in Healthcare Based on Association Rules. *IEEE Access, 8*, 129002–129011.

# 2 An Overview of Big Data Applications in Healthcare

## *Opportunities and Challenges*

*Akansha Gautam and Indranath Chatterjee*

**CONTENTS**



## 2.1 INTRODUCTION

### 2.1.1 WHAT IS DATA?

Data are an assemblage of different symbols, words, measurements, and numbers. A single value denoting a single variable refers to 'datum'. *Data* means "things given" in Latin, although we tend to use it as a mass noun in English as if it denotes a substance. Ultimately, almost all useful data are given to us either by nature, as a reward for careful observation of physical processes, or by other people, usually inadvertently (Jacobs 2009). It is a term used for raw facts and figures that can be processed to retrieve information systematically in a useful and meaningful manner. Data are measured, collected, reported, and analyzed, and after that, they can be visualized using charts, graphs, maps, images, and other various tools that can perform analysis. It is considered meaningless until correctly manipulated,

DOI: 10.1201/9781003142751-3

processed, and analyzed. Processed and meaningful data are known as information that is accountable for executing relatable actions using it. Information is dependent on data. Data become information only after certain operations are performed. Data can be categorized into primary data and secondary data during its collection. Primary data are the new data, referred to as 'pure', which is collected directly or for the first time by an investigator to achieve a particular goal. Secondary data are nothing but the facts and figures assembled or recorded before the project. It is preexisting data. Mainly, data can be bisected into two categories: qualitative data and quantitative data. Qualitative data, nonnumerical data, provide descriptive information, whereas quantitative data deal in numerical form. Quantitative data, also known as numerical data, referring to a number, can be further bifurcated into discrete data and continuous data. Discrete data can be computed and counted, whereas continuous data can only be measured. Qualitative data, referred to as categorical data, are further fractionated into nominal data and ordinal data. Nominal means names. Ordinal data follow some kind of implied order. Both humans and machines can generate data.

In computer science, data are raw pieces of information presented in a messy form, usually translated into an understandable and well-structured form that is later on processed and managed to extract substantial knowledge and create value out of it in an efficient way. Data can be called information once it is relevant enough to be used to make smart decisions when analyzed and processed in some fashion. Data can be collected using various ways such as surveys, social media monitoring, online tracking, online market analytics, transactional data tracking, interviews, documents, health records, sensors, and mobile phones (Sagiroglu et al. 2013). Data are a hot topic these days. They manifest the status of past activities and allow us to make better decisions in the future.

When organized in a well-structured way and performed the analysis, data can be act as a fuel for decision-making, and they help obtain the expected results. However, the size of data is not restricted to a certain fixed number of megabytes or gigabytes. Data are present in enormous sizes. Interestingly, it will continue to grow exponentially with every moment that passes by. Digital data pay a significant contribution to the escalation of data size. This inevitable quantity of data requires a modern approach to managing, analyzing, and monetizing the data instead of using the conventional approach. Traditional technologies have limited storage capacity and rigid management tools and are expensive. They lack scalability, flexibility, and performance needed in the big data context (Oussous et al. 2018).

Data are everywhere. It can be present around us in different forms. Such existing forms are structured, unstructured, and semistructured. Structured data means organized data that can be stored, accessed, and processed directly and attractively in the form of a table. It follows a consistent order. A human or computer can easily access this type of data. A database containing a student table is a classic example of structured data. The unstructured data do not follow a predefined structure, such as collecting videos, audio, text files, and images. Semistructured data do not comply with the data models tabular structure but still, have some organizational properties. JSON documents, NoSQL databases, CSV files, XML files are some examples of semistructured and schema-less databases. Interestingly, the amount of data present in an unstructured format is much larger than data in a structured format. There are many data to be handled other than just the structure type of data that we normally consider.

In previous times, data used to be measured in megabytes or gigabytes. Nevertheless, nowadays, a considerable amount of data being generated from different sources is stored in petabytes or zettabytes. This is just the beginning. It will keep on increasing exponentially with every moment that passes. This constant and exponential growth of data in an enormous amount refers to the term *big data*. Surprisingly, if $x$ amount of data exists today in this world, then 90 percent of this entire $x$ amount of data is generated in the last two years. This increasingly massive proportion of data is nothing but termed as 'big data'. It is considered an explosion of available information (Fan et al. 2014). All the credit for this sudden escalation in data growth in the past 10 years goes to a new digital media era and our day-to-day reliability on technology and science.

Big data is an amalgamation of different types of granular data (Yaqoob et al. 2016). It requires a different approach to deal with these larger datasets. It is a great challenge to capture, store, understand, and analyze an incredible amount of information using traditional software such as RDBMS. Conventional data processing application software cannot manage, process, and leverage such large and complex datasets as it demands a properly structured model to manipulate further and fetch the data (Gautam et al. 2020). For instance, has anyone ever wondered how Facebook or Google can deal with such large quantities of information quickly? One's Facebook profile page contains data in different forms, such as texts, images, and videos. Apart from the personal data like name and age stored in an individual Facebook profile, even each picture that is liked and reacted to by a person also generates data. This shows that data are not only getting generated at a larger scale but also in various forms. Such a combination of data cannot be stored in a single structured table. Big data technologies can only handle it. Smartphones are the basic need of every individual today, but the point to be highlighted here is that have we ever given a thought to the amount of data a smartphone generates in the form of texts, calls, emails, images, videos, Internet searches, and audio. Approximately, data are produced in terms of exabytes every month by a single smartphone user. Now imagine this number getting multiplied with billions of smartphone users. This massive amount of data is vast and complex for traditional computing systems to handle. Hence, we require another approach to deal with unstructured or semistructured data and to scale the database regularly to handle the data as the data will keep growing systematically. Meaningful insights can be extracted from such datasets, which can provide significant assistance to official surveys (George et al. 2014).

Data in the digitalization era went from scanty, costly, and inaccessible to superabundant and cheap. That is when the concept of big data emerged. There is plenty of data available from different sources and platforms that can lead to the generation of vast big data. There are multiple sources from which big data can be collected. Some of them include online transactions, health records, mobile applications, social media platforms, education, banking, videos, text documents, audios, images, customer databases, and so on. A massive chunk of data is generated when performing these activities. The percentage of this uncertain data production from such various sources will keep increasing without pause. We call it uncertain data because it is available in an unorganized format. All these data are precious. Big data trends are getting admired in the healthcare sector as they help improve and target customers and public health.

Big data is often combined with machine learning and artificial intelligence technologies for better decision-making and predictive analytics that bring the information's value in the limelight.

### 2.1.2 What is Big Data?

Big data, the term itself, says a lot about the size or amount of data it presents. Its importance revolves around the dataset size and helps in analyzing and efficiently retrieving concrete and crucial information. Big data is data only but of a larger size. It is a myth that big data is only about the size or volume of produced data but refers to the large amount of data collected from various data sources and has distinct formats. In a single minute, a quintillion bytes of data are generated by humans every day, and this pace is accelerating. Classical database management systems are not capable of handling such vast and varied datasets. The advancement in computer systems and Internet technologies has led to the overall development of computing hardware. However, the problem of managing large-scale and complex data is still there when we are entering the era of big data (Tsai et al. 2015). Big data is an asset that can be used for many benefits. Big data tends to exist everywhere. This constant generation of data using audio, images, videos, social media, telecom, business applications, and various other domains led us to the emergence of big data.

Big data includes seven core characteristics: volume, variety, velocity, variability, veracity, visualization, and value (Figure 2.1). All these 7Vs are interconnected. We can classify any data as big data with the concept of these 7Vs. Volume depicts the unimaginable amount or size of information produced every single second from various data sources. The second characteristic, variety, refers to the multiple varieties of the data in which it can be available such as structured, unstructured, or semi-structured. The third characteristic is velocity. The term *velocity* itself indicates its implicit meaning. It states the speed of something in a given direction. There is no point in ending up waiting for the data if we invest time and resources. So this aspect

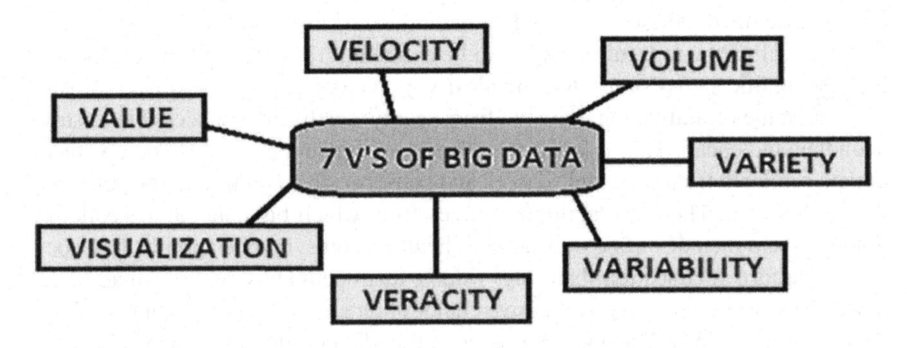


**FIGURE 2.1** Characteristics of Big Data.

of big data is to provide data whenever demanded and at a faster pace. Precisely, velocity cites the rate at which data are being generated every day. The fourth characteristic, which is variability, focuses on the exact and implicit meaning of the raw data as per its context. Veracity, the fifth characteristic, represents clear and consistent states of data available to us. It shows the degree of reliability that data offer. The term *data visualization* comes into play to bring the obscure raw details of data into the glance. Data visualization converts raw data into graphs, images, charts, and even videos that allow us to acquire knowledge from them, discover new patterns, and spot trends. The last characteristic is a value that makes sure that the processed data provide an immense value. It refers to the valuable, reliable, and authentic data that need to be analyzed to find better apprehension.

The term *big data* has gained a critical value at present. We all used to store data in a CD or a floppy in the early 21st century. The utilization of files, hard discs, floppy discs, and manual paper records has become obsolete. People have replaced them with relational database systems for storing exponentially growing data. Moreover, curiosity for accessing new technologies, inventions, and applications with fast response times and the Internet's evolution have led to continuous and enormous amounts of data. Due to the technology evolution, data have also been evolved into big data. Before, we had landline phones, but now every person carries an Android or IOS smartphone that makes our lives smarter. Every action we perform on our smartphones leads to the generation of many data. Apart from that, we were using bulky desktops managing data in megabytes only and then floppies and, later on, hard disks to store terabytes of data. Now we can store our data in clouds. Social media is also considered a significant factor in the evolution of big data. The Internet of Things (IoT) is also accountable for producing large datasets. The IoT accumulates data from various sources in order to make our physical devices act smarter. This enhancement of technology is implicitly responsible for the generation of big data. Google and Facebook are considered Internet giants that collect and store an enormous amount of data daily. For instance, Google can keep track of various information such as user preferences, location, Internet browsing history, emails, contacts, and so on (Dash et al. 2019).

Similarly, Facebook stores many user-generated data (Dash et al. 2019). The only point of concern is that the extensive amount of data is very complex. Moreover, it is not in a format that traditional database systems can handle. Previously, companies could not store their available archives in vast amounts for an extended period because it was too inefficient as these traditional technologies have a specific storage limit, follow rigidity, and are expensive at the same time (Oussous et al. 2018).

Big data analysis illustrates patterns and connections that will tremendously improve most human activities. Big data has a few problems, such as storing large and exponentially growing datasets; processing complex datasets from different sources in distinct formats, including structured, unstructured, and semistructured; and the speed at which we perform the computation of a huge amount of data (Kumar et al. 2016). To resolve these issues and store and process this big data efficiently, different software frameworks, algorithms, and techniques such as Cassandra, Hive, Hadoop, Kafka, and Apache Spark, perform the job required by a particular organization. Hadoop is an example to show how it helps in performing the job. Hadoop has mainly two components: HDFS and MapReduce. The Hadoop Distributed File System (HDFS) component stores big data, and the MapReduce technique is used to process the data. Since all the data will not fit in one single processor any longer, we have to distribute it. *Distribute* here means breaking the extensive amount of data into smaller chunks and storing them on various machines rather than accumulating them on one centralized machine, which happens in the traditional approach. So if one machine fails, data are safe on another machine. Distributing data across various nodes can help in computation that is thousands of times faster. We can run the data in parallel, which is impossible with conventional systems because storage will be limited to one system. That is why Hadoop is known for processing huge datasets in parallel and distributed fashion, which helps in easy and faster processing. Also, various copies are being made of each broken file that goes into different nodes.

After storing and processing our big data, these data can be analyzed to retrieve better future results for many various applications. In many e-commerce sites, designers analyze user data to understand customers' likes and dislikes when purchasing the items. They analyze customer buying habits using their purchasing history. This will help them suggest new products that the patrons might be interested in trying. These overall analyses will help them improve their services and enhance and personalize their customer's experience.

Similarly, in many games, designers analyze user data to determine at which stage most users get stuck or stop playing the game. These analyses can help them improve the game's storyline and enhance the users' experience, which helps increase the customer access rate. This is known as 'big data analytics'. Big data analytics is widely used in the business sector to smooth their continuous growth and development. In analyzing a particular extensive dataset, various data mining algorithms are applied, which will help them with better decision-making. Valuable insights are extracted by doing a hypothesis formulation, which is mainly based on conjectures assembled from experience and discovering correlations among different variables (Rajaraman 2016). Hundreds of servers running enormously parallel software must analyze big data (Sagiroglu et al. 2013). The technological developments in big data infrastructure, analytics, and services allow firms to transform themselves into data-driven organizations (Lee 2017).

Big data is considered the most powerful fuel capable of running huge information technology (IT) industries and firms of the 21st century. There has been immense growth in the operation of big data analysis across the world. Big data is observed to be a widely spread technology used in every possible sector these days. Some of them include the following:

a. The entertainment industry such as Netflix, YouTube, Instagram, Facebook, and many other different websites uses big data.
b. Insurance companies use big data to predict illnesses and accidents to tag their products' prices as per the predicted data incurred.
c. Driverless cars require a huge amount of data for their successful execution.
d. The education sector is also choosing big data technology as a learning tool instead of traditional lecture methods.

e. Big data can also be used in automobiles. Driverless cars have thousands of sensors fitted into their engines to record every tiny detail about their operations, such as the size of an obstacle, distance from the obstacle, and how to react or find out the best course of action situation. A large amount of data is being generated for each kilometer that a person drives on that car.
f. The travel and tourism sector is considered the biggest user of big data technologies. It allows predicting travel facilities requirements at different places.
g. The financial and banking sectors massively use big data technology. Customer behavior can be predicted using various patterns such as shopping trends, investments, and fashion.
h. Big data has started creating a major difference in the healthcare sector. Predicted outputs can help medical facilitators to provide the best services to each of their patients.
i. The telecommunication and media sector is one of the primary users of big data. Data are generated in terms of zettabytes every day. To handle such largely scaled data, big data technologies are indeed required.
j. Government and military are also extensive users of big data. A single jet-fighter plane requires processing petabytes of data during its flight.
k. Big data technologies can be used in politics to analyze patterns, influence, and predict a win.

All the points mentioned earlier ensure that big data technologies have a vast scope in every field and sector. The job opportunities in big data, including openings for prominent data engineers, big data analysts, big data solution architecture, and many more, are currently in tremendous demand. Also, there is a rising demand for analytics professionals and data scientists. Many companies hire big data developers. A few of them are Oracle, Amazon, SAP, Dell, Cognizant, Capgemini, Accenture, and many more.

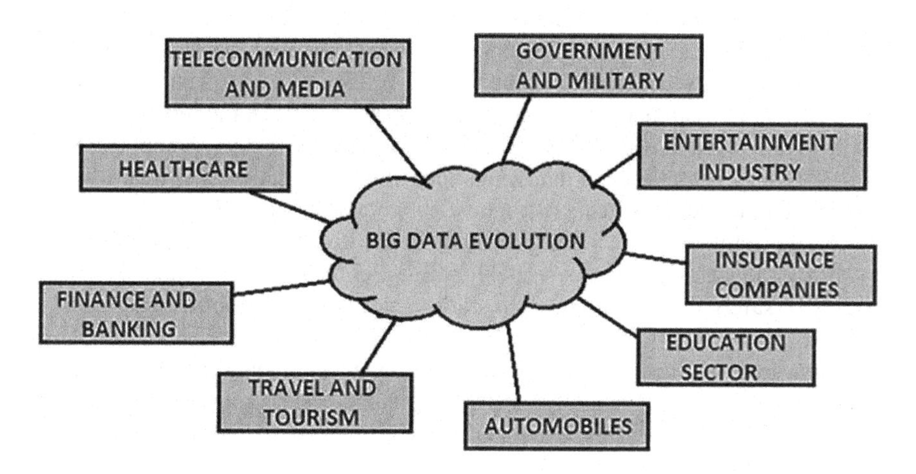


**FIGURE 2.2** Big Data Evolution and Its Various Applications.

There are numerous significant advantages of big data that can help in the continuous growth and development of various organizations. Some of them include predictive analysis, which reduces a risk factor in an organization and businesses' growth by analyzing each customer's needs. Multimedia platforms like YouTube, Instagram, Facebook, and many more can share data due to developing big data technologies. The medical and healthcare sectors can also keep their patients under constant observation using big data tricks. Big data has an excellent capability to escalate the growth of customer-based companies on a broader scale.

### 2.1.3 Big Data and the Healthcare System: An Introduction

The healthcare system is one of the essential areas that play a vital role in every individual's life. Without healthcare facilities, we humans cannot even imagine ostensibly living our lives. To live healthy and long lives, humans require healthcare facilities to cure their illnesses whenever required. The healthcare industry is enormous. Different health professionals are associated with various health sectors such as medicine, nursing, dentistry, psychology, physiotherapy, and many others (Dash et al. 2019). A massive amount of data is regularly produced in healthcare areas. Big data and the healthcare sector can be connected to bring a great revolution to the healthcare sector. It can transform the healthcare industry. Data are fetched, processed, and converted into actionable information using big data technologies to improve healthcare services.

Healthcare professionals can fetch this vast amount of medical data constructed by acquiring digital technologies and compute needful information by using the best strategies. Analyzing a large amount of medical data from various locations and census will concede to regulate which conditions improve a particular treatment's effectiveness and which do not work. Algorithms can be developed to detect and treat diseases. The plentiful sources of big data in healthcare include patient portals, research studies, electronic health records (EHRs), search engine data, insurance records, government agencies, smartphones, wearable devices, generic databases, payer records, pharmacy prescription, and medical imaging data (Chatterjee et al. 2019). The EHR itself could be considered big data (Ross et al. 2014). Various EHRs include billing data, laboratory data, and medication records (Manogaran et al. 2017). Medical researchers use medical data present in big data to recognize all the risk factors associated with different kinds of diseases to avoid any complications in the near time. Doctors use big data to comprehensively diagnose and treat illnesses occurring in individual patients to improve public health comprehensively.

Big data technology offers many different applicative areas in the healthcare sector, such as advanced quality care, public health, clinical decision support, disease scrutiny or safety inspection, research studies, and predictive modelling (Lee et al. 2017).

Healthcare organizations and government agencies receive information every moment regarding any contagious disease outbreaks by accessing the data collected from electronic health records, social media platforms, the web, or any other sources.

Also, hospitals and clinics generate massive volumes of data across the world, in exabytes, annually in various forms such as patient history and current records and

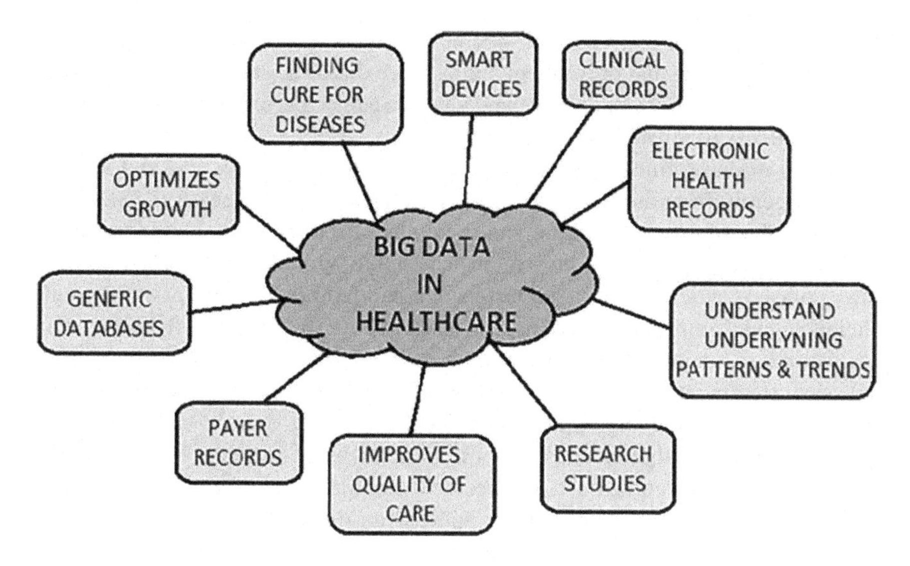


**FIGURE 2.3** Big Data in Healthcare.

test results. Especially amid the COVID-19 pandemic, a massive amount of valuable data is coming out of healthcare at a continuous pace in the entire world. Data derived from healthcare records of COVID-19 patients all over the world deliver up-to-the-minute information to the World Health Organization (WHO), various medical researchers, and government agencies, which can help them in the diagnosis of the novel coronavirus and help in inventing vaccine and eventually saving lives of many people across the world. These healthcare data are produced at a very high speed, showing the velocity attribute of the big data. Healthcare data can be present in Excel records, log files, and X-ray images. Analyzing all these data will profit the medical sector by allowing the early detection of diseases faster, revealing better treatment procedures, and reducing the expenditure associated with it.

With the help of predictive analyses, healthcare professionals can provide personal healthcare services to an individual patient.

## 2.2 BIG DATA ANALYTICS IN THE HEALTHCARE INDUSTRY

This section addresses the need for big data techniques and data analysis in the healthcare area and why and how it can resolve various requirements in a much more innovative and efficient way. Big data technologies in the healthcare industry have constructive and lifesaving consequences. It can influence the delivery of the entire healthcare industry around the world. A vast amount of data assembled by digitizing every possible thing that brings convenience to people has led to an upsurge in data growth.

Similarly, health data available in large and entangled forms are processed to bring out a positive output, monitor hospitals' performance, and improve patient care, which cannot be tackled by traditional computing software. In simple words, data should be processed and analyzed to maximize benefits and minimize the risk factors for people. The analyses of a specific amount of health data of a sample population will help detect and prevent epidemics, cure diseases, and reduce costs associated with it. Data mining techniques and analysis can be performed to determine the causes of illness to inflate diagnostic services.

Today's massive amount of medical data requires consolidating and examining intelligently to assist better healthcare delivery (Chatterjee et al. 2020). Big data is capable of creating unique and innovative networks of knowledge-sharing systems. Data comparison is now more accessible by measuring and monitoring data digitally. Critical insights gained from it can facilitate workflows and greater efficiencies and improve patient health. Patterns can be detected from extensive data using a systematic analysis approach so that clinicians can inform about the necessary treatment to individuals and predict its outcomes accordingly. Health data processing using big data technologies can help deliver context-relevant clinical information that enables better holistic decision-making. Making healthcare structured, relevant, smart, and accessible could benefit from big data trends. Progress in using big data in treating and curing diseases increases the benefits associated with it. Every person around us today has some device that can collect data, such as a fitness tracker, different applications residing on our mobile phones that measure heart rate. These devices can measure our oxygen level or body temperature. Such devices also contribute to the production of health data. A learning healthcare system by practicing medical data can be developed to predict the interaction between the human body and any particular disease and make it relevant. Big data techniques can predict the cause of various diseases based on the medical records before they start emerging (Senthilkumar et al. 2018; Mohamed et al. 2020).

Big data analytics is applied to healthcare to detect serious illness signs in an individual's life at a very early stage. Doctors need to understand their patients better to deal with the medical situations that may arrive in their lives and their adverse repercussions. Early detection and treatment are much better and simpler in terms of complications and risks at an early stage. Moreover, it is less expensive. We all have heard "Prevention is better than cure". Its exact meaning can be delivered in real time by analyzing vast medical datasets in this big data era.

There are few major problems associated with modern healthcare, including inflated costs, irrelevant waste, and low quality. Big data could prove to offer quality care delivery and at a reasonable cost.

In big data, we talk about different Vs depicting its characteristics. Healthcare explains these Vs in their terms. A large volume of medical data, such as genome data and medical imaging data, is regularly generated in healthcare. Healthcare also creates a much different kind of information available in different formats, such as clinical information, including patients' diagnostic information, demographics, procedure, medication, lab test results, and clinical prescriptions. Various medical devices used for the improvement of a patient's health also produce many data. Healthcare generates big real-time data such as oxygen level, heart rate, body temperature, blood pressure measures, and sugar level, which depicts the velocity attributes of big data. There can be much noise or many inconsistencies present in healthcare showing missing values, errors, etc. Veracity is considered a big challenge in healthcare data analytics. Variability in the health system can be defined as the delivery of data such as EHRs, and files remain consistent every time they are accessed. It can create an immense impact on data homogeneity if the data are continuously changing. Data visualization has a different place in the medical sector. It can help find patterns or trends and compile them together in the medical history of a patient to predict future outcomes. If processed, managed, and monitored in a well-defined manner, medical data can add tremendous value to the entire healthcare system. Healthcare data's value can be translated into understanding new diseases and therapies, predicting outcomes at earlier stages, making real-time decisions, promoting patients' health, enhancing medicine, reducing cost, and improving healthcare value and quality (Asri et al. 2015).

In healthcare, an extensive amount of unstructured medical data is present in various healthcare organizations (Chatterjee 2018). This data can act as a resource for acquiring relevant critical and actionable insights for enhancing care delivery services and reducing costs. Big data applications in the healthcare industry include discovering new treatments and medicines by medical researchers, genetic and lifestyle data analyses to determine the primary root cause of a disease and its prevention. Big data analytics could revolutionize healthcare by improving hospital care quality and patient safety in intensive care units (ICUs). Big data analytics is becoming crucial in ICUs, where patients' safety and quality care play a vital role. Data analysis and visualization of various biomarkers, including blood pressure, cholesterol, and the like, could help predict the likelihood of heart disease.

## 2.3 BIG DATA IN CLINICAL DECISION SUPPORT SYSTEMS

The healthcare sector's sole aim is to prevent, diagnose, and provide treatment for any health-related problems occurring in human beings. Analyses of large-scale medical data can identify new and different underlying outlines and inclinations in the data that show the way to scientific discoveries in pathogenesis, classification, diagnosis, treatment, and disease progression (Shilo et al. 2020). Clinical health data is mainly composed of data being collected from imaging, laboratory examinations, demographic information, diagnosis, prescription drugs, treatment, progress notes, and hospitalization. Using big data analytics, we can make smarter and cost-effective decisions to effectively and efficiently deal with health data. Artificial intelligence and machine learning techniques are used to enable intelligent decision-making. The acquisition of functioning magnetic resonance imaging (fMRI) data is a complex process that generates vast volumes of data. Knowledge extraction from these data involves several steps, including preprocessing, feature reduction, and modelling, often using machine learning techniques (Chatterjee et al. 2018; Chatterjee et al. 2020).

Clinical decision support tools embedded in EHR systems hold the potential to support complex clinical reasoning, thus improving patient safety (Roosan et al. 2016). Computer-based programs are developed to make an intelligent clinical decision support system using big data analytics to handle medical history data, imaging data, omics data, and the continuity to monitor data.

## 2.4 BIG DATA IN SMART HEALTHCARE

In this digitalized world, every person is obsessed with tracking their health statistics and achieve fitness goals. Portable and wearable devices play a vital role here. A single smart device can also contribute a lot to big data. Each day a considerable amount of data is being created by a single smart device user. This combination of big data and smart devices leads to another new wave of transformation. With the increase in technology dependency, people are using many smart devices that directly or indirectly generate data in massive quantities. In healthcare, smart devices, such as fitness trackers, smartphones containing different healthcare applications, wearable devices, tablets, and the like, are responsible for data production. With an app serving many healthcare needs such as diet plans, medicine knowledge, measuring blood pressure, and counting one's steps in a day, big data is the next wave in medical science. These smart devices and wearable sensor devices contribute a colossal amount of data at a continuous pace.

Figure 2.4 describes how data are collected and go into the next state to be retrieved for understanding the dataset's underlying patterns. Health data are first collected from various smart devices. This state is known as 'data collection'. Then, collected data is further processed using different approaches, and the output is termed as 'big data'. This state of data is known as 'data processing'. In the next step, big data is analyzed by deploying various algorithms and patterns are detected within it to fulfill the required goals such as advance care quality, finding a cure for a disease, enhancing operational efficiency, and predicting whether a patient can develop a symptom by observing their past treatment history.

Many different areas of healthcare such as pharmacies, hospitals, physicians, and insurance companies are trying to search various paths to get a better understanding of big data and its applications in order to work toward the direction of cost reductions, the enhancement in services, and the like (Pramanik et al. 2017).

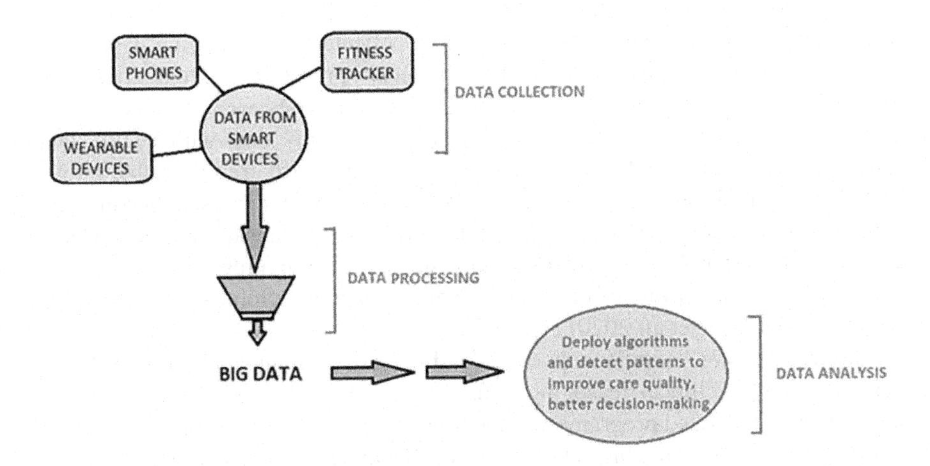


**FIGURE 2.4** Data Processing From Smart Devices.

## 2.5 COMPUTATIONAL TOOLS USED IN HEALTHCARE

Various computational tools or software are used to deal with healthcare data and perform its analysis:

a. EHR Software: It is one of the most common healthcare software used by clinics and hospitals. It collects all the data related to patients, for example, the medication a patient is taking, surgeries or any procedure a patient has undergone in his medical history, recommendations provided by a doctor, and so on.
b. Hospital management software: It helps the administrative area of a hospital perform day-to-day activities or operations.
c. Medical imaging software: In order to process MRI, computerized tomography (CT) scans, and positron-emission tomography (PET) scans primarily, visualization and medical imaging software are used.
d. Medical database software stores all the patient's medical history and treatment plans a patient has undergone. This software assists doctors in better decision-making and educates themselves by scrutinizing clinical cases of a given disease.
e. Hadoop/MapReduce: Open-source platforms such as Hadoop/MapReduce, available from the cloud, can handle big data analytics in the healthcare sector.

## 2.6 OPPORTUNITIES OF BIG DATA IN HEALTHCARE

Significant data trends in healthcare are offering many opportunities. All biomedicine areas, including genetics and genomics, health systems, population, and public health, can benefit from big data and all the other associated technologies (Martin-Sanchez et al. 2014). There can be various benefits of big data in healthcare, which includes evidence-based care, reduced cost of healthcare, increases in the participation rate of patients in the care process, improvements in public health surveillance, reductions in mortality rate, increases in communication between healthcare providers and patients, improvements in the quality of care, and early detection of security threats and fraud in healthcare (Olaronke et al. 2016). Moreover, big data analytics should also be outlined to be easy to use and learn for all.

Big data offers many opportunities in the field of healthcare. Some of them include the following:

1. Advanced quality care: Analysis performed in healthcare can help health professionals predict a patient's condition to enhance care quality. Moreover, it will help in improving all healthcare services by making optimized decisions for every individual patient.
2. Discovering treatment for diseases: A research study can be done by analyzing the required information retrieved from various sources. A cure might be found on an individual or general level.
3. Reduction in costs: Productively accessing technologies and financial resources can help crosscut the healthcare sector's cost. Also, diseases being detected at an early stage can reduce cost and times of wait.

4. Convenience for patients: Big data in healthcare can benefit patients to make the right decision promptly. If required, proactive care can be identified on time to avoid degradation in his health.
5. Advantage to researchers: Researchers and doctors can study new diseases and improve treatments by deploying new algorithms and healthcare techniques.
6. Declination in fraud: Fraud can happen from medical records and hospital visits. Proper analysis can prevent such misuses.
7. Follow-up on wearable smart devices: Big data has provided exposure for collecting and analyzing data using different phone applications or smart wearables. It can give a continuous follow-up on a patient's health condition.

## 2.7 CHALLENGES AND FUTURE SCOPE

The colossal and complex datasets offer significant challenges in the analyses. If poorly implemented, it can cause harm to a patient's health, and only a rigorous approach to perform evaluation and implementation can alleviate this risk (Sanchez-Pinto et al. 2018). Various algorithms and technologies should be compared and evaluated to implement the best big data solutions in the health system. Patient privacy and security also become an essential concern while adopting big data analytics in healthcare. Many other challenges of big data can be outlined in healthcare, such as data integrity and accuracy, should be maintained in order to resist any kind of changes, difficulty in integrating fragmented and dispersed data being stored in proprietary heterogeneous systems in different healthcare organizations, ethical challenges such as access control to patient's information, confidentiality, data privacy, the proliferation of healthcare standards, security, and privacy issues (Olaronke et al. 2016). The heterogeneous nature of data is also considered another challenge in healthcare. Data assembled from clinical notes, imaging formats, laboratory results, fitness gadgets, operational machines, and the like make it more complicated. Storing a massive amount of data is also a challenge in itself. Cloud-based storage can serve as a great help to deal with this issue. Data cleaning needs to be performed before retrieving something relevant from the dataset. It can be performed either manually or using automatized approaches. Machine learning algorithms can be used to reduce costs and time. Big data also delivers a huge risk in the replacement of medical staff. Computers and technology can diminish the number of doctors in healthcare. Apart from this, a big investment is required initially to adopt big data techniques and build a perfect healthcare system.

## 2.8 CONCLUSION

The increase in availability and scalability of healthcare data demands new and different strategies to process and manage the data beyond existing information systems' limits. These needs are still to be achieved. Big data can predict results at a very early stage in the healthcare system and optimize costs associated with it. Using the right approach, big data analytics can deal efficiently with healthcare data that are collected, analyzed, and managed. The main goal is to achieve a seamless connection between healthcare providers and patients. It will enhance the advanced quality care and facilitate early detection of diseases and security threads and frauds. Embedding these cognitive strategies in the future can help in building a robust and optimized healthcare system.

## REFERENCES


Asri, H., Mousannif, H., Al Moatassime, H., & Noel, T. (2015, June). Big data in healthcare: Challenges and opportunities. In *2015 International Conference on Cloud Technologies and Applications (CloudTech), Marrakech, Morocco* (pp. 1–7). IEEE. https://doi.org/10.1109/CloudTech.2015.7337020

Chatterjee, I. (2018). Mean deviation based identification of activated voxels from time-series fMRI data of schizophrenia patients. *F1000Research*, 7(1615). https://doi.org/10.12688/f1000research.16405.2

Chatterjee, I., Agarwal, M., Rana, B., Lakhyani, N., & Kumar, N. (2018). Bi-objective approach for computer-aided diagnosis of schizophrenia patients using fMRI data. *Multimedia Tools and Applications*, 77(20), 26991–27015. https://doi.org/10.1007/s11042-018-5901-0

Chatterjee, I., Kumar, V., Rana, B., Agarwal, M., & Kumar, N. (2020). Impact of ageing on the brain regions of the schizophrenia patients: An fMRI study using evolutionary approach. *Multimedia Tools and Applications*, 79(33), 24757–24779. https://doi.org/10.1007/s11042-020-09183-z

Chatterjee, I., Kumar, V., Sharma, S., Dhingra, D., Rana, B., Agarwal, M., & Kumar, N. (2019). Identification of brain regions associated with working memory deficit in schizophrenia. *F1000Research*, 8(124). https://doi.org/10.12688/f1000research.17731.1

Chatterjee, I., & Mittal, K. (2020). A concise study of schizophrenia and resting-state fMRI data analysis. *Qeios*, 414 (599711), 2. https://doi.org/10.32388/599711.2

Dash, S., Shakyawar, S. K., Sharma, M., & Kaushik, S. (2019). Big data in healthcare: Management, analysis and future prospects. *Journal of Big Data*, 6(1), 54. https://doi.org/10.1186/s40537-019-0217-0

Fan, J., Han, F., & Liu, H. (2014). Challenges of big data analysis. *National Science Review*, 1(2), 293–314.

Gautam, A., & Chatterjee, I. (2020). Big data and cloud computing: A critical review. *International Journal of Operations Research and Information Systems (IJORIS)*, 11(3), 19–38. https://doi.org/10.4018/IJORIS.2020070102

George, G., Haas, M. R., & Pentland, A. (2014). Big data and management. *AMJ*, 57, 321–326, https://doi.org/10.5465/amj.2014.4002.

Jacobs, A. (2009). The pathologies of big data. *Communications of the ACM*, 52(8), 36–44.

Kumar, A., & Chatterjee, I. (2016). Data mining: An experimental approach with WEKA on UCI Dataset. *International Journal of Computer Applications*, 138(13). https://doi.org/10.5120/ijca2016909050

Lee, C. H., & Yoon, H. J. (2017). Medical big data: Promise and challenges. *Kidney Research and Clinical Practice*, 36(1), 3.

Lee, I. (2017). Big data: Dimensions, evolution, impacts, and challenges. *Business Horizons*, 60(3), 293–303. http://doi.org/10.1016/j.bushor.2017.01.004

Manogaran, G., Thota, C., Lopez, D., Vijayakumar, V., Abbas, K. M., & Sundarsekar, R. (2017). Big data knowledge system in healthcare. In *Internet of Things and Big Data Technologies for Next-Generation Healthcare* (pp. 133–157). Cham: Springer. https://doi.org/10.1007/978-3-319-49736-5_7

Martin-Sanchez, F., & Verspoor, K. (2014). Big data in medicine is driving big changes. *Yearbook of Medical Informatics*, 9(1), 14.

Mohamed, W. M., Chatterjee, I., & Kamal, M. A. (2020). Integrated neural technologies: Solutions beyond tomorrow. *Neuroscience Research Notes*, 3(5), 1–3. https://doi.org/10.31117/neuroscirn.v3i5.59

Olaronke, I., & Oluwaseun, O. (2016, December). Big data in healthcare: Prospects, challenges and resolutions. In *2016 Future Technologies Conference (FTC), San Francisco, USA* (pp. 1152–1157). IEEE.

Oussous, A., Benjelloun, F. Z., Lahcen, A. A., & Belfkih, S. (2018). Big data technologies: A survey. *Journal of King Saud University-Computer and Information Sciences*, 30(4), 431–448.

Pramanik, M. I., Lau, R. Y., Demirkan, H., & Azad, M. A. K. (2017). Smart health: Big data enabled health paradigm within smart cities. *Expert Systems with Applications*, 87, 370–383.

Rajaraman, V. (2016). Big data analytics. *Resonance*, 21(8), 695–716.

Roosan, D., Samore, M., Jones, M., Livnat, Y., & Clutter, J. (2016, October). Big-data based decision-support systems to improve clinicians' cognition. In *2016 IEEE International Conference on Healthcare Informatics (ICHI), Chicago, IL, USA* (pp. 285–288). IEEE. http://doi.org/10.1109/ICHI.2016.39

Ross, M. K., Wei, W., & Ohno-Machado, L. (2014). "Big data" and the electronic health record. *Yearbook of Medical Informatics*, 9(1), 97.

Sagiroglu, S., & Sinanc, D. (2013, May). Big data: A review. In *2013 International Conference on Collaboration Technologies and Systems (CTS), San Diego, CA, USA* (pp. 42–47). IEEE. https://doi.org/10.1109/CTS.2013.6567202

Sanchez-Pinto, L. N., Luo, Y., & Churpek, M. M. (2018). Big data and data science in critical care. *Chest*, 154(5), 1239–1248.

Senthilkumar, S. A., Rai, B. K., Meshram, A. A., Gunasekaran, A., & Chandrakumarmangalam, S. (2018). Big data in healthcare management: A review of literature. *American Journal of Theoretical and Applied Business*, 4(2), 57–69. http://doi.org/10.11648/j.ajtab.20180402.14

Shilo, S., Rossman, H., & Segal, E. (2020). Axes of a revolution: Challenges and promises of big data in healthcare. *Nature Medicine*, 26(1), 29–38.

Tsai, C. W., Lai, C. F., Chao, H. C., & Vasilakos, A. V. (2015). Big data analytics: A survey. *Journal of Big Data*, 2(1), 1–32. https://doi.org/10.1186/s40537-015-0030-3

Yaqoob, I., Hashem, I. A. T., Gani, A., Mokhtar, S., Ahmed, E., Anuar, N. B., & Vasilakos, A. V. (2016). Big data: From beginning to future. *International Journal of Information Management*, 36(6), 1231–1247.

# 3 Clinical Decision Support Systems and Computational Intelligence for Healthcare Industries

*Swapnil Singh, Ameyaa Biwalkar, and Vidhi Vazirani*

## CONTENTS



## 3.1 INTRODUCTION: BACKGROUND AND DRIVING FORCES

The field of medicine has undergone a significant revolution in the last few decades, and technology has played a vital role. Doctors rely on technological devices and methods to detect and treat diseases in an efficient manner. In recent times, people have required medical aid at younger ages due to a complete change in their lifestyles. With such an increase in demand for healthcare facilities comes a need to adopt foolproof methods to detect chronic diseases at an early stage. While most of these diseases cannot be cured completely, some can be controlled if treated at the initial stages. This chapter shows how machine learning and deep learning algorithms are used for the early detection of heart diseases, pneumonia, diabetes mellitus, and brain tumors. Most models discussed are updated versions of traditional algorithms to maximize their accuracy score and overall efficiency. Since these models are used for the detection of diseases, errors are a cause of worry.

DOI: 10.1201/9781003142751-4

There are two types of errors: Type 1 error (a patient who does not have a disease tests positive) and Type 2 error (a patient who has a disease tests negative). Type 2 errors are dangerous and need to be minimized. Thus, data that are used to train these models undergo rigorous preprocessing. Before feeding a model data, the data should be cleaned to get rid of noise or outliers, missing values should be filled or removed, and any kind of imbalance in the data should be eliminated. Knowledge discovery in databases (KDD) and principal component analysis (PCA) are some of the most common preprocessing methods. The KDD process includes phases like data selection, preprocessing, transformation, classification, and finally evaluation. PCA is a method for reducing a dataset's dimensionality while retaining the information it tries to portray.

## 3.2 MACHINE LEARNING

Machine learning is a way of capturing data and turning it into useful information. It uses computational algorithms to learn information from data. Algorithms adaptively improve their performance as the number of samples available for learning increases. Machine learning algorithms find natural patterns in data that generate insights and help make better results. Machine learning is subdivided into three subcategories of supervised, unsupervised, and semi-supervised learning. Supervised learning is further divided into regression and classification. Classification is used to predict what class something belongs to. Classification is applied in the health sector to classify and predict diseases. Machine learning can detect anomalies in various diseases like heart diseases, diabetes, pneumonia, brain tumors, and cancer. Various attributes are taken into consideration while predicting these diseases. For example, in heart diseases, attributes such as age, blood pressure, and electrocardiogram (ECG) data values are collected. In the case of diabetes, we consider the insulin levels and the body mass index of an individual. The previously stated are specific examples of the datasets taken into account for prediction. Machine learning has revolutionized the diagnosis of diseases. One can get the results almost instantly, thus reducing the diagnosis time, thereby providing better treatment, hence better chances of surviving in cases of deadly diseases like the coronavirus and Zika virus. In this chapter, the prediction of these diseases at an early stage using trivial machine learning algorithms is discussed.

## 3.3 DEEP LEARNING

Deep learning (DL) is a subsection of machine learning that deals with artificial neural networks (ANNs). Neural networks mimic the structure and function of our nervous system. It resembles a human brain in two ways. It acquires knowledge through learning and stores this within the interconnections. ANNs are built from neurons. Neurons consist of input, node, weight, and output. ANNs are of six major types: modular neural networks, feed-forward neural networks, convolutional neural networks, radial basis function neural networks, recurrent neural networks, and Kohonen self-organizing neural networks. For medical diagnosis, convolutional neural networks and feed-forward neural networks are most commonly used. For detecting anomalies in images for medical diagnosis, convolutional neural networks are used. They filter and transform images so that they can be used for training and testing the neural network. Some medical application areas concerning DL techniques are discussed ahead in the chapter.

We discuss more DL applications in the medical field later in the chapter.

## 3.4 APPLICATION OF DL TECHNIQUES ON HEART DISEASE DATA SETS

According to the World Health Organization (WHO) statistics, around 17.9 million people die every year due to heart diseases accounting for 31% of all the world's deaths (Mattingly, 2020). Therefore, there must be early and accurate detection of heart diseases. According to statistics, four out of five patients suffering from heart diseases die due to a heart attack or a stroke, and one-third of the deaths are prematurely in patients younger than 70. Patients with a high risk of heart diseases demonstrate high blood pressure, glucose, and lipid, along with overweight and obesity. Lifestyle and eating habits play an essential factor, which could lead to heart diseases. The identification of people with such a lifestyle and displaying the earlier-stated symptoms, after that providing them with appropriate treatment, could prevent premature deaths.

The advent of technology in the field of medicine has revolutionized medical diagnosis. These technologies can be used to detect and cure heart diseases. Blood tests, two-dimensional echoes, ECGs, cardiac computerized tomography (CT) scans, cardiac magnetic resonance imaging (MRI) Holter monitoring, stress tests, and cardiac catheterization are some of the top ways to detect heart diseases. Angioplasty, that is, placing a pacemaker in the patient's heart, is the most common way to cure the most prevalent heart diseases. Regular monitoring of the enzyme levels in the blood, blood pressure, serum cholesterol, and blood glucose can lead to the early detection of heart diseases.

A. Sankari Karthiga et al. (2017) compared decision trees (DTs) and naive Bayes (NB) algorithms for the early prediction of heart-related diseases. After performing basic preprocessing on the data, the data was fed to both these models, and the model was evaluated using accuracy, specificity, and sensitivity. The DT gave a better overall performance. The DT model's specificity and sensitivity were 97.79 and 95.45, respectively, whereas for the NB algorithm, it was 70.94 and 85.53, respectively. Thus, it was concluded that the DT is more preferred for the University of California, Irvine (UCI) dataset. This model's future scope is to include text mining to process big and unstructured data.

Seyedamin Pouriyeh et al. (2017) investigated and compared different data mining techniques. The authors have applied ensemble techniques for the prediction of heart diseases. Ensemble in machine learning means combining machine learning algorithms and then taking the majority vote of these algorithms combined as output. For example, if we combine three algorithms for detecting an object, say, of class A. If two of these algorithms say that it is of class A whereas the third algorithm says that it belongs to class B, we consider the maximum vote and say that the object is from class A. This is also called max-voting. Ensemble not only increases the accuracy of the model but also decreases variance (Singh, 2018). The authors applied these techniques to the Cleveland dataset for heart diseases, which had 303 instances. They employed a 10-fold cross fold technique as they had a small dataset and then applied algorithms such as single conjunctive rule learner (SCRL), multilayer perceptron (MLP), radial basis function (RBF), NB, support vector machine (SVM), K-nearest neighbor (KNN), and DT. They also applied techniques such as bagging, boosting, and stacking on the dataset for ensemble predictions. KNN was applied with four values of K, namely, 1, 3, 9, and 15. K in KNN is the number of nearest neighbors to be considered in most voting processes. In general practice, K's value is chosen to be the square root of N, where N is the number of data points in the training dataset. They achieved the highest accuracy when the value of K was 9. DT had an accuracy of 77.55% accuracy, NB had an accuracy of 83.49%, KNN had an accuracy of 83.16%, MLP had an accuracy of 82.83%, RBF had an accuracy of 83.82%, SCRL had an accuracy of 69.96%, and accuracy of SVM was achieved as 84.16%. The greatest accuracy was given by SVM while RBF, NB, and KNN are in the next places. The next part of their implementation was using ensemble methods of bagging, boosting, and stacking. When they used bagging, DT improved from 77.55% to 78.54%, and more significantly, SCRL increased to 80.52% from the initial 69.96%. For SVM, the accuracy remained unchanged. Upon applying boosting, the DT's accuracy increased from 77.55% to 82.17% and of the SCRL from 69.96% to 81.18%. The accuracy of other algorithms remained approximately the same. Stacking was done by the combination of

- SVM, NB, RBF, KNN, MLP;
- SCRL, NB, SVM, MLP, RBF;
- SVM, NB, MLP, RBF; and
- MLP, SVM, DT, MP, SVM,

achieving an accuracy of 78.54%, 82.17%, 82.17%, 83.16%, and 84.15% respectively. As it was observed, the combination of SVM and MLP gave the highest accuracy of 84.15%. They concluded that SVM outperformed all algorithms, even the ensemble algorithm.

Amanda H. Gonsalves et al. (2019) proposed the use of J48, NB, and SVM to predict coronary heart disease (CHD). The South African Heart Disease Dataset was extracted from KEEL (Knowledge Extraction from Evolutionary Learning) to carry out the research. The dataset has 462 instances; however, there is an imbalance in the classification. Three hundred two individuals do not suffer from CHD, and only 160 patients suffer from CHD. Data were analyzed using WEKA. For enhancing and boosting the model's performance, cross-validation was used. The number of folds was kept at ten, and it remained constant for all models. For evaluating the performance, measures like accuracy, error rate, specificity, and sensitivity were calculated. Although all three models had an accuracy greater than 70%, NB performed the best, with comparatively higher accuracy measures. J48 and SVM gave lower sensitivity while NB gave a lower specificity. Nonetheless, it was concluded that NB works best for the given data.

Mythili T., Dev Mukherji, Nikita Padalia, and Abhiram Naidu (2013) proposed a rule-based model to compare logistic regression and SVM's accuracy on the Cleveland Heart Disease Dataset to find the most accurate model of the two for predicting heart diseases. The dataset consisted of 13 attributes and 303 data points. In the preprocessing, they removed the data points with missing attributes. The cleaned dataset was then applied to logistic regression, DT, and SVM. The node-splitting criterion for the DT was the Gini coefficient. Gini coefficient is the most commonly used splitting criterion that works on the population diversity of attributes. SVM was trained over 10-fold cross-validation. DT is a rule-based algorithm; rule-based algorithms use a small number of attributes to make decisions. Various rules can be used depending on the attributes of the dataset. The rules are as follows similarity or prototype-based rules (P-rules), classical propositional logic (C-rules), threshold rules (T-rules), association rules (A-rules), M-of-N, and fuzzy logic (F-rules). The authors recommend the use of C-rules in this model. C-rules are in if–else ladder form, providing a comprehensible and straightforward way of expressing knowledge. They provide high specificity and sensitivity, but the accuracy achieved would be on the lower side. After training the proposed models, the sensitivity, specificity, and accuracy were compared, and the most weighed model was found. Their future work is to develop a tool for predicting the risk of disease of a prospective patient. Furthermore, they plan to work on neural networks and ensemble algorithms. Ensemble algorithms like bagging, boosting, and stacking provide better accuracy than trivial machine learning algorithms. While neural networks are better learners and resemble the human neural network, they provide better accuracy than trivial machine learning algorithms. Thus, their future work may provide better performance.

Dr. K. Usha Rani (2011) analyzed a heart disease dataset using the neural network approach. A neural network is made up of neurons. Neurons are the building blocks of neural networks. A neuron is a function that acts on the input with a bias and weight. The neuron's output is the summation of the bias and the multiplication of the input and weight. In a neural network, we work on optimizing the weight and the bias. The weights assigned to the inputs make neural networks robust. The same Cleveland dataset was used in this experiment. Since there were 13 features in the dataset, the input layer of the dataset had 13 neurons. The output layer had two neurons. The neural networks were made with and without hidden layers. There were various combinations of testing and training samples used over the neural network. For the training sample, being 100 and testing samples being 300, the accuracy for single layer was 76%, and that of multilayer was 82%; similarly, when the training sample was 150 and the test samples were 200, the accuracy of the single layer was 76% and for multilayer was 83%. For the training samples being 250 and the testing samples being 150, the accuracy achieved was 86.2% and 89.3% for single- and multilayer neural networks, and when the training samples were 350 and when the test samples were 100, the accuracy achieved was 90.6% for a single-layer neural network and 94% accuracy for multilayer neurons. Parallelism is implemented at each neuron in all hidden and output layers to speed up the learning process.

C. Sowmiya and Dr. P. Sumitra (2018) evaluated the potential of nine classification techniques, namely, Bayesian neural network, SVM, ANN, KNN, and the algorithm proposed by the authors of Apriori algorithm and SVM. Apriori algorithm is a frequent item mining algorithm. The frequent item sets help us find the association rules. Association rules help us determine the general patterns in the dataset. According to the author's algorithm, they first read the data, removed noise from the data, and then peeked the data. These data were then conditionally read and checked the data, the values were calculated, and then the time domain, frequency domain, and the nonlinear results were analyzed. The future work was to experiment with another algorithm for the Apriori algorithm.

Deepali Chandna (2014) applied data mining techniques to extract the hidden patterns, important for heart diseases, from the dataset gathered by the International Cardiovascular Hospital. The proposed study portrayed a system using adaptive neuro-fuzzy inference systems and information gain, for diagnosis of heart disease. Using entropy, we can find information gain. It is a ranking measure for attributes, being used in the DT (ID3 algorithm) to decide the splitting attribute. The ensemble of fuzzy inference system and neural networks gives us the adaptive neuro-fuzzy inference system. This study joined information gain with adaptive neuro-fuzzy systems. Information gain was used to select quality attributes, and these attributes were then given as input to the adaptive neuro-fuzzy inference system. Adaptive neuro-fuzzy inference systems then train and test using these attributes. The proposed system was evaluated by finding the error rate and accuracy. The average number of misclassified data points divided by the number of data points is called the error rate, whereas one minus the error rate is called the accuracy. The model proposed by the authors was done using two popular machine learning tools, namely, WEKA and MATLAB. It was found that the accuracy was 98.24%, which was higher than that of existing models. This study's future work is to focus on classification speed and computational cost and broaden the scope to other diseases.

Senthilkumar Mohan, Chandrasegar Thirumalai, and Gautam Srivastava (2019) proposed a novel method of applying machine learning techniques using significant features to improve the accuracy of predicting cardiovascular diseases. Various known models and many combinations of features were introduced to the prediction model. With the combination of mixed random forest (RF) and linear models (HRFLM), they enhanced the model accuracy to 88.7%. The dataset used in this study was the UCI heart disease dataset. In HRFLM, a computational approach of three computational rules, namely, Apriori, predictive, and Tertius, is used to find heart disease factors on the UCI dataset. HRFLM makes use of an ANN with 13 clinical features as its input. The obtained results were analyzed against trivial machine learning algorithms, namely, KNN, DT, support network machine, and neural network. The dataset was first preprocessed, followed by a feature selection using the DT entropy, followed by modelling. The feature selection and modelling process are repeated a various number of times for various combinations. During the modelling phase, the best model was selected based on the low error rate. These models were applied to clusters, which were made based on DT features and variables. The model applied were DT, language model, SVM, RF, NB, neural network, and KNN. After that, the HRFLM model was applied to these results. By combining the characteristics of RF and the linear method, HRFLM gave the best results.

Shamsher Bahadur Patel, Pramod Kumar Yadav, and Dr. D. P. Shukla (2013) applied three classifiers—classification by clustering, NB, and DT—on the UCI database for heart diseases. Classification via clustering is classifying clusters. To avoid high-value attributes from dominating low-value attributes, all attributes were normalized, before applying clustering. Out of the 14 attributes, only six were selected for the classification phase using feature selection. The attributes were maximum heart rate achieved, resting blood pressure, exercise-induced angina, old peak, number of significant vessels colored, and chest pain type. The previously stated attributes were used to train the earlier-stated algorithms. These models were analyzed over the accuracy, model construction time, and mean absolute error. DT gave the highest accuracy of 99.2% and a low mean absolute error of 0.00016, although it took the highest model construction time of 0.09 s. The future work of the study included using fuzzy learning models to evaluate the intensity of cardiac diseases.

Refer to Table 3.1 for a comparison of various models as stated in the previous study and the accuracy of the models are visualized in Figure 3.1.

## 3.5 APPLICATION OF MACHINE LEARNING TECHNIQUES ON PNEUMONIA DATA SETS

Pneumonia is a form of acute respiratory lung infection, and there is alveolus in the lungs, which are small sacs; when one breathes, they fill up with air. When someone is infected by pneumonia, these sacs fill up with fluid and pus, hindering oxygen intake and making breathing painful. There are three types of pneumonia, viral, bacterial, and fungal pneumonia. Patients with viral and bacterial pneumonia have similar symptoms, although viral pneumonia has more symptoms than bacterial pneumonia. Difficulty breathing and/or coughing, with or without fever, are the symptoms shown by children younger than 5. To diagnose pneumonia, one must observe the following symptoms: lower chest all indrawing, the chest retracting inward while breathing, or the presence of fast breathing. One of the common symptoms of viral pneumonia is sneezing. Unconsciousness, convulsions, and hyperthermia can be experienced in severely ill infants; they may be unable to eat or drink.

Pneumonia is the single largest cause of death in children worldwide. According to the WHO, pneumonia killed 8,08,694 children younger than 5 in 2017, accounting for 15% of children's deaths younger than 5 (*Pneumonia*, 2019). Pneumonia is a common disease in sub-Saharan Africa and South Asia. It can be cured and prevented with low-cost, low-tech medication and care. Transmission of the disease occurs in multiple ways. It can be airborne through the droplets released during sneezing and coughing; it also spreads through blood, especially after birth in infants. Children have these viruses and bacteria in their trachea, which, when inhaled, could lead to pneumonia. Parental smoking, in-house pollution, and living in a crowded house are the leading environmental reasons that cause pneumonia. Pneumonia can be cured easily if there is efficient and early detection. There are more than 10 million cases each year alone in India; considering the large numbers, there is a need for a faster and efficient way to detect pneumonia. Pneumonia has also been one of the early symptoms of numerous pandemics; thus, efficient and faster detection would help in defeating the pandemic.

Xin Li, Fan Chen, Haijiang Hao, and Mengting Li (2020) proposed an improvised CNN (convolutional neural network) model for detecting pneumonia using

**TABLE 3.1**
**F1 Score and Summary for Heart Disease Prediction**

| Paper | Dataset | Decision Tree | Naive Bayes | Multilayer Perceptron | Single Layer Perceptron | K-Nearest Neighbour | Single Conjunctive Rule Learner | Radial Basis Function | Logistic Regression | Support Vector Machine | Apriori | Classification Using Clustering |
|---|---|---|---|---|---|---|---|---|---|---|---|---|
| Karthiga et al. (2017) | UCI | ✓ | ✓ | | | | | | | | | |
| (Pouriyeh et al. (2017) | Cleveland | 0.801 | 0.851 | 0.824 | | 0.824 | 0.718 | 0.853 | | 0.860 | | |
| Gonsalves et al. (2019) | South African Heart Disease Dataset | ✓ | ✓ | | | | | | | ✓ | | |
| T. et al. (2013) | Cleveland | ✓ | | | | | | | ✓ | ✓ | | |
| Rani (2011) | Cleveland | | | ✓ | ✓ | | | | | | | |
| Sowmiya and Sumitra (2018) | Cleveland | | | ✓ | | ✓ | | | | ✓ | ✓ | |
| Chandna (2014) | Cleveland | | | | | ✓ | | | | | | |
| Mohan et al. (2019) | Cleveland | 0.918 | 0.845 | 0.926 | | | | | 0.902 | 0.925 | | |
| Bahadur et al. (2013) | Cleveland | ✓ | ✓ | | | | | | | | | ✓ |

*Source:* Based on Bahadur et al. (2013), Chandna (2014), Gonsalves et al. (2019), Karthiga et al. (2017), Mohan et al. (2019), Pouriyeh et al. (2017), Rani (2011), Sowmiya and Sumitra (2018); T. et al. (2013).

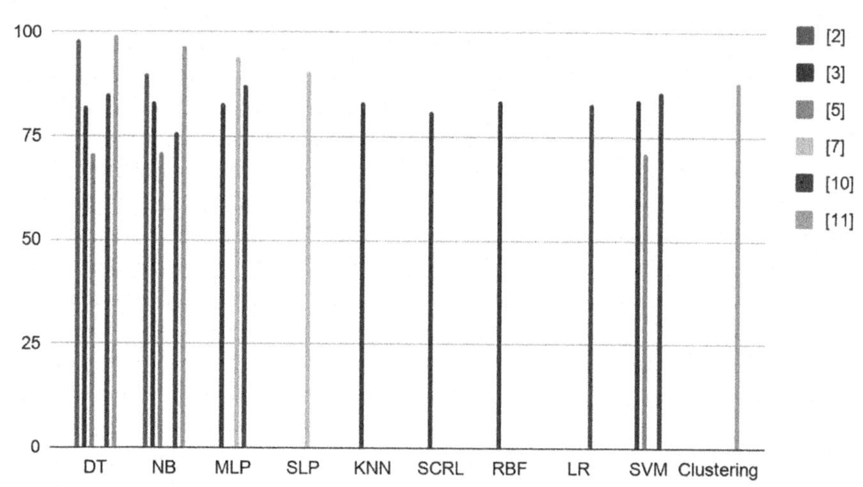


**FIGURE 3.1** Accuracy Comparison—Heart Diseases.

*Source:* Based on Bahadur et al. (2013), Chandna (2014), Gonsalves et al. (2019), Karthiga et al. (2017), Mohan et al. (2019); Pouriyeh et al. (2017), Rani (2011), Sowmiya and Sumitra (2018), and T. et al. (2013).

DLmodels. On the Lenet-5 model, they added pooling layers and convolutional layers; for an optimal model for better classification, they added a feature integration layer. Their proposed method uses fewer features than traditional CNN, hence avoiding complex features. Their first dataset was first used in 2018 by Daniel Kermany et al. at the University of California, San Diego. This dataset had two categories that were pneumonia and normal. There were 1,575 normal and 4,265 images. As the dataset was small, they augmented the dataset and increased the size to eight times its original size by changing brightness, contrast, sharpness, and Gaussian blur. Finally, the dataset was divided into a 7:3 ratio for training and testing. The other dataset taken into account for the study was provided in the 2018 medical image pneumonia recognition contest, which was sponsored by the Radiology Society of North America (RSNA). The dataset consisted of 4,300 pneumonia samples and 12,500 normal samples. The images are of uniform size, that is, 64 × 64. They fused the two datasets and got 14,075 normal images and 8,565 pneumonia images. The fused dataset was also divided into 7:3 train–test ratios. They applied the existing convolutional models, namely, transfer learning, Google InceptionV3 + Data Augmentation (GIV + DA), and Google Inception V3 + Random Forest (GIV3 + RF) along with the improved Lenet. After a comparison of accuracy, it was proved that their improved Lenet performed the best. The training accuracy of the improved Lenet on dataset 1 was 98.83% and that on dataset 2 was 98.44%, while the testing accuracy was 97.26% on dataset 1 and 91.41% on dataset 2. Their future work includes detecting the type of pneumonia and using a convolutional layer to segment the image to localize the lesion. Upon the application of their future work, the detection time for pneumonia would decrease significantly, leading to a decrease in the mortality rate.

Okeke Stephen, Mangal Sain, Uchenna Joseph Maduh, and Do-Un Jeong (2019) proposed a custom convolutional neural network for the classification and detection of pneumonia using a collection of chest X-ray image samples. This study used a custom convolutional neural network for extracting features of the chest X-ray images and classifying them to determine if the patient had pneumonia or not. As the dataset was limited, they applied data augmentation to increase the dataset. Data augmentation is a way where the dataset undergoes specific changes such that various variations of the images are formed, for example, rotation, increasing or decreasing brightness. The dataset had 5,856 anterior–posterior chest X-ray images chosen from pediatric patients younger than 5. For validating, 2,134 images were used, and 3,722 images were used for training. For augmenting the data, they used a rotation of the images to 40 degrees, width shifts by 0.2%, height shifts by 0.2%, shear range of 0.2% clips the image angles in a counterclockwise direction, zooming images with the ratio of 0.2%, and finally flipping the image horizontally. The convolutional neural network built had four feature extraction layers, 32, 64, 128, and 128 neurons. There are 2 × 2 max-pooling layers and Rectified Linear Unit (ReLU) activators between these layers. The convolutional layers' output is passed through a flattened layer to convert it into a one-dimensional feature vector. A sigmoid activation function was used for classification; there were two dense layers with 1 and 512 neurons, respectively, and a dropout layer with 50% dropout, and the ReLU used the activation function for layers other than output layer. The results obtained were a training loss of 0.1288, a training accuracy of 0.9531, a validation loss was 0.1835, and a validation accuracy was 0.9373 when the data size was kept as 200. The future work includes classifying X-ray images into lung cancer and pneumonia, therefore broadening the scope of their model.

Lin Li et al. (2020) developed an automatic framework for detecting COVID-19 with the help of chest X-ray images, and then they evaluated its performance. The dataset used for the study contained other non-pneumonia, COVID-19-positive, and community-acquired pneumonia CT scans. The dataset was gathered from six hospitals and consists of 4,356 CT exams of 3,322 patients. The COVNet model was developed by the authors of this paper using a three-dimensional DL framework. The COVNet framework uses RestNet50 as its backbone. It is first preprocessed for a three-dimensional CT exam, and the lung is extracted as the region of interest using U-Net-based segmentation. This image is then sent to the COVNet for classification. The training accuracy of the model was 90%. The area under the curve was found to be 0.95.

Khalid El Asnaoui, Youness Chawki, and Ali Idri (2020) presented a study for automatic binary classification of various pneumonia images using trained versions of Xception, VGG16, MobileNet_v2, VGG19, Resnet50, DenseNet201, Inception_V3, and Inception_ResNet_V2. They used an X-ray and CT dataset containing 5,856 images, of which 4,273 images had pneumonia and 1,583 images were normal. As the dataset was limited, they augmented the dataset by rescaling the images by dividing the images by 255, rotating the image 90 degrees, and horizontally and vertically shifting the range to 0.2; the shear and zoom range was 0.2; and there were horizontal flips. The batch size was taken as 32, and the training was done over 300 epochs. The validation and training samples were, respectively, set to 109 and 159. The learning rate was reduced to 0.000001 from the initial 0.00001, and Adam with beta 1 as 0.9 and beta 2 as 0.999 was chosen as the optimization. The models were evaluated using F1 score, accuracy, precision, sensitivity, and septicity. Inception_Resnet_V2, MobileNet_V2, and Resnet50 gave high accuracies compared with other models. The future work includes feature extraction using U-Net and You-Only-Look-Once for segmentation. Pinpointing the infected region and finding out the extent of the infection would be done with the help of segmentation; thus, the appropriate treatment could be provided to the patient.

Pranav Rajpurkar et al. (2017) developed an algorithm that can detect pneumonia from chest X-rays at a level exceeding practicing radiologists. Their algorithm, CheXNet, is a 121-layer convolutional layer neural network trained on chest X-ray 14, the largest publicly available chest X-ray dataset, containing more than 100,000 frontal-view X-ray images with 14 diseases. The model's test performance was compared with the help of the chest X-ray dataset annotated by four working radiologists. CheXNet is a DenseNet. DenseNet improves the flow of information and gradients through the networks, optimizing very deep networks tractable. The final layer was replaced with a single neuron layer to provide a single output with the sigmoid activation function. The model was trained using the batch size of 16, and the initial learning rate was taken to be 0.001, which decayed by a factor of 10 each time the validation loss plateaued after an epoch and picked the model with the lowest validation loss. The average F1 score of the radiologists was found to be 0.387, whereas the F1 score of CheXNet was better, proving it to be better than practicing radiologists. The limitation to the model is that only frontal X-rays were used for training, 15% of the accurate predictions required lateral X-rays, and neither the model nor the radiologists were provided with the patient history. The future scope includes extending the model to detect the other diseases in the dataset used for this experiment.

Xianghong Gu, Liyan Pan, Huiying Liang, and Ran Yang (2018) proposed a system to detect viral and bacterial pneumonia. The model was trained over the open JSRT (Japanese Society of Radiological Technology) database consisting of 241 images and the MC (Montgomery County, Maryland), constituting 138 images. The left and right lung images were segmented using a fully convolutional neural network. These images were then given as input to a convolutional neural network with many layers. Features of the target lung regions were extracted on the basis of the convolutional neural network model; with the help of manual features, the performance of the model was evaluated. Finally, the manual and extracted features are merged and then inputted to SVM for binary classification. The proposed model was then evaluated over a dataset containing 4,513 chest x-rays of pediatric patients, collected from Guangzhou Women and Children's Medical Center, China. The performance was evaluated by calculating the area under the curve, accuracy, specificity, precision, and sensitivity. The experimental results show the precision of 0.8886, the sensitivity as 0.7755, specificity as 0.9267, and the area under the curve as 0.8234. The future work includes tuning the deep convolutional neural network with chest radiography and feature extraction and studying more universal features in the future.

Tawsifur Rahman et al. (2020) aims to provide three types of comparisons, namely, pneumonia versus normal, viral versus bacterial pneumonia, and bacterial, normal, and viral pneumonia, by using some already trained (convolutional neural network), namely, SqueezeNet, AlexNet, DenseNet, and ResNet18, for transfer learning. The dataset consisted of 5,347 chest X-ray images. These images were preprocessed and trained for transfer learning-based classification tasks. During preprocessing, the images were augmented to increase the size of the dataset. The classification accuracy of pneumonia versus normal was 0.98, the accuracy of bacterial, normal, and viral pneumonia was 0.933, and the accuracy for viral and bacterial pneumonia classification was 0.95; all these accuracies were provided by DenseNet201. DenseNet201 exhibited a marvelous performance by effectively training over a small collection of complex data-like images; it also reduced bias and had a higher generalization.

Naseem Ansari, Ahmed Rimaz Faizabadi, S. M. A. Motakabber, and Muhammad Ibn Ibrahimy (2020) used ResNet50 along with transferred learning on two different datasets to detect pneumonia. The datasets used for the study were RSNA dataset from the Radiological Society from North America consisting of 26,684 samples consisting of three classes, which are lung opacity (31% of the dataset), no lung opacity/not normal (40% of the dataset), and normal (29% of the dataset). The other dataset was the CHEST X-Ray Image Dataset consisting of 5,856 pediatric chest X-ray images; this dataset was gathered from the Guangzhou Women and Children Hospital and Children's Medical Centre Guangzhou, China. The dataset had two classes, which were normal and pneumonia. They performed preprocessing on the dataset to add, remove, and transform attributes. For the training phase, a pretrained ResNet 50 model was used. This model had been trained over the ImageNet dataset containing 1,000 classes. The model was tuned before training it with the two datasets. For the CXI dataset, the optimizer was stochastic gradient descent, loss function as binary cross-entropy, learning rate as 0.01, trained over four epochs, with a batch size of 16.

Similarly, the parameters were tuned before training the model with the RSNA dataset, the only difference was that the optimizer used was Adam, and the batch size was increased to 20. The best accuracy was found for the train test split being in the 80:20 ratio. The future work of this paper includes the use of deeper neural networks like ResNet-101 and ResNet-152. They also aim to classify the CXR dataset for lung cancer and pneumonia.

Yan Han et al. (2021) experimented with the RSNA dataset. They performed the experiment using ResNet, ResNetAtt, ResNetRadi, and ResNetAttRadi. The models were trained for 200 epochs and with a batch size of 64. The optimizer was set to stochastic gradient descent, and the initial learning rate was set to 0.1. They used accuracy, F1 score, and area under the curve (AUC) as their evaluation parameters. ResNetAtt and ResNetAttRadi gave better regions of interest.

Refer to Table 3.2 for a comparison of various models as stated in the previously mentioned study and the accuracy of the models are visualized in Figure. 3.2.

**TABLE 3.2**
**F1 Score and Summary for Pneumonia Detection**

| Paper | Dataset | Transfer Learning | GIV + DA | GIV3 + RF | Improved Lenet | Custom Convolutional Neural Network | Res Net50 | VG G16 | VG G19 | Dense Net201 | Inception_ ResNet_V2 | Incept- ion_V3 | Mobile Net_V2 | Xcept- ion | Res Net18 | Alex Net | Squeeze Net |
|---|---|---|---|---|---|---|---|---|---|---|---|---|---|---|---|---|---|
| X. Li et al. (2020) | RSNA, chest X-ray images | ✓ | ✓ | ✓ | ✓ | | | | | | | | | | | | |
| Stephen et al. (2019) | CXI | | | | | ✓ | | | | | | | | | | | |
| L. Li et al. (2020) | Custom | | | | | | ✓ | | | | | | | | | | |
| El Asnaoui et al. (2020) | CXI | | | | | 0.857 | 0.967 | 0.865 | 0.851 | 0.939 | 0.962 | 0.945 | 0.963 | 0.850 | | | |
| Rajpurkar et al. (2017) | ChestX-ray14 | | | | | 0.435 | | | | | | | | | | | |
| Gu et al. (2018) | JSRT | | | | | ✓ | | | | | | | | | | | |
| Rahman et al. (2020) | CXI | | | | | | | | | 0.935 | | | | | 0.909 | 0.885 | 0.865 |
| Ansari et al. (2020) | CXI, RSNA | | | | | | ✓ | | | | | | | | | | |
| Han et al. (2021) | RSNA | | | | | | 0.795 | | | | | | | | | | |

*Source:* Based on Ansari et al. (2020), El Asnaoui et al. (2020), Gu et al. (2018), Han et al. (2021), L. Li et al. (2020), X. Li et al. (2020), Rahman et al. (2020), Rajpurkar et al. (2017), and Stephen et al. (2019).

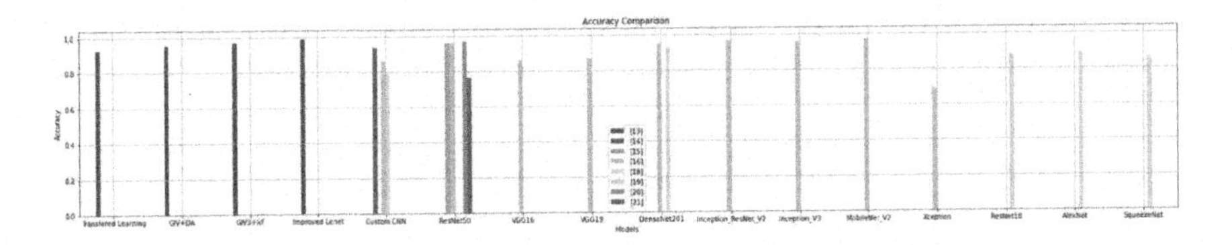

**FIGURE 3.2** Accuracy Comparison—Pneumonia.

*Source:* Based on Ansari et al. (2020), El Asnaoui et al. (2020), Gu et al. (2018), Han et al. (2021), L. Li et al. (2020), X. Li et al. (2020), Rahman et al. (2020), Rajpurkar et al. (2017), and Stephen et al. (2019).

## 3.6 APPLICATION OF MACHINE LEARNING TECHNIQUES ON DIABETES DATA SETS

Diabetes is a chronic disease caused by high levels of sugar in the blood. There are three types of diabetes: type 1, type 2, and gestation. Type 1 diabetes is caused due to absence or lower amounts of insulin production in the body. This type usually affects young children. Type 2 diabetes, usually found in older persons, is caused due to lack of insulin in the body. The major cause of this type is obesity. Typically found in older people, this type of diabetes is found in the younger generation as well, mainly due to a change in lifestyle habits. The last type, gestational diabetes, is observed in pregnant ladies and is usually cured naturally after the pregnancy. However, such women are at a higher risk of acquiring type 2 diabetes at later stages in life.

Diabetes mellitus or type 2 diabetes increases the risk of heart-related diseases by two to four times. The number of people that get affected by this disease is increasing each year exponentially. Diabetes cannot be completely cured but can be controlled if detected at an early stage. Thus, there is a demand for an efficient system that helps doctors accurately detect diabetes at an early stage by looking at a few medical records. Machine Learning algorithms play a key role here. To detect diabetes, we use the supervised learning technique of classification.

Aishwarya Jakka and Vakula Rani J (2019) wrote a paper that compared the various machine learning algorithms. The dataset selected for this study is the PIMA Indians Diabetes Dataset of the National Institute of Diabetes and Digestive and Kidney Diseases. This dataset contains the data of female diabetic patients. The data include the medical parameters of the patients. The clean dataset was fed to different models, and the results were compared. The models compared were KNN, DT, NB, SVM, logistic regression, and RFs. While logistic regression gave the highest accuracy of about 77.6%, the SVM was the least accurate, with a measure of 65.62%. Other evaluation parameters, like precision, recall, F1 score, misclassification rate, and receiver operating characteristics (ROC)–AUC score, were also calculated. The logistic regression model was found as the most efficient one after a complete evaluation was performed. To understand the comparisons in a more effective way, refer to Table 3.3.

Veena Vijayan V. and Anjali C. (2016) suggest the use of Adaboost to enhance the accuracy of these traditional models. Adaboost increases the efficiency of a model by assigning weights to the samples at each stage. Initially, all samples have the same

**TABLE 3.3**
**Comparing Different ML Algorithms and Their Performance on the PIMA Indians Diabetes Dataset**

| Classifier | ROC–AUC Score |
|---|---|
| KNN | 69.8 |
| DT | 69.2 |
| NB | 70.2 |
| SVM | 60.5 |
| LR | 73.6 |
| RF | 70.1 |

*Source:* Jakka and J (2019).

**TABLE 3.4**
**AdaBoost**

| Algorithm | Accuracy before AdaBoost | Accuracy after AdaBoost |
|---|---|---|
| Decision Tree | 76.000 | 77.600 |
| Naive Bayes | 78.100 | 79.687 |
| Support Vector Machines | 79.680 | 79.687 |
| Decision Stump | 74.470 | 80.720 |

*Source:* Vijayan and Anjali (2016)

weight (usually 1/N, where N is the number of samples). After the training of the model is complete, the weight of each sample is updated. If the sample is correctly classified, its weight decreases, and in case the sample is incorrectly classified, the weight would increase. The new weights are calculated using a series of formulae:

Calculate the error, € = Count of incorrectly classified sample/count of total samples
If € > 0.5,

Weight for base learner, αt = (½)*ln(1 – €t)/ €t. (3.1)
If classification was wrong, Dnew = Dold/zj*et. (3.2)
If classification was right, Dnew = Dold/zj*e – t. (3.3)

(zj is the normalization factor.)

This AdaBoost algorithm was applied to base classifiers like DTs, NB, SVMs, and decision stumps. A comparison of the accuracy measures before and after applying the AdaBoost algorithm is shown in Table 3.4. The authors used other evaluation methods like error rate, sensitivity, and specificity as well. While SVM did not show any significant changes, decision stumps showed improvement after AdaBoost was used.

Uswa Ali Zia and Dr. Naeem Khan (2019) focused on increasing generic classifier models' accuracy by using resampling techniques on the benchmark dataset. The data were acquired for the PIMA dataset and were preprocessed using principal component analysis (PCA). The resampling technique is applied to the preprocessed dataset. A supervised resampling filter generates a subsample of the dataset with or without replacement. While doing so, there is an assurance that the class distribution is even promising the best results. Resample filter is a must when the entire dataset does not have an even distribution of the output class. This technique is fast and efficient. Uswa Ali Zia and Dr. Naeem Khan chose the bootstrap method of resampling, which obtains a random sample with replacement. The sample generated is then fed to the classifier models, namely, DT (J48 and J48graft), NB, and KNN (k = 1 and k = 3). While both the DT models gave the highest accuracy of 94.44%, NB gave the lowest accuracy of 74.89%. The KNN gave an accuracy of 93.79% when the value of k was 1 and fell to 76.79% when the value of k was set to 3. The main focus of this study was the bootstrap dataset. Instead of testing these data on traditional models, it can be used in combination with upgraded models to boost accuracy further.

Minyechil Alehegn and Rahul Joshi (2017) proposed a hybrid model to improve individual classifier models' performance. After processing the dataset to retrieve acceptable data, the training set is produced. The data are then fed to various classifier models like RF, KNN, NB, and DT (J48). After the models are trained, new data are introduced to these models. Once the prediction of all these models is produced, these individual classifiers' results are then combined using assemble methods. This is known as the meta-classifier technique (stacking and voting). The results of these models are combined to yield one output. This is done by taking the majority class in the case of classification. Thus, if most of the models yield the result that the patient has diabetes, the output is 'Diabetic'; otherwise, the output is 'Not Diabetic.' The authors also confirm that this classifier method yields better results as compared to individual classifiers.

Aishwarya, Gayatri, and Jaisankar (2013) proposed using a modified algorithm of the SVM to help detect diabetes. They started preprocessing their dataset—the PIMA Indians Dataset—using PCA. The dataset has 769 records and eight attributes. PCA was used to maximize the effect of preprocessing with minimum loss in data. The SVM algorithm is efficient when the classification happens within two output classes (namely, diabetes present and diabetes absent). The modified model, in practice, divides the main problem into smaller subproblems and then limits the performance with every subproblem. They use incremental and decremental procedures to learn and unlearn. This increases the performance of the model by considerable amounts. While the incremental procedure keeps checking on changes in the marginal coefficient values and the margin values placed in equilibrium, the decremental model leaves out the values and looks for new values. The SVM system they proposed yields an accuracy of 95%, which is higher than the accuracy of the existing system, that is, 89.07%. A deeper performance analysis is done using the ROC curve. Using the confusion matrix values, other performance parameters like F1 score, precision, and recall were calculated. The performance observed is better when compared to the traditional model. To further enhance this model, features to detect the possible complications in the future can be added. This would help in not only detecting diabetes but also how it will adversely affect the health of the patient at later stages.

Kumari and Chitra (2018) came up with another modification for the traditional SVM model. They use the RBF kernel of SVM as it can handle higher dimensionality. The kernel's output depends on the Euclidean distance between the testing data point and the support vector. The support vector is placed in the center of the RBF kernel, and an RBF parameter determines the area of impact this support vector will have over the data space.

The function is defined as

$$K\,(xi, xj) = \exp(-\boldsymbol{\gamma} \mid xi - xj \mid 2), \boldsymbol{\gamma} > 0. \quad (3.4)$$

Here,

$\boldsymbol{\gamma}$ is the RBF parameter,
xi is the training vector, and
xj is the testing data point.

A dataset with 400 samples is split. Two hundred points are used for training, while 260 points are used for testing the model. The dataset had eight attributes and two output classes (diabetic and nondiabetic). Using the confusion matrix, the accuracy was calculated and estimated to be around 78.2%, while the sensitivity and specificity stood at 80% and 76.5%, respectively.

Huma Naz and Sachin Ahuja (2020) proposed a comparative analysis of ANNs, NB, DT, and DL. They approved the algorithms mentioned above on the PIMA dataset and achieved accuracy in the range of 90% to 98%. The PIMA dataset had 768 instances and eight attributes. The dataset was split into testing and training in a 20:80 ratio. The dataset needs to be properly balanced between the training and testing dataset; for this, we use sampling. The process of extracting parameters and characteristics from a representative portion of a large dataset consistently is called sampling. This helps divide the dataset into training and testing sets, with the same proportion of classes as in the original dataset. A few ways to sample the dataset are automatic sampling, linear sampling, stratified sampling, and shuffled sampling. In linear sampling, a dataset is linearly divided into proportions as a dataset representative, whereas in shuffled sampling, subsets of the dataset are built by random splits in the dataset. In stratified sampling, the dataset is split arbitrarily, and then it rebuilds to form subsets, whereas in automatic sampling, stratified sampling is used as the default technique for building subsets. A DL model was a multilayer perceptron with an input layer of 8 neurons, two hidden layers with 50 neurons each, and an output layer with two neurons and the activation function used was softmax. The precision and recall for DL were found to be 95.22%, 98.46%, respectively, whereas the precision and recall for DT were found to be 94.02%, 95.45%, respectively. In the case of ANN, the precision and recall were found to be 88.05%, 83.09%, respectively, and the precision and recall were found to be 59.07%, 64.51%, respectively for NB. The future work includes building an app or a website where the DL model could be deployed for the early detection of diabetes. The DL model, although quite efficient, still needs improvement before deployment. The model gives an error of 1.93%, which is dangerous when it comes to medical applications.

Rajeeb Dey et al. (2008) used an ANN on a custom-made dataset. This dataset was gathered from Sikkim Manipal Institute of Medical Sciences hospital, Gangtok, Sikkim. The data consisted of readings from 530 patients, out of which 249 were diabetic whereas the others were nondiabetic. The dataset had parameters like occupation, fasting and random blood sugar, sex, blood plasma, age, and blood sugar. The occupation was included in determining the stress levels. It was subdivided into working class, homemakers or retired persons, students, and labor class. Before training, the data was normalized using the log sigmoid function, which transformed the training data in the range of 0 to 1.350 samples (141 diabetic and 209 nondiabetics) were used for training the model. Two neural networks were trained. One network had the architecture of 6–10–1 (6 input neurons, 10 neurons in one hidden layer, 1 output neuron) and 6–14–14–1 (6 input neurons, 14 neurons in each of the two hidden layers, 1 output neuron). A nonlinear log sigmoid function was applied to the hidden layers, and a smooth linear activation was applied to the input layer of the model. Momentum was 0.3, and the learning rate was fixed at 0.1 for the two networks. Two hundred test samples were used for a performance evaluation of the networks. Both networks made 185 correct classifications.

Refer to Table 3.5 for a comparison of various models as stated in the preceding study and the accuracy of the models are visualized in Figure 3.3.

## 3.7 APPLICATION OF DL TECHNIQUES ON BRAIN TUMOR DATA SETS

A brain tumor is a mass of abnormal cells in the brain. The growth of these cells is restricted in the skull; this may lead to serious brain damage. Based on the tumor's nature, they are of two types: malignant (cancerous) and benign (noncancerous). As these tumors increase in size, the pressure on the brain increases, leading to life-threatening conditions. Based on the origin, brain tumors are of two types: primary and secondary. Primary tumors, majorly benign, originate in the brain, whereas secondary tumors are malignant because of the spread of cancerous cells from other parts of the body like the lungs and the breast.

The major symptom of a brain tumor is a headache. This headache increases while sleeping and is the most painful in the early mornings. This is accompanied by vomiting, blurred vision, weakness in limbs, memory loss, and many other symptoms. Tumors are diagnosed with the help of CT scans, MRI, angiography, and other such procedures. The proposed treatment for brain tumors includes brain surgery, radiation therapy, chemotherapy, and newly developed hormone therapy. If brain tumors are detected at an early stage, then risky brain surgery can be prevented. However, due to a lack of medical professionals and medical experts, we need something to automate the detection process; machine learning and DL come into the picture.

Mohammad Shahjahan Majid, T. M. Shahriar Sazzad, and Md. MahbuburRahman (2020) proposed a segmentation and classification model for the detection of brain tumors. They used image-processing techniques to enhance the image and then used morphological operations to get the region of interest. The dataset used for

**TABLE 3.5**
**F1 Score and Summary for Type 2 Diabetes Detection**

| Paper | Dataset | K Nearest Neighbours | Decision Tree | Support Vector Machine (SVM) | Logistic Regression | Naive Bayes | Random Forests | Modified SVM | Decision Stumps | Deep Learning | Artificial Neural Network |
|---|---|---|---|---|---|---|---|---|---|---|---|
| Jakka and J (2019) | PIMA | 69 | 72 | 59 | 75 | 74 | 69 | | | | |
| Vijayan and Anjali (2016) | PIMA | | ✓ | ✓ | | ✓ | | | ✓ | | |
| Zia and Khan (2019) | PIMA | ✓ | ✓ | | | ✓ | | | | | |
| Alehegn et al. (2017) | PIMA | ✓ | ✓ | | | ✓ | ✓ | | | | |
| Aishwarya et al. (2013) | PIMA | | | | | | | ✓ | | | |
| Kumari andChitra (2018) | PIMA | | | | | | | ✓ | | | |
| Naz and Ahuja (2020) | PIMA | | 94.72 | | | 61.67 | | | | 96.81 | 85.98 |
| Dey et al. (2008) | Custom dataset | | | | | | | | | | ✓ |

*Source:* Based on Aishwarya et al. (2013), Alehegn et al. (2017), Kumari and Chitra (2018), Dey et al. (2008), Jakka and J (2019), Naz and Ahuja (2020), Vijayan and Anjali (2016), and Zia and Khan (2019).

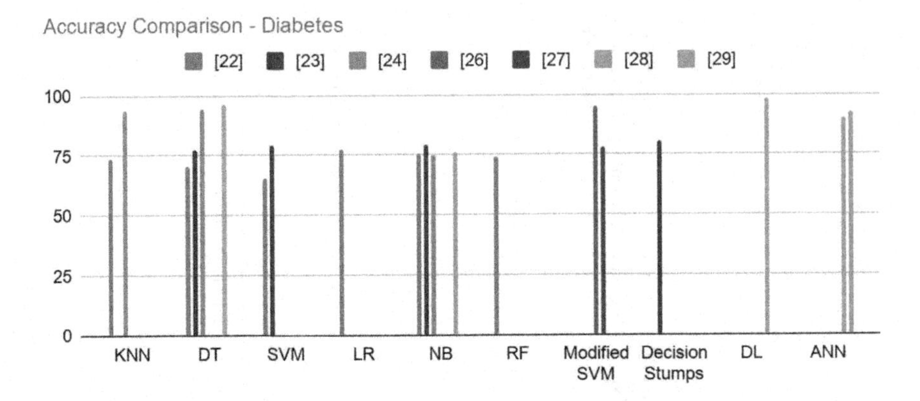


**FIGURE 3.3** Accuracy Comparison—Diabetes.

*Source:* Based on Aishwarya et al. (2013), Alehegn et al. (2017), Kumari and Chitra (2018), Dey et al. (2008), Jakka and J (2019), Naz and Ahuja (2020), Vijayan and Anjali (2016), and Zia and Khan (2019).

the framework had images containing glioma, meningioma, and pituitary tumor. The images were of size 350 × 350 and were divided into batches of 120 images. As stated earlier, these images were enhanced by first converting the image to red, green, and blue (RGB) model and then to hue, saturation, value (HSV); the S value was not considered, whereas the V number was enhanced by histogram equalization, and then the V channel median filter was applied to preserve the edges of the image. Then this image was again converted to RGB. For better results, the morphological operation of erosion and dilation was applied to this image. Finally, threshold-based Otsu's segmentation was implemented on the filtered image. Hole filling is applied to the images to remove unwanted black regions in the tumor. The region of interest is identified by a semi-automated manner using the size and shape features. We then applied classification using the SVM classifier. The limitation to this study is the fewer number of images taken in the dataset for classification. Future work could include deep learning algorithms like convolutional neural networks and gated recurrent units.

B. V. Kiranmayee, Dr. T. V. Rajinikanth, and S. Nagini (2016) performed a classification of brain tumor images to classify them as positive (with a brain tumor) and negative (without brain tumor). The images are preprocessed, segmentation, and feature extraction. These extracted features are used for classification using the ID3 algorithm. The algorithm was coded using java.

Mohammadreza Soltaninejad et al. (2017) worked on two datasets, one created by them and the other BRATS 2012 dataset. The dataset consisted of FLAIR (fluid-attenuated inversion recovery) MRI. They used the superpixel classification technique. Feature extraction was done to extract Gabor textons, fractal analysis, and curvatures of each superpixel of the entire brain area. A simple linear iterative clustering algorithm does superpixel segmentation. After segmentation, feature extraction is done to extract intensity statistics, textons, and curvature features. Texton features are extracted with the help of Gabor filters and k-means clustering.

Minimum redundancy maximum relevance is applied to the extracted features to select the best suitable features for classification. Extremely randomized trees are used for classification, using the minimum redundancy maximum relevance algorithm feature. They also performed a comparison between scalar vector machines and extremely random trees, proving extremely random trees to be more efficient in the classification of brain tumor images. The precision and sensitivity for the support vector machine were found to be 83.59% and 87.82%, respectively, whereas the accuracy and the sensitivity for extremely random trees were found to be 87.86% and 89.48%, respectively. The major limitation to the study was the insufficient number of FLAIR MRI images in the dataset.

Dr. Chinta Someswararao, R Shiva Shankar, Sangapu Venkata Appaji, and VMNSSVKR Gupta (2020) proposed a model to detect brain tumors using conventional neural networks. They worked on the UCI Dataset containing two classes: images with brain tumors and images without brain tumors. In the data cleaning process, the images were resized to (224, 224). This dataset was then split into a 75:25 ratio for training and testing. They then used this dataset to train and test their convolutional neural network and used accuracy as the evaluation metrics. With the testing data, they achieved 100% accuracy.

Suchita Goswami and Lalit Kumar P. Bhaiya (2013) developed an unsupervised learning-based neural network for a brain tumor detection model. Their dataset consisted of MRI images of the brain. They performed edge detection on the images, followed by histogram equalization and thresholding. Next, features were extracted using feature extraction. These features were contrast, correlation, energy, and homogeneity. Using these features, they developed a self-organizing map for the classification of the images. Self-organizing maps is an ANN based on competitive unsupervised learning. The idea is that the output neurons compete to get fired. The output neuron that gets activated is known as the winner takes all neurons. The model gave 96.80% specificity and 100% sensitivity, thus outperforming other DL algorithms. The limitation of this study was that the dataset consisted of merely 70 images.

Deipali Vikram Gore and Vivek Deshpande (2020) performed an analysis of the various techniques using DL for brain tumor detection. They summarized various convolutional neural networks used for brain tumor detection. In the end, they identified various research gaps:

- MRI images can be enhanced and could be applied to convolutional neural networks
- Developing quality models for medical determination, presence forecast, and visualization
- Computerized therapeutic leading to early tumor detention and hence lower mortality
- Feature extraction for reducing the computational costs
- Consideration of various types of datasets for classification to increase the scope of the developed model

G. Hemanth, M. Janardhan, and L. Sujihelen (2019) explored machine learning and DL to compare these algorithms' effectiveness on brain tumor classification. They used the brain tumor dataset from Kaggle, consisting of two classes, that is, positive (with the tumor) and negative (without tumor). This dataset was then preprocessed to eliminate noise and improvise the quality of the image. The cleaned images undergo average filtering. Average filtering removes salt-and-pepper noise. This cleaned dataset was used to train and test the convolutional neural network. The same dataset was used to train conditional random fields, SVM, and genetic algorithm. After training convolutional neural network was 92.7% efficient, the conditional random field was 87.5% efficient, the SVM was 90.3% efficient, and the genetic algorithm was 84.78% efficient; therefore, they concluded that the convolutional neural network outperformed all other models.

Shubham Kumar Baranwal et al. (2020) compared convolutional neural networks' performance and SVMs on brain tumor images. The dataset used by them was the Brain Tumor Dataset, consisting of 3,064 MRI scan images of 233 patients of Nanfang Hospital and General Hospital. There were 708 meningioma images, 1,426 glioma images, and 930 pituitary tumor images. These images were resized to 384 × 384 and then split into training and testing containing 2,451 and 631 images, respectively. The convolutional layer consisted of five convolutional layers, five max-pooling layers, one dropout layer, a flatten layer, a dense layer of 1024 neurons, and an output layer with three neurons for the three classes. The SVM framework included a two-dimensional Gabor filter, statistical feature extraction, and, finally, the SVM classifier. The statistical features included contrast, correlation, energy, homogeneity, and dissimilarity. After training and testing the model, they achieved 0.94 precision, 0.94 recall, and 0.94 F1 scores for the convolutional neural network; 0.80 precision, 0.81 recall, and 0.81 F1 score for the linear SVM; and 0.82 precision, 0.82 recall, and 0.82 F1 scores for polynomial SVM; thus, they concluded that the convolutional neural network outperformed the linear and polynomial SVMs. The study's limitation was the fewer images in the dataset, and the future scope includes grading the tumors based on their size.

Refer to Table 3.6 for a comparison of various models as stated in the above study and the accuracy of the models are visualized in Figure. 3.4.

## 3.8 CONCLUSION

Machine learning is taking data, applying certain algorithms to these data, and deriving insights from them. In contrast, deep learning is a branch of machine learning that majorly deals with neural networks. This chapter dealt with the applications of machine learning and DL in the field of medical diagnosis. The detection of heart diseases, pneumonia, diabetes, and brain tumor detection has been portrayed in this chapter. Certain diagnosis systems for the mentioned diseases have been highlighted. This chapter also portrays comparisons between the performance of trivial machine learning and DL algorithms.

## 3.9 IMPLICATIONS

Proper clinical diagnosis and treatment are required to automate the health systems. This can be achieved with different data mining techniques, machine learning, and

**TABLE 3.6**
**F1 Score and Summary for Brain Tumor Detection**

| Paper | Dataset | Support Vector Machine | ID3 | Extremely Random Tree | Convolutional Neural Network | Self-Organizing Map | Conditional Random Field | Genetic Algorithm |
|---|---|---|---|---|---|---|---|---|
| Majib et al. (2020) | Brian Tumor Dataset, Figshare | ✓ | | | | | | |
| Kiranmayee et al. (2016) | Brain Tumor Dataset, Kaggle | | ✓ | | | | | |
| (Soltaninejad et al., 2017) | Custom, BRATS 2012 | ✓ | | ✓ | | | | |
| Someswararao et al. (2020) | UCI | | | | ✓ | | | |
| Goswami and Bhaiya (2013) | – | | | | | ✓ | | |
| Hemanth et al. (2019) | Brain Tumor Dataset, Kaggle | ✓ | | | ✓ | | ✓ | ✓ |
| Baranwal et al. (2020) | Brain Tumor Dataset | ✓ | | | ✓ | | | |

*Source:* Based on Baranwal et al. (2020), Gore and Deshpande (2020), Goswami and Bhaiya (2013), Hemanth et al. (2019), Kiranmayee et al. (2016), Majib et al. (2020), Soltaninejad et al. (2017), and Someswararao et al., 2020).

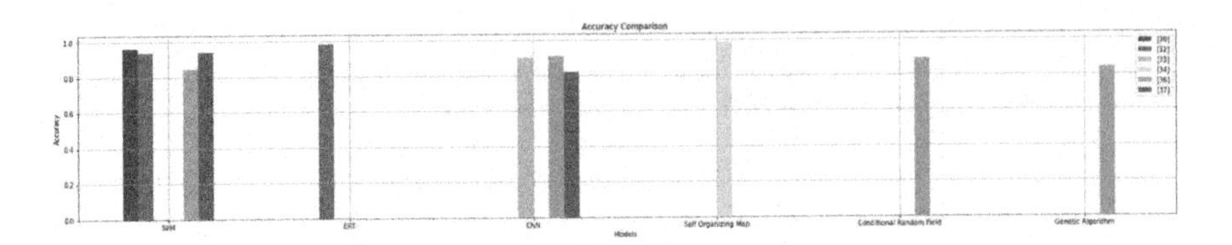


**FIGURE 3.4** Accuracy Comparison – Brain Tumor.

*Source:* Based on Baranwal et al. (2020), Gore and Deshpande (2020), Goswami and Bhaiya (2013), Hemanth et al. (2019), Kiranmayee et al. (2016), Majib et al. (2020), Soltaninejad et al. (2017), and Someswararao et al. (2020).

the recent deep learning models to increase the decision support systems' efficiency. These models use the raw medical data for analysis and predict the details and heath reports for individuals. This helps the physicians in their decision-making process, healthcare institutions, insurance fraud detection, emergency health record analysis, and much more. All these practices help improve the treatment process and reduce the overall costs. Computational intelligence and analytics are a must for the healthcare industry to automate systems and increase healthcare practices. The previously discussed models provide detailed analysis for associations between the patients, medicines as well as their outcomes. The pros and cons of the types of data, algorithms, and methodologies have been crafted in a tabular format. Major challenges to be faced by these support systems are scalability, human intervention, patterns based on historical knowledge, and complex data types. By taking care of privacy concerns and providing regulated surveillance, these models can be used well for the healthcare industry.

## REFERENCES

Aishwarya, R., Gayathri, P., & Jaisankar, N. (2013). A method for classification using machine learning technique for diabetes. *International Journal of Engineering and Technology*, *5*(3), 2903–2908.

Alehegn, M., & Joshi, R. (2017). Analysis and prediction of diabetes diseases using machine learning algorithm: Ensemble approach. *International Research Journal of Engineering and Technology*, *4*(10), 426–436. www.irjet.net

Ansari, N., Faizabadi, A. R., Motakabber, S. M. A., & Ibrahimy, M. I. (2020). Effective Pneumonia Detection using Res Net based Transfer Learning, *Test Engineering and Management*, *82*, 15146–15153.

Bahadur, S., Yadav, P. K., & Shukla, D. P. (2013). Predict the diagnosis of heart disease patients using classification mining techniques. *IOSR Journal of Agriculture and Veterinary Science*, *4*(2), 60–64. https://doi.org/10.9790/2380-0426164

Baranwal, S. K., Jaiswal, K., Vaibhav, K., Kumar, A., & Srikantaswamy, R. (2020). Performance analysis of brain tumour image classification using CNN and SVM. In *Proceedings of the 2nd International Conference on Inventive Research in Computing Applications, ICIRCA 2020*, IEEE, 537–542. https://doi.org/10.1109/ICIRCA48905.2020.9183023

Chandna, D. (2014). Diagnosis of heart disease using data mining algorithm. *International Journal of Computer Science and Information Technologies*, *5*(2), 1678–1680.

Dey, R., Bajpai, V., Gandhi, G., & Dey, B. (2008). Application of Artificial Neural Network (ANN) technique for diagnosing diabetes mellitus. In *IEEE Region 10 Colloquium and 3rd International Conference on Industrial and Information Systems, ICIIS 2008*, IEEE, 155, 50–53. https://doi.org/10.1109/ICIINFS.2008.4798367

El Asnaoui, K., Chawki, Y., & Idri, A. (2020). Automated methods for detection and classification pneumonia based on X-ray images using deep learning. *ArXiv.*

Gonsalves, A. H., Thabtah, F., Mohammad, R. M. A., & Singh, G. (2019). Prediction of coronary heart disease using machine learning: An experimental analysis. In *ACM International Conference Proceeding Series.* Association for Computing Machinery, New York, NY, USA, 51–56. https://doi.org/10.1145/3342999.3343015

Gore, D. V., & Deshpande, V. (2020). Comparative study of various techniques using deep learning for brain tumor detection. In *2020 International Conference for Emerging Technology, INCET 2020*, IEEE, 1–6. https://doi.org/10.1109/INCET49848.2020.9154030

Goswami, M. S., & Bhaiya, M. L. K. P. (2013). Brain tumour detection using unsupervised learning based neural network. In *Proceedings – 2013 International Conference on Communication Systems and Network Technologies, CSNT 2013*, IEEE, 573–577. https://doi.org/10.1109/CSNT.2013.123

Gu, X., Pan, L., Liang, H., & Yang, R. (2018). Classification of bacterial and viral childhood pneumonia using deep learning in chest radiography. In *ACM International Conference Proceeding Series*, Association for Computing Machinery, New York, NY, USA, 88–93. https://doi.org/10.1145/3195588.3195597

Han, Y., Chen, C., Tewfik, A. H., Ding, Y., & Peng, Y. (2021). Pneumonia detection on chest X-ray using radiomic features and contrastive learning. In *Proceedings—2021 IEEE 18th International Symposium on Biomedical Imaging, ISBI*, IEEE, 247–251. http://doi.org/10.1109/ISBI48211.2021.9433853

Hemanth, G., Janardhan, M., & Sujihelen, L. (2019). Design and implementing brain tumor detection using machine learning approach. In *Proceedings of the International Conference on Trends in Electronics and Informatics, ICOEI 2019, 2019-April (ICOEI), IEEE*, 1289–1294. https://doi.org/10.1109/icoei.2019.8862553

Jakka, A., & J, V. R. (2019). Performance assessment of machine learning based models for diabetes prediction. In *2019 IEEE Healthcare Innovations and Point of Care Technologies, HI-POCT 2019*, IEEE, 11, 147–150. https://doi.org/10.1109/HI-POCT45284.2019.8962811

Karthiga, A. S., Mary, M. S., & Yogasini, M. (2017). Early prediction of heart disease using decision tree algorithm. *International Jounral of Advanced Research in Basic Engineering Sciences and Technology*, *3*(3), 1–16.

Kiranmayee, B. V., Rajinikanth, T. V., & Nagini, S. (2016). A novel data mining approach for brain tumour detection. In *Proceedings of the 2016 2nd International Conference on Contemporary Computing and Informatics, IC3I 2016*, IEEE, 46–50. https://doi.org/10.1109/IC3I.2016.7917933

Kumari, V. A., & Chitra, R. (2018). Classification of diabetes disease using support vector machine. *International Journal of Engineering Research and Applications*, *3*(2), 1797–1801. www.researchgate.net/publication/320395340

Li, L., Qin, L., Xu, Z., Yin, Y., Wang, X., Kong, B., Bai, J., Lu, Y., Fang, Z., Song, Q., Cao, K., Liu, D., Wang, G., Xu, Q., Fang, X., Zhang, S., Xia, J., & Xia, J. (2020). Artificial intelligence distinguishes COVID-19 from community acquired pneumonia on chest CT. *Radiology.* http://arxiv.org/abs/2003.13865

Li, X., Chen, F., Hao, H., & Li, M. (2020). A Pneumonia detection method based on improved convolutional neural network. In *Proceedings – 2020 International Workshop on Electronic Communication and Artificial Intelligence, IWECAI 2020*, IEEE, 4, 488–493. https://doi.org/10.1109/IWECAI50956.2020.00028

Majib, M. S., Shahriar Sazzad, T. M., & Rahman, M. M. (2020). A framework to detect brain tumor cells using MRI images. In *HORA 2020–2nd International Congress on Human-Computer Interaction, Optimization and Robotic Applications, Proceedings*, IEEE. https://doi.org/10.1109/HORA49412.2020.9152893

Mattingly, Q. (2020). *Cardiovascular Diseases*. World Health Organization. www.who.int/health-topics/cardiovascular-diseases/#tab=tab_1

Mohan, S., Thirumalai, C., & Srivastava, G. (2019). Effective heart disease prediction using hybrid machine learning techniques. *IEEE Access*, *7*, 81542–81554. https://doi.org/10.1109/ACCESS.2019.2923707

Naz, H., & Ahuja, S. (2020). Deep learning approach for diabetes prediction using PIMA Indian dataset. *Journal of Diabetes and Metabolic Disorders*, *19*(1), 391–403. https://doi.org/10.1007/s40200-020-00520-5

*Pneumonia*. (2019). World Health Organization. www.who.int/news-room/fact-sheets/detail/pneumonia

Pouriyeh, S., Vahid, S., Sannino, G., De Pietro, G., Arabnia, H., & Gutierrez, J. (2017). A comprehensive investigation and comparison of Machine Learning Techniques in the domain of heart disease. In *Proceedings – IEEE Symposium on Computers and Communications, Iscc*, IEEE, 204–207. https://doi.org/10.1109/ISCC.2017.8024530

Rahman, T., Chowdhury, M. E. H., Khandakar, A., Islam, K. R., Islam, K. F., Mahbub, Z. B., Kadir, M. A., & Kashem, S. (2020). Transfer learning with deep Convolutional Neural Network (CNN) for pneumonia detection using chest X-ray. *Applied Sciences (Switzerland)*, *10*(9). https://doi.org/10.3390/app10093233

Rajpurkar, P., Irvin, J., Zhu, K., Yang, B., Mehta, H., Duan, T., Ding, D., Bagul, A., Ball, R. L., Langlotz, C., Shpanskaya, K., Lungren, M. P., & Ng, A. Y. (2017). CheXNet: Radiologist-level pneumonia detection on chest X-rays with deep learning. *ArXiv*, 3–9.

Rani, K. U. (2011). Analysis of heart diseases dataset using neural network approach. *International Journal of Data Mining & Knowledge Management Process*, *1*(5), 1–8.

Singh, A. (2018). *A Comprehensive Guide to Ensemble Learning (with Python Codes)*. Analytics Vidhya. www.analyticsvidhya.com/blog/2018/06/comprehensive-guide-for-ensemble-models/

Soltaninejad, M., Yang, G., Lambrou, T., Allinson, N., Jones, T. L., Barrick, T. R., Howe, F. A., & Ye, X. (2017). Automated brain tumour detection and segmentation using superpixel-based extremely randomized trees in FLAIR MRI. *International Journal of Computer Assisted Radiology and Surgery*, *12*(2), 183–203. https://doi.org/10.1007/s11548-016-1483-3

Someswararao, C., Shankar, R. S., Appaji, S. V., & Gupta, V. (2020). Brain tumor detection model from MR images using convolutional neural network. In *2020 International Conference on System, Computation, Automation and Networking, ICSCAN 2020*, IEEE, 1–4. https://doi.org/10.1109/ICSCAN49426.2020.9262373

Sowmiya, C., & Sumitra, P. (2018). Analytical study of heart disease diagnosis using classification techniques. In *Proceedings of the 2017 IEEE International Conference on Intelligent Techniques in Control, Optimization and Signal Processing, INCOS 2017*, IEEE, 2018-*February*, 1–5. https://doi.org/10.1109/ITCOSP.2017.8303115

Stephen, O., Sain, M., Maduh, U. J., & Jeong, D. U. (2019). An efficient deep learning approach to pneumonia classification in healthcare. *Journal of Healthcare Engineering*, *2019*. https://doi.org/10.1155/2019/4180949

T, M., Mukherji, D., Padalia, N., & Naidu, A. (2013). A heart disease prediction model using SVM-Decision Trees-Logistic regression (SDL). *International Journal of Computer Applications*, *68*(16), 11–15. https://doi.org/10.5120/11662-7250

Vijayan, V. V., & Anjali, C. (2016). Prediction and diagnosis of diabetes mellitus: A machine learning approach. In *2015 IEEE Recent Advances in Intelligent Computational Systems, RAICS 2015, December*, IEEE, 122–127. https://doi.org/10.1109/RAICS.2015.7488400

Zia, U. A., & Khan, N. (2019). Predicting diabetes in medical datasets using machine learning techniques. *International Journal of Scientific Research & Engineering Trends, 5*(2), 1538–1551. www.ijser.org

# 4 Proposed Intelligent Software System for Healthcare Systems Using Machine Learning

*Manu Banga*

**CONTENTS**



## 4.1 INTRODUCTION

'Prevention is better than cure'. Benjamin Franklin's famous aphorism is worth recalling at this troubling time for healthcare and the economy, especially during the COVID-19 pandemic (Abdel-Basset et al., 2021). There are various schemes sponsored by the Indian government under the flagship of Ayushman Bharat National Health Protection Mission (AB-NHPM) hybrid of two major health initiatives, namely, health and wellness centers (HWC) and the National Health Protection Scheme (NHPM). This research explores the effectiveness of preventive care models using an intelligent healthcare system (ISS) to predict early hospital readmission in the Indian healthcare system. In this research study, a novel ISS for preventive care is proposed for reducing hospital readmission by using machine learning analytics techniques on patients of high- and low-risk categories using just-in-time (JIT) deduction. JIT analytics not only ensure best readmission quality but also handle patients' readmission by identifying and prioritizing

DOI: 10.1201/9781003142751-5

them according to their medical aid needed so patients with higher comorbidities can be given better medical facility based on their past medical history, like lab tests and medicines, thereby enhancing the reliability of the intelligent software system. An ISS comprises programs, methodologies, rules and related documentation and research that empower the client to collaborate with a computer and its equipment, that is, hardware, or perform errands. It comprises four Vs and suffices the first V (Volume) of big data. The speed of healthcare data created from patient encounters and patient monitors is increasing in and out of the clinic—the second V (Velocity). More than 80 percent of medical data resides in unstructured formats, such as doctors' notes, images and charts from monitoring instruments. The third V (Variety) and the fourth V (Veracity) deal with unsure or vague data. Most healthcare data from clinic and hospital records are afflicted with errors, as while entering data, technicians frequently attach research to the wrong person's record or copy research incorrectly. This section provides the context of the study research and its aim and objectives. It then demonstrates the significance of this research. In today's digital world, life without a computer is not possible. Nearly everything around us is on an intelligent system, helping us lead life in a better way. The revolution brought by smart systems has increased global productivity and has benefited the world. Intelligent systems are useless without software. For excellence in intelligent systems, there is a strong need to develop more and more reliable software systems for healthcare. Thus, the role played by the intelligent system is appreciated in the healthcare industry as ISSs can handle huge volumes of data, the amount of research with respect to the capacity for its storage and management in patient readmission cases.

### 4.1.1 Motivation of Work

The Indian healthcare system needs major reinvention based on income levels difference, aging population, rising health awareness and positive outlook toward preventive healthcare are expected to increase the demand of healthcare services in the near future (Abdelrahman, 2020) thus making the healthcare industry a critical and fast-growing industry in India, with expectations of US$280 billion in 2025. With the dawning of the digital era and massive growth in healthcare, a vast amount of data can be anticipated from different health science data sources, including data from patient electronic records, claims systems, lab test results, pharmacies, social media, drug research, gene sequencing, home monitoring mobile apps and so on. These data are called big data, and big data analytics can possibly change the way healthcare providers utilize modern innovations to extract knowledge from their medical data repositories and settle on educated choices. The new trend of medical data digitization is leading to an optimum model change in the healthcare industry. As a result, the healthcare industry is experiencing an increment in the sheer volume of data regarding unpredictability, timeliness and diversity. Successfully acquiring and effectively analyzing a variety of healthcare data for a long period of time can shed light on a significant number of approaching healthcare challenges. Figure 4.1 shows the four dimensions of big data in designing an ISS for Indian healthcare. The human body is a big source of big data. Data increase and move

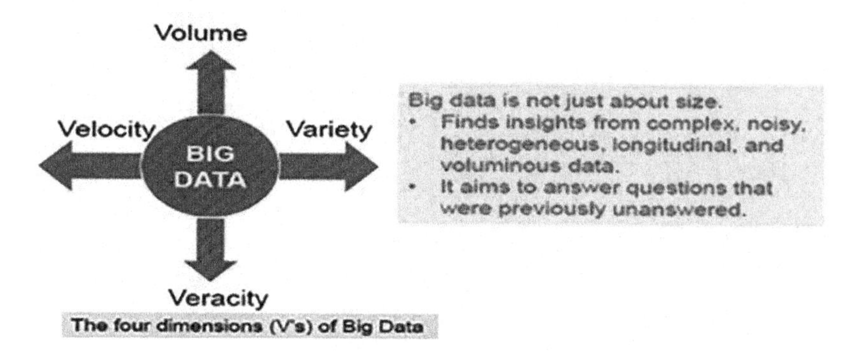


**FIGURE 4.1** The Four Dimensions of Big Data.

faster than healthcare organizations can ingest them. Collecting this important data and analyzing it for clinical and advanced investigation is critical to enhancing healthcare and outcomes. Instead of looking at "bigness", look for "smartness" in data. In India, even if the digitization of medical records is in its infancy (Forestiero et al., 2018), in today's research age, healthcare is transforming from assessment-based decisions to informed decisions based on data and analytics (Gravili et al., 2018). Figure 4.2 depicts the human body as big data. The current research study deals with the volume, variety and veracity of data. Due to privacy issues, this study could not access millions of instances, but still, thousands of real instances make this dataset reliable and efficient, with noisy, inconsistent and missing values in heterogeneous data. India, a developing country with a mammoth population, faces various problems in the field of healthcare in the form of huge expenditures, meeting the needs of the poor people, accessibility to hospitals and medical research, especially when an epidemic, such as dengue or malaria, spreads (Khanra et al., 2020). Operational efficiencies in Indian healthcare can be improved by capturing every single detail about a patient to form a complete view that will help predict and prevent an existing condition or disease from escalating and plan responses to disease epidemics. Most Indian healthcare organizations are currently setting out on the analytics journey. Research management systems and electronic medical records (EMRs) have been integrated by certain tertiary care hospitals to create a central repository of historical data in the form of data warehouses and consequently use it to mine data and do research and analytics to make smarter decisions for the enhanced quality of healthcare. Big data analytics is opening many avenues and opportunities in the Indian healthcare system. Unraveling the "big data"–related complexities can give useful research in designing an intelligent healthcare system by using data efficiently and effectively; it can provide some quick returns with respect to predicting the spread of epidemic preventive healthcare and fraud management (Babu et al., 2017), which will significantly lower healthcare costs. Thus, this study has performed various preprocessing techniques on the data to make it more accurate and reliable. Figure 4.3 shows the use of big data in an Indian healthcare scenario.

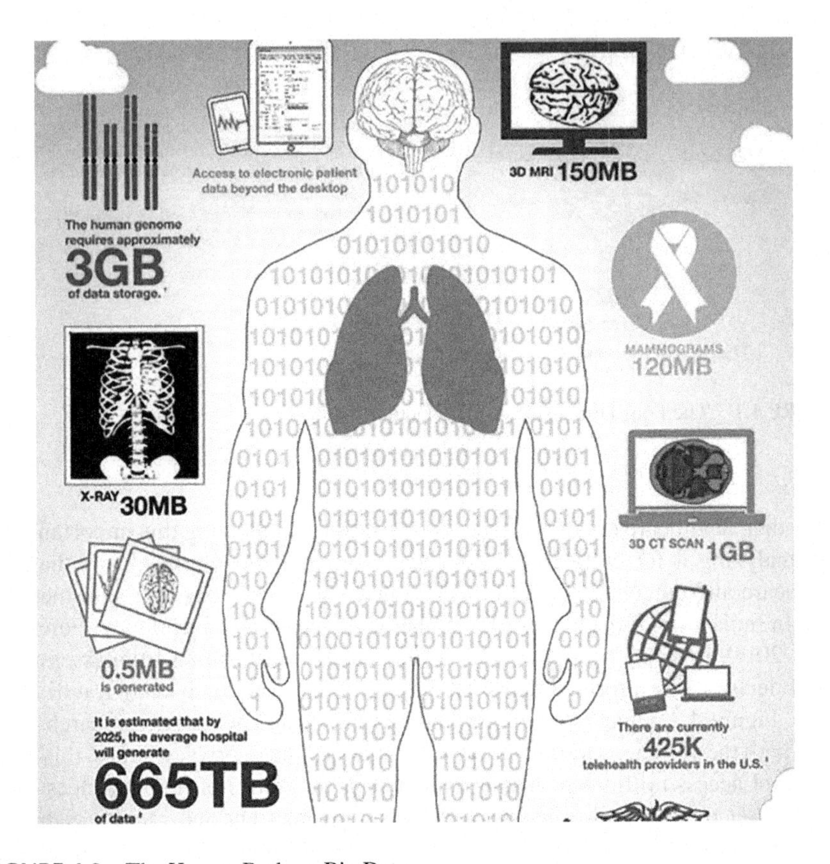


FIGURE 4.2 The Human Body as Big Data.

FIGURE 4.3 Use of Big Data in Indian Heathcare Scenario.

## 4.2 LITERATURE REVIEW

An intelligent healthcare system discovers trends in readmission data of various departments, thereby predicting and assessing readmission rates. Organizations utilize multiple methods to examine their data to predict future events. An intelligent healthcare system is a combination of statistical analysis and various data mining techniques, such as association, classification, clustering and pattern matching. It comprises the exploration and preparation of data, defining an intelligent system, and follows its process. Figure 4.4 shows stages in developing an intelligent system. In the development of an intelligent healthcare system, various authors have carried out research primarily on prescriptive analytics using descriptive analytics, diagnostic analytics and predictive analytics and have proposed a framework for the healthcare industry of real-time patients using support vector machines. Good accuracy was achieved, but biased datasets resulted in false predictions (Abdel-Basset et al., 2021). For dealing with ambiguous, biased datasets, a framework for COVID-19 prediction was proposed using personality traits. Researchers conducted a comprehensive study of Qatar and accessed various healthcare records arising from COVID-19 (Abdelrahman, 2020). Eminent researchers proposed a system for a hospital infrastructure management system. They conducted a survey for designing a comprehensive system covering multispecialty domains (Babar et al. 2016). Some researchers designed diagnostic analytics system for heart failure readmission cases for heart patients' readmission based on pulmonary infection using naïve Bayes theorem. They achieved 71% accuracy in probability assessment (Babu et al., 2017). Researchers carried out an extrinsic survey on various

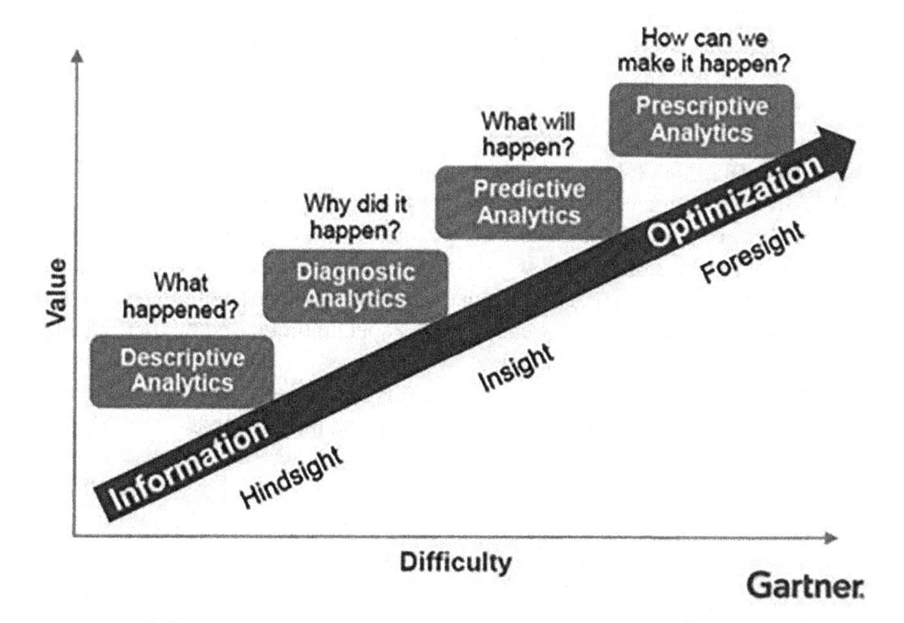


**FIGURE 4.4** Stages in Developing Intelligent System.

reasons for hospital readmission between 2016 to 2020, and a novel method based on a hybrid approach of extracting relevant feature selection was proposed for recurring deterioration of patients with comorbidities (Bossen et al., 2020). After following the sequence of readmission, a real-time system for tracking nosocomial disease was proposed by Cheng et al. (2018), who intelligently collaborated with doctors on call with a patient message. Researchers extensively applied big data analytics on diabetes patients and achieved, invariably, the same accuracy on multiple datasets (De Silva et al., 2015). As data are highly skewed, some researchers applied distributed and parallel algorithms to the healthcare system and predicted that various department operations are in parallel with their individual classification (Forestiero et al., 2018). For accessing operation and maintenance, researchers reviewed big data analytics in healthcare (Galetsi et al., 2020). In many diseases, patient readmission required using statistical modelling approaches for finding factors that affect early readmission, but it is not easy to predict each specific event that has same characteristics for a new patient (Gowsalya et al., 2014). Some researchers designed a specialized software system to help hospitals with elderly patients readmission, thus developing a ubiquitous environment for old-aged persons (Gravili et al. 2018). Researchers designed a trustworthy system for heath analytics using big data for the identification of predictor variables in readmission (Jin et al., 2016). Some researchers worked on the risk involved in predicting the readmission of patients with a chronic medical history (Kamble et al., 2018). For further assessment, researchers worked on the low to high risk categorization of patients having suffering chronic diseases (Khanra et al., 2020). Utilizing needs-based patient–doctor connections, researchers developed an intelligent healthcare system using mobile computing, thus accessing patient and doctor records in an online platform (Ma et al., 2018). An assessment of patient and doctor on call an online system was designed for trustworthy data processing of lab records, test results in healthcare analytics (Moutselos et al., 2018). Researchers developed an application for the online assessment of patient records by hospital administration sorted by doctor and department name (Navaz et al., 2018). Authors studied using deep learning methodologies for finding pattern distribution and classification of healthcare datasets in identification of readmission for patients with comorbidities (Sabharwal et al., 2016). Researchers conducted a survey on persons with diabetes for the need, application and characteristics of big data analytics in healthcare industry (Salomi et al., 2016) and designed a mobile application predicting patients requiring special treatment, an attendant and a special diet based on their past history through an e-commerce application (Wu et al., 2017). So implementing appropriate plans to prevent readmission and identifying patients who are at a greater risk of hospitalization is very important. People with uncontrollable comorbidities often become acutely ill and enter a cycle of hospital admission and readmissions. Patients with diabetes who are hospitalized are at greater risk of getting readmitted compared to patients without commodities (Jin et al., 2016). Besides this, intelligent systems based on a specific patient subpopulation categorized by disease type are of more value than the models based on entire cohort. So this work has been limited to diabetic patient readmission as diabetes is one of the most prevalent diseases and a major health hazard in developing countries like India as 17.1 percent of the adult population suffers from several heath issues as per Indian Council of Medical Research data (Gravili et al., 2018).

## 4.3 RESEARCH METHODOLOGY

The research methodology is a conventional collection of different techniques for examining a certain problem. These practices are continuous in nature and help in the acquisition of new knowledge by modifying earlier knowledge. The research methodology starts with interpreting the events happening in the surrounding world; developing myths or hypotheses by interrogating how a certain the process takes place, usually making way for predictions; and validating these hypotheses using different statistical tests thus generated from a controlled investigation that generate the empirical data, thus adding on to the previous results and making us alter, refine or reject our myths or hypotheses. Thus, the research methodology used for designing an ISS for readmission prediction uses a developmental design based on an incremental approach for imbalanced learning using the evaluation of previous datasets using statistical measures. Using machine learning for accessing readmission of a patient using an ISS helps reduce medical costs, thereby increasing the quality of healthcare in an optimal manner by formularizing the patient identification problem in a count of monthly discharges, assessing various steps involved in readmission risk prediction by exploring the EMR of tests like a nerve conduction test (NCV), electromyography (EMG), and magnetic resonance imaging (MRI) causing readmission related attributes, thereby extracting relevant features for predicting readmission of patients (Jin et al., 2016). Thereafter, applying data preprocessing techniques on readmissions data for efficient prediction using machining learning techniques and how can this prediction aid in reducing cost and improving the overall quality of healthcare in hospitals involved using optimal sets of patients who are expected to be readmitted after immediate discharge from an Indian hospital to study the attributes related to cause readmission of different patients by accessing the features that can be used in predicting readmission of a discharged patient with health issues within a stipulated time by diagnosis and interventions from clinical notes to enhance predictor variables for the model using data mining techniques with the importance of each predictor variable with respect to its ability to predict readmissions by the generation of frequent patterns that implicate readmission risk using various supervised learning algorithms in predicting the risk of readmissions. A best intelligent system is built and validated by statistical techniques, and using cost-sensitive analysis on savings can be done if the model is implemented with the average in the time lapse between discharge and readmission corresponding to different diseases (Navaz et al., 2018), thereby reducing and replacing missing values in the dataset and then normalizing the dataset values. The normalized datasets' relevant features influencing patient readmission are extracted using feature subset selections and then are overlapped with these subsets to complete a dataset in a bottom-up manner without redundancy after obtaining an optimized dataset of relevant features; then classification is done to identify patients from most likely to least likely for readmission, and after that, the model is evaluated on real hospital dataset. Figure 4.5 depicts the methodology adopted for data preprocessing.

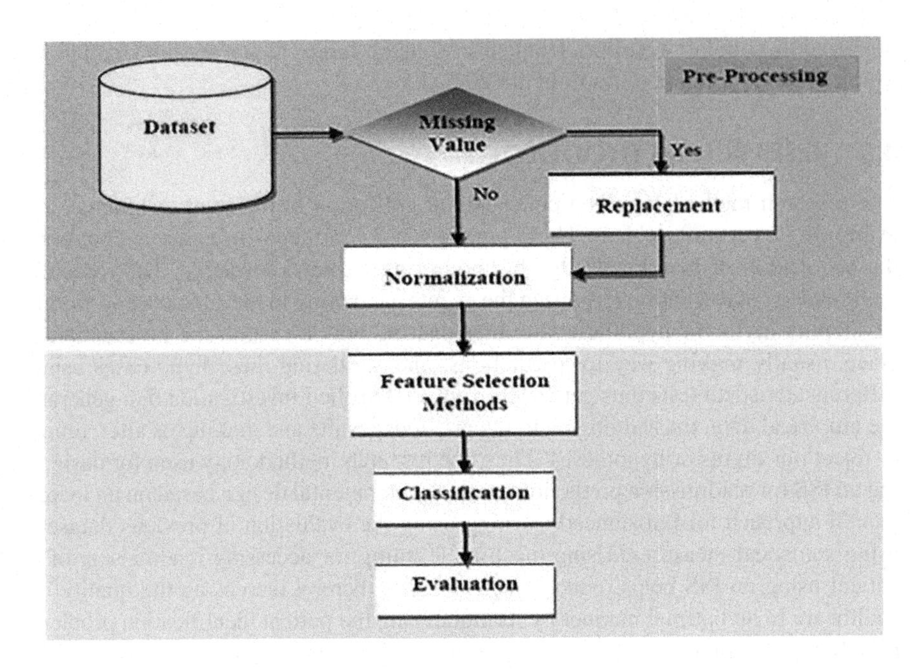


**FIGURE 4.5** Data Preprocessing Methodology.

## 4.4 PROPOSED APPROACH

Designing an ISS by automated feature selection by using relevant features of data exploration using data collection, acquiring domain knowledge and describing and exploring datasets thus resolves the class imbalance problem in an iterative manner by data preprocessing using support vector machines to define class labels as risk class and non-risk class directly to the features of the raw data, as constructed based on the history of the patient. So most of the features are extracted from individual patient admission records, but some features are aggregated across multiple admission records of the same patient and for multiple records using discretization and frequency distribution on prediction modelling and evaluating the model accuracy based on factors like precision, recall, and area under the curve. Figure 4.6 depicts the model prediction for the healthcare system.

### 4.4.1 Datasets Used

National Health Systems Resource Centre, Ministry of Health and Family Welfare of various districts in accordance with primary health centers and secondary and tertiary health centers provided by specialized hospitals. The dataset for this research was obtained from an Indian hospital ministry. A total of 58,625 inpatient admission records were explored from 1 April 2016 to 31 December 2020. Out of these, 9,381 records were diabetic patient encounters that were considered for modelling purposes. Each admission record included demographic research (e.g., name, gender,

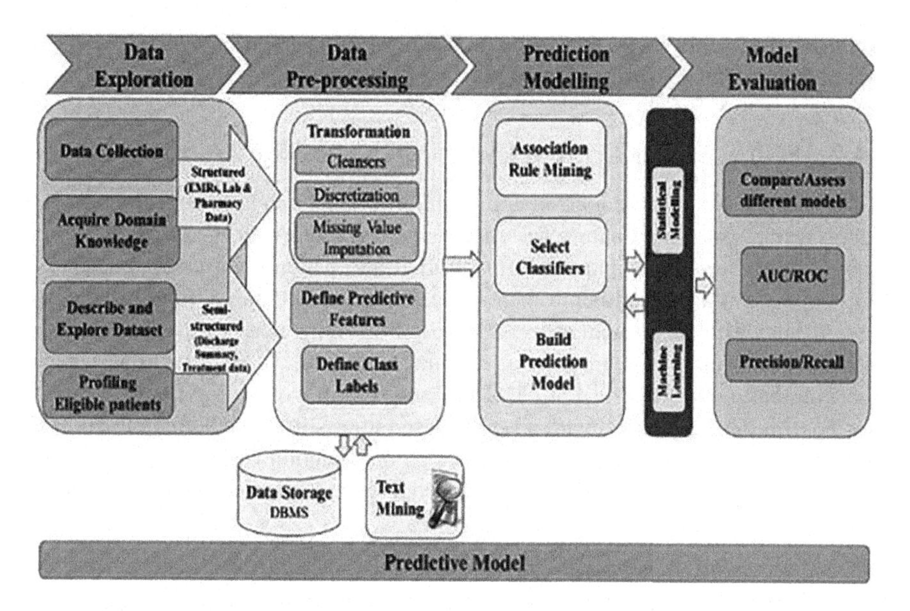


**FIGURE 4.6** Model Prediction for the Healthcare System.

**TABLE 4.1**
**Ranking of Relevant Attributes for Patients with Comorbidities**

| Attributes | Criteria |
|---|---|
| Sugar level | 1 |
| Hemoglobin level | 1 |
| Blood pressure | 1 |
| Varicose veins | 2 |
| Nerve palsy test: high level/low level | 3 |
| Number of surgeries in last 5 years | 4 |
| Past history: Nonunion of factures | 5 |
| Gastroenteritis | 5 |
| Period of stay in hospital | 5 |

age), clinical research (e.g., diagnosis, treatment, lab tests, medication data), administrative data (e.g., length of stay) and billing research (e.g., charge amount). After preprocessing, a subset of features was considered as shown in Table 4.1. Some of them are derived features known as aggregation of multiple datasets as the presence of these features leads to a particular illness. Discretization and frequency distribution were done using MATLAB tool, with analysis of 9,496 records and criteria for individual attribute as shown in the table.

In the proposed approach, the hybrid algorithm based on PSO-MGA (particle swarm optimization–modified genetic algorithm) is used to improve the accuracy of the intelligent healthcare system estimation using feature selection by PSO-MGA for reducing noise and selecting relevant attributes.

### 4.4.2 Feature Selection Using Hybrid MGA and PSO Algorithm

In this section, we propose a novel approach for an intelligent readmission risk prediction framework using feature selection using PSO-MGA for the efficient categorization of a class of patients to be readmitted and patients not to be readmitted. Wu et al. (2017) used healthcare data and categorized structured and unstructured datasets but with a class imbalance problem. For resolving overtraining and the class imbalance problem, the extraction of relevant features is needed; for that, a novel algorithm based on a hybrid PSO-MGA is proposed for obtaining the best feature sets. For this dataset, loaded with a kernel-based fitness function defined in an iterative manner, it is updated based on the best global function value of PSO or MGA, and after, the best sets of features are obtained. Figure 4.7 depicts the working of the proposed algorithm.

### 4.4.3 Proposed Approach—An Intelligent Healthcare Framework Using PSO-MGA Feature Selection with Bagging

First, the unstructured and structured dataset is loaded, and features are extracted using PSO-MGA; for extracting further relevant features, it is iteratively scanned, and a new dataset with the most relevant feature dataset is obtained. After obtaining the most relevant feature bagging technique and avoiding overtraining in the PSO-MGA hybrid algorithm, thus resolving class imbalance problem of readmission class or non-readmission class, the feature subset optimization was strengthened to obtain an optimized value of feature sets with the highest accuracy. Figure 4.8 depicts the working of a hybrid algorithm.

### 4.4.4 Performance Measures for Classification Problem

Dealing with the classification problem of predictive performance is an important task with the F-measure (Babu et al., 2017). So an F-measure is used for predictive performance using confusion for classification performance measurement. This represents the counts of the four possible outcomes as shown in Table 4.2.

In a binary classifier, there are four possible outcomes:

- True Positive (TP): Both the prediction outcome and the actual value are true.
- False Positive (FP): The prediction outcome is true, but the actual value is false.
- True Negative (TN): Both the prediction outcome and the actual value are false.
- False Negative (FN): The prediction outcome is false but the actual value is true.

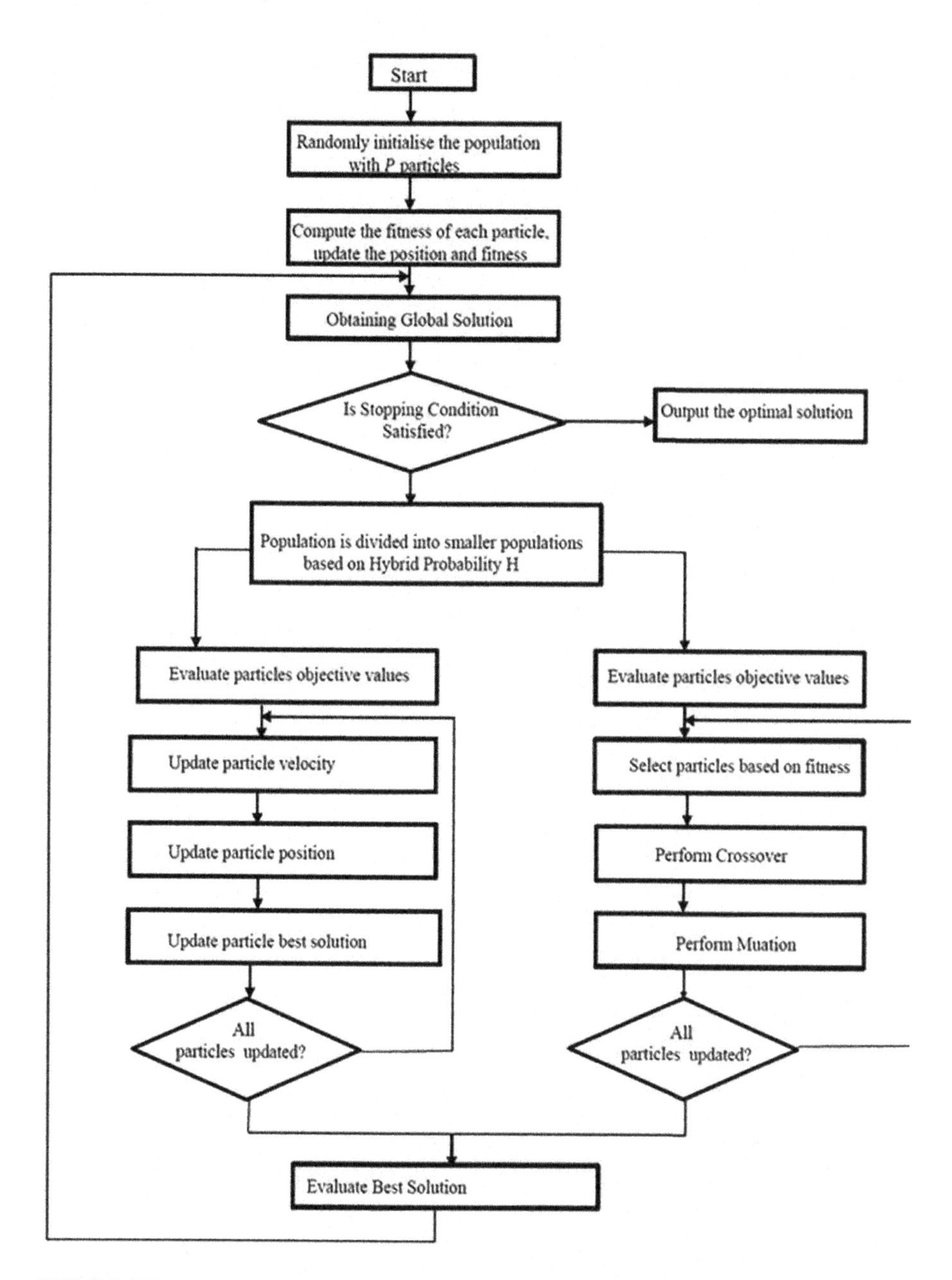


**FIGURE 4.7** Feature Selection Using PSO-MGA.

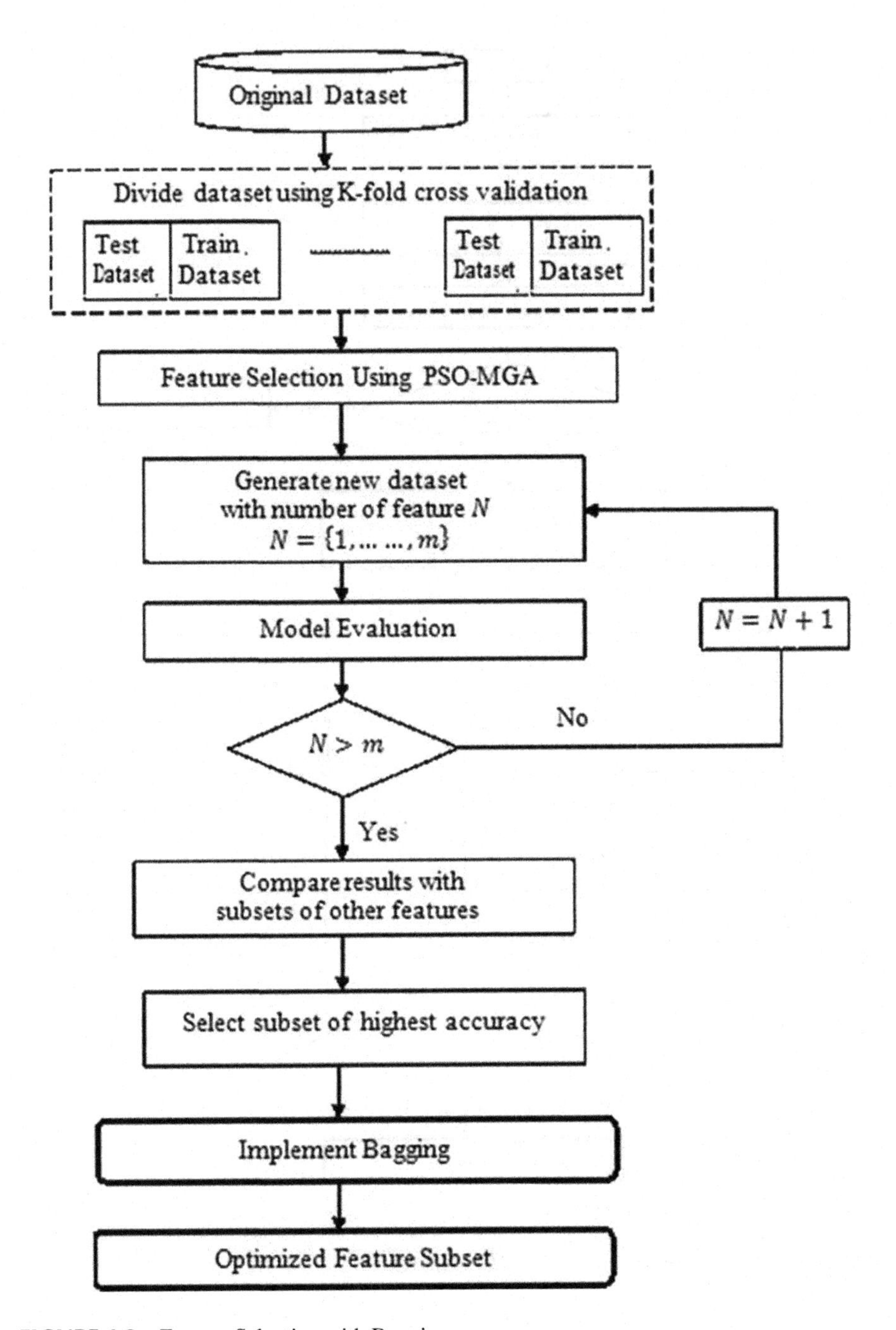


**FIGURE 4.8** Feature Selection with Bagging.

**TABLE 4.2**
**Confusion Matrix**

| | Actual Positive | Actual Negative |
|---|---|---|
| Predict Positive | TP | FP |
| Predict Negative | FN | TN |

$$\text{True Positive Rate}(t_p) = \frac{\text{Positive correctly classified}}{\text{Total Negative}} = \frac{TP + TN}{TP + TN + FP + FN} \quad (4.1)$$

$$\text{True Negative Rate}(f_p) = \frac{\text{Negative correctly classified}}{\text{Total Negative}} = \frac{FP + FN}{TP + TN + FP + FN} \quad (4.2)$$

$$\text{Sensitivity} = \text{recall} = \frac{\text{Total Positive classified}}{\text{Total samples}} = \frac{TP}{TP + FN} \quad (4.3)$$

$$\text{Specificity} = \frac{\text{Total Negative classified}}{\text{Total samples}} = \frac{TN}{TP + FN}$$

The F-measure is used for assessing the entire confusion matrix for evaluating a specific classifier as it properly takes into account all true and negative cases for interpretation. The accuracy of the classifiers are defined as a percentage of positively predicted over all the predicted values:

$$\text{Accuracy} = \frac{TP + TN}{TP + TN + FP + FN}; \quad (4.4)$$

$$\text{Precision} = \frac{TP}{TP + FP}; \quad (4.5)$$

$$F - \text{measure} = 2.\frac{\text{Precision Recall}}{\text{Precision} + \text{Recall}} \quad (4.6)$$

## 4.5 RESULTS AND DISCUSSION

In this research, a comprehensive experiment was conducted for exploring the basic characteristics of this problem, the effect of imbalanced learning and its interactions with data imbalance, type of classifier, input metrics and imbalanced learning method. Table 4.3 depicts feature analysis age on readmission. Table 4.4 depicts feature analysis gender on readmission. Table 4.5 depicts feature analysis admission source on readmission. Table 4.6 depicts feature analysis length of stay in days on readmission. Table 4.7 depicts feature analysis length of number of inpatient visits on readmission. Table 4.8 depicts feature analysis glucose serum test result visits on readmission. Table 4.9 depicts feature analysis HbA1c test result visits on readmission. Table 4.10 depicts feature analysis HbA1c test result visits on readmission. Table 4.11 depicts feature analysis number of medications on readmission. Table 4.12 depicts feature analysis number of lab procedures on readmission.

## TABLE 4.3
## Feature Analysis Age on Readmission

| | Total (9,476) | | Readmitted (1,306) | | Not Readmitted (10,782) | | |
|---|---|---|---|---|---|---|---|
| Feature Name | Frequency | Percentage | Frequency | Percentage | Frequency | Percentage | *p* -Value |
| **Age** | | | | | | | |
| <= 25 | 32 | 0.3 | 5 | 0.4 | 27 | 0.3 | |
| 26–30 | 33 | 0.4 | 5 | 0.4 | 28 | 0.3 | |
| 31–35 | 66 | 0.7 | 4 | 0.3 | 62 | 0.8 | |
| 36–40 | 93 | 1 | 11 | 0.9 | 82 | 1 | |
| 41–45 | 252 | 2.7 | 37 | 3.1 | 215 | 2.6 | |
| 46–50 | 434 | 4.6 | 32 | 2.6 | 402 | 4.9 | |
| 51–55 | 916 | 9.8 | 1 | 10.7 | 786 | 9.6 | 0.012 |
| 56–60 | 19 | 14 | 155 | 12.8 | 1154 | 14.1 | |
| 61–65 | 1759 | 18.8 | 264 | 21.8 | 1495 | 18.3 | |
| 66–70 | 1760 | 18.8 | 216 | 17.8 | 1544 | 18.9 | |
| 71–75 | 1310 | 14 | 164 | 13.5 | 1146 | 14 | |
| 76–80 | 780 | 8.3 | 112 | 9.2 | 668 | 8.2 | |
| 81–85 | 447 | 4.8 | 56 | 4.6 | 391 | 4.8 | |
| 86–90 | 143 | 1.5 | 15 | 1.2 | 128 | 1.6 | |
| 91+ | 47 | 0.5 | 5 | 0.4 | 42 | 0.5 | |

## TABLE 4.4
## Feature Analysis Gender on Readmission

| | Total (9,476) | | Readmitted (1,306) | | Not Readmitted (10,782) | | |
|---|---|---|---|---|---|---|---|
| Feature Name | Frequency | Percentage | Frequency | Percentage | Frequency | Percentage | *p* -Value |
| **Gender** | | | | | | | |
| Female | 3,606 | 38.4 | 23 | 34.9 | 3,183 | 39 | .007 |
| Male | 5,775 | 61.6 | 88 | 65.1 | 4,987 | 61 | |

## TABLE 4.5
## Feature Analysis Admission Source on Readmission

| | Total (9,476) | | Readmitted (1,306) | | Not Readmitted (10,782) | | |
|---|---|---|---|---|---|---|---|
| Feature Name | Frequency | Percentage | Frequency | Percentage | Frequency | Percentage | *p* -Value |
| **Admission Source** | | | | | | | |
| Emergency | 3,694 | 39.4 | 524 | 43.3 | 3,170 | 38.8 | .003 |
| Planned | 5,687 | 60.6 | 687 | 56.7 | 5,000 | 61.2 | |

**TABLE 4.6**
**Feature Analysis Length of Stay in Days on Readmission**

| | Total (9,476) | | Readmitted (1,306) | | Not Readmitted (10,782) | | |
|---|---|---|---|---|---|---|---|
| Feature Name | Frequency | Percentage | Frequency | Percentage | Frequency | Percentage | p -Value |
| **Length of Stay in Days** | | | | | | | |
| 1 | 1,559 | 16.6 | 208 | 17.2 | 1,351 | 16.5 | |
| 2 | 1,509 | 16.1 | 176 | 14.5 | 1,333 | 16.3 | |
| 3 | 1,064 | 11.3 | 138 | 11.4 | 926 | 11.3 | 0.001 |
| 4 | 822 | 8.8 | 94 | 7.8 | 728 | 8.9 | |
| 5 | 674 | 7.2 | 86 | 7.1 | 588 | 7.2 | |
| 6 | 528 | 5.6 | 75 | 6.2 | 453 | 5.5 | |
| 7 | 831 | 8.9 | 72 | 5.9 | 759 | 9.3 | |
| 8 | 720 | 7.7 | 68 | 5.6 | 652 | 8 | |
| 9 | 373 | 4 | 41 | 3.4 | 332 | 4.1 | |
| 10 | 312 | 3.3 | 64 | 5.3 | 248 | 3 | |
| 11 | 209 | 2.2 | 39 | 3.2 | 170 | 2.1 | |
| 12 | 143 | 1.5 | 22 | 1.8 | 121 | 1.5 | |
| 13+ | 637 | 6.8 | 128 | 10.6 | 509 | 6.2 | |

**TABLE 4.7**
**Feature Analysis Length of Number of Inpatient Visits on Readmission**

| | Total (9,476) | | Readmitted (1,306) | | Not Readmitted (10,782) | | |
|---|---|---|---|---|---|---|---|
| Feature Name | Frequency | Percentage | Frequency | Percentage | Frequency | Percentage | p -Value |
| **No. of Inpatient Visits** | | | | | | | |
| 0 | 6,928 | 73.9 | 676 | 55.8 | 6,252 | 76.5 | |
| 1 | 1,431 | 15.3 | 223 | 18.4 | 1,208 | 14.8 | |
| 2 | 492 | 5.2 | 109 | 9 | 383 | 4.7 | |
| 3 | 225 | 2.4 | 82 | 6.8 | 143 | 1.8 | 0.001 |
| 4 | 125 | 1.3 | 37 | 3.1 | 88 | 1.1 | |
| 5 | 65 | 0.7 | 22 | 1.8 | 43 | 0.5 | |
| 6 | 34 | 0.4 | 16 | 1.3 | 18 | 0.2 | |
| 7 | 23 | 0.2 | 9 | 0.7 | 14 | 0.2 | |
| 8 | 12 | 0.1 | 4 | 0.3 | 8 | 0.1 | |
| 9 | 4 | 0.3 | 3 | 0.2 | 1 | 0.1 | |
| 10+ | 42 | 0.4 | 13 | 2.5 | 12 | 0.1 | |

**TABLE 4.8**
**Feature Analysis Glucose Serum Test Result Visits on Readmission**

| | Total (9,476) | | Readmitted (1,306) | | Not Readmitted (10,782) | | |
|---|---|---|---|---|---|---|---|
| Feature Name | Frequency | Percentage | Frequency | Percentage | Frequency | Percentage | *p* -Value |
| Glucose Serum Test Result | | | | | | | |
| >200 | 845 | 9 | 87 | 7.2 | 758 | 9.3 | |
| >300 | 355 | 3.3 | 41 | 3.4 | 264 | 3.2 | .0.17 |
| Normal | 1,927 | 20.5 | 226 | 18.7 | 1,701 | 20.8 | |
| None | 6,334 | 67.2 | 857 | 70.8 | 5,447 | 66.7 | |

**TABLE 4.9**
**Feature Analysis HbA1c Test Result Visits on Readmission**

| | Total (9,476) | | Readmitted (1,306) | | Not Readmitted (10,782) | | |
|---|---|---|---|---|---|---|---|
| Feature Name | Frequency | Percentage | Frequency | Percentage | Frequency | Percentage | *p* -Value |
| HbA1c Test Result | | | | | | | |
| >6 | 391 | 4.2 | 41 | 3.4 | 350 | 4.3 | 0.162 |
| >7 | 1,198 | 12.8 | 139 | 11.5 | 1,059 | 13 | |
| Normal | 186 | 2 | 28 | 2.3 | 158 | 1.9 | |
| None | 7,606 | 81.1 | 1,003 | 82.8 | 6,603 | 80.8 | |

**TABLE 4.10**
**Feature Analysis HbA1c Test Result Visits on Readmission**

| | Total (9,476) | | Readmitted (1,306) | | Not Readmitted (10,782) | | |
|---|---|---|---|---|---|---|---|
| Feature Name | Frequency | Percentage | Frequency | Percentage | Frequency | Percentage | *p* -Value |
| Number of Diagnosis | | | | | | | |
| <= 1 | 117 | 1.2 | 12 | 1 | 105 | 1.3 | |
| 2 | 599 | 6.4 | 150 | 12.4 | 449 | 5.5 | |
| 3 | 1,591 | 17 | 185 | 15.3 | 1,406 | 17.2 | |
| 4 | 1,207 | 12.9 | 135 | 11.1 | 1,072 | 13.1 | 0.001 |
| 5 | 1,622 | 17.4 | 151 | 12.5 | 1,479 | 18.1 | |
| 6 | 1,395 | 14.9 | 144 | 11.9 | 1,251 | 15.3 | |
| 7 | 1,038 | 11.1 | 122 | 10.1 | 916 | 11.2 | |
| 8 | 720 | 7.7 | 114 | 9.4 | 606 | 7.4 | |
| 9 | 445 | 4.7 | 73 | 6 | 372 | 4.6 | |
| 10 | 253 | 2.7 | 46 | 3.8 | 207 | 2.5 | |
| 11 | 137 | 1.5 | 28 | 2.3 | 109 | 1.3 | |
| 12 | 89 | 0.9 | 10 | 0.8 | 79 | 1 | |
| 13+ | 160 | 1.7 | 41 | 3.4 | 119 | 1.5 | |

**TABLE 4.11**
**Feature Analysis Number of Medications on Readmission**

| | Total (9,476) | | Readmitted (1,306) | | Not Readmitted (10,782) | | |
|---|---|---|---|---|---|---|---|
| Feature Name | Frequency | Percentage | Frequency | Percentage | Frequency | Percentage | *p*-Value |
| **Number of Medications** | | | | | | | |
| <= 25 | 1,399 | 14.9 | 199 | 16.4 | 1,200 | 14.7 | |
| 26–50 | 26 | 28 | 346 | 25.3 | 2,324 | 28.4 | |
| 51–75 | 1,542 | 16.4 | 201 | 16.6 | 1,341 | 16.4 | |
| 76–100 | 965 | 10.3 | 141 | 11.6 | 824 | 10.1 | |
| 101–125 | 1,041 | 11.1 | 108 | 8.9 | 933 | 11.4 | |
| 126–150 | 685 | 7.3 | 76 | 6.3 | 609 | 7.5 | 0.001 |
| 151–175 | 360 | 3.8 | 56 | 4.6 | 304 | 3.7 | |
| 176–200 | 199 | 2.1 | 29 | 2.4 | 170 | 2.1 | |
| 201–225 | 137 | 1.5 | 19 | 1.6 | 118 | 1.4 | |
| 226–250 | 87 | 0.9 | 12 | 1 | 75 | 0.9 | |
| 251+ | 336 | 3.6 | 64 | 5.3 | 272 | 3.3 | |

**TABLE 4.12**
**Feature Analysis Number of Lab Procedures on Readmission**

| | Total (9,476) | | Readmitted (1,306) | | Not Readmitted (10,782) | | |
|---|---|---|---|---|---|---|---|
| Feature Name | Frequency | Percentage | Frequency | Percentage | Frequency | Percentage | *p*-Value |
| **No. of Lab Procedures** | | | | | | | |
| <= 25 | 1,427 | 15.2 | 216 | 17.8 | 1,306 | 14.8 | |
| 26–50 | 1,129 | 12 | 122 | 10.1 | 1,007 | 12.3 | |
| 51–75 | 1,286 | 13.7 | 131 | 10.8 | 1,155 | 14.1 | 0.001 |
| 76–100 | 1,248 | 13.3 | 110 | 9.1 | 1,138 | 13.9 | |
| 101–125 | 1,099 | 11.7 | 111 | 9.2 | 988 | 12.1 | |
| 126–150 | 784 | 8.4 | 102 | 8.4 | 682 | 8.3 | |
| 151–175 | 575 | 6.1 | 73 | 6 | 502 | 6.1 | |
| 176–200 | 376 | 4 | 67 | 5.5 | 329 | 3.8 | |
| 201–225 | 293 | 3.1 | 46 | 3.8 | 247 | 3 | |
| 226–250 | 221 | 2.4 | 47 | 3.9 | 174 | 2.1 | |
| 251+ | 943 | 10.1 | 186 | 15.4 | 757 | 9.3 | |

Based on preceding tables, *p*-values, removing imbalance in the data, are useful for readmission prediction compared with the usual learning methods and imbalanced learning methods pairwise on the same base classifier, metrics and datasets. From this, we can compute the difference between predicting defects with and without an imbalanced learner for each dataset as follows from the repeated measure design of the experiment in performance without imbalanced learning under differing levels of data imbalance, causing negative effect as depicted in Figures 4.9 through Figure 4.12.

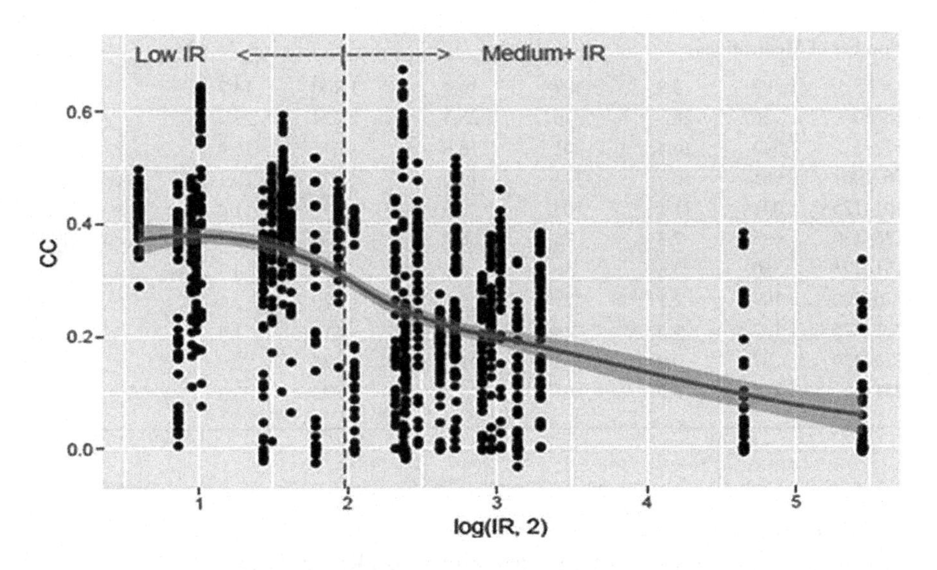


**FIGURE 4.9** Effect of Class Imbalance on Learning Ability.

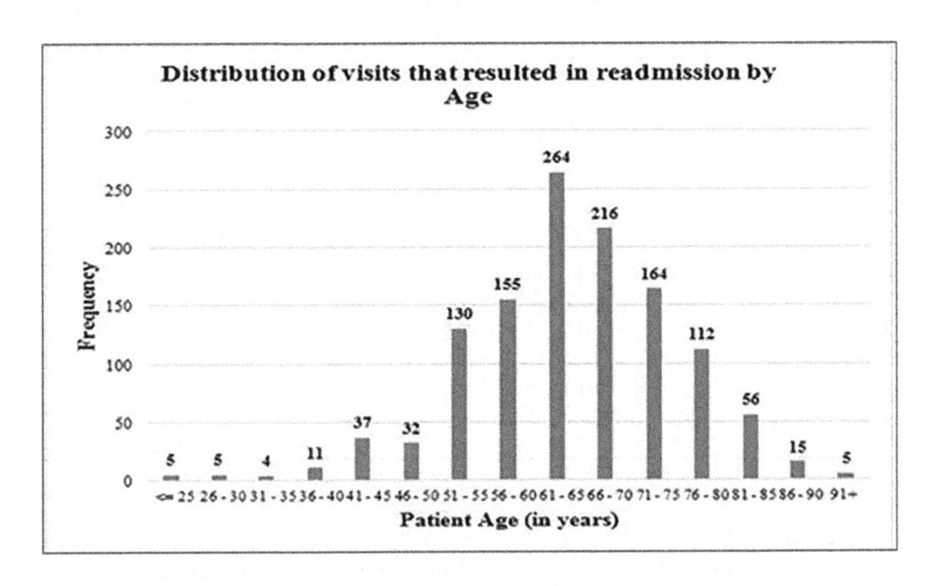


**FIGURE 4.10** Distribution of Visits That Resulted in Readmission by Age.

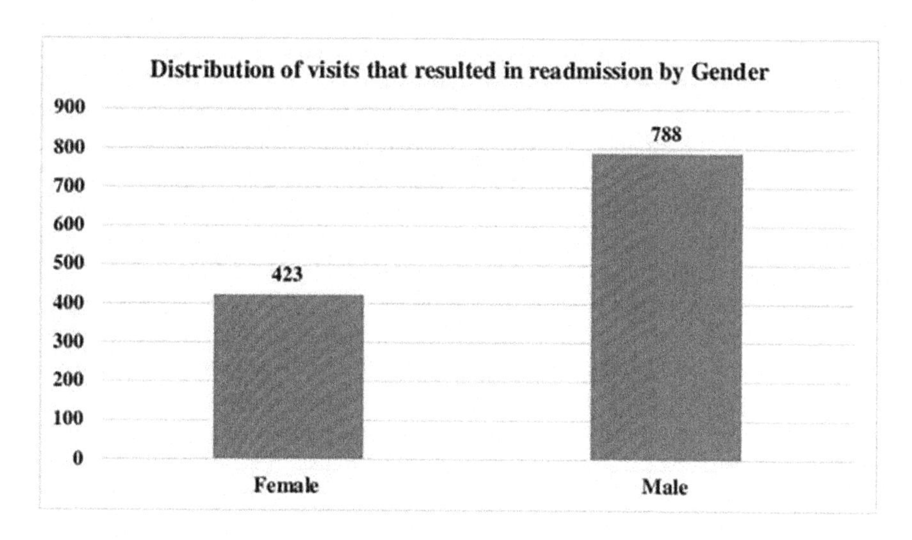


FIGURE 4.11 Distribution of Visits That Resulted in Readmission by Gender.

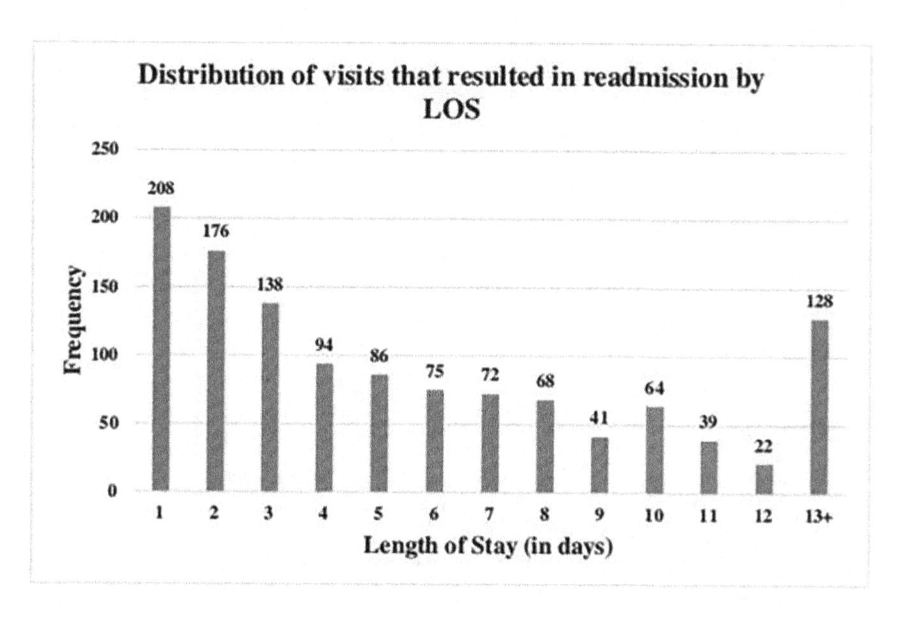


FIGURE 4.12 Distribution of Visits That Resulted in Readmission by Length of Stay.

## 4.6 CONCLUSION AND FUTURE WORK

In this research study, the real-world hospital data of beneficiaries under the AB-PMJAY having different comorbidities were analyzed and assessed with the objective of building an intelligent system to help identify patients who are prone of being readmitted. Thus, among 10 crores family, approximately 50 crores members' beneficiaries, this study classifies the patients into different risk groups of readmission as high, moderate, low, and very low after discharge based on patients' characteristics using their previous clinical and lab test data from an Indian hospital, mainly finding patient demographic characteristics divided among clusters of patients suffering from similar illnesses and then using an overlapping method for extracting repeated patients in two clusters using their diagnosis, lab reports, past history, medications used by them for specific period and treatment given to a particular class label admission. So, the prediction of readmissions is studied using different characteristics based on age, as senior citizen patients prone to sugar; type 2 diabetes; gender; admission source as outpatient or inpatient; department or medical specialty availed with length of stay and number of inpatient visits; number of lab procedures; number of diagnoses; number of medications; HbA1c test results; triglycerides test results; hypertension glucose serum test results; fasting glucose serum test results; hemoglobin, creatinine test results; albumin test results; low-density/high-density lipoprotein ratio; renal test results; diabetes medications; retinopathy; neuropathy; diabetic foot problems; troponin-I test results; ketones test results; Clexane test results; heart problems; gastropathy; obesity; potassium test results; amylase test results; and lipase test results. This study has assessed various feature selection techniques and proposed an optimal solution by avoiding recursion in an optimal subsets merger, thereby avoiding the class balancing that hampers accurate results of prediction modelling of the readmission for patients. Based on this empirical study on predicting risk of readmission, hospital readmission expense can be significantly reduced, which is greatly required for the Indian healthcare system. Further studies can be conducted in rural and urban community settings in multiple states across India by designing a specialized healthcare system as this allows determining how various readmission factors differ based on patient geographical location. Moreover, this would also strengthen both urban and rural hospitalization based on age categorization with various health conditions, such as heart disease and kidney disease, among others, in emergency readmissions.

## REFERENCES

Abdel-Basset, M., Chang, V., & Nabeeh, N. A. (2021). "An intelligent framework using disruptive technologies for COVID-19 analysis". *Technological Forecasting and Social Change, 163*, p. 120431.

Abdelrahman, M. (2020). "Personality traits, risk perception, and protective behaviors of Arab residents of Qatar during the COVID-19 pandemic". *International Journal of Mental Health and Addiction*, pp. 1–12.

Babar, M. I., Jehanzeb, M., Ghazali, M., Jawawi, D. N., Sher, F., & Ghayyur, S. A. K. (2016, October). "Big data survey in healthcare and a proposal for intelligent data diagnosis framework". In *2016 2nd IEEE International Conference on Computer and Communications (ICCC)*, pp. 7–12. Chengdu, China, IEEE Xplore.

Babu, S. K., Vasavi, S., & Nagarjuna, K. (2017, January). "Framework for predictive analytics as a service using ensemble model". In *2017 IEEE 7th International Advance Computing Conference (IACC)* pp. 121–128. Hyderabad, India, IEEE Xplore.

Bossen, C., & Piras, E. M. (2020). "Introduction to the special issue on information infrastructures in healthcare: Governance, quality improvement and service efficiency". *Computer Supported Cooperative Work (CSCW), 29*(4), pp. 381–386.

Cheng, C. H., Kuo, Y. H., & Zhou, Z. (2018). "Tracking nosocomial diseases at individual level with a real-time indoor positioning system". *Journal of medical systems, 42*(11), pp. 1–21.

De Silva, D., Burstein, F., Jelinek, H. F., & Stranieri, A. (2015). "Addressing the complexities of big data analytics in healthcare: The diabetes screening case". *Australasian Journal of Information Systems, 19.*

Forestiero, A., & Papuzzo, G. (2018, December). "Distributed algorithm for big data analytics in healthcare". In *2018 IEEE/WIC/ACM International Conference on Web Intelligence (WI)* (pp. 776–779). Santiago, Chile, IEEE Computer Society.

Galetsi, P., & Katsaliaki, K. (2020). "A review of the literature on big data analytics in healthcare". *Journal of the Operational Research Society, 71*(10), 1511–1529.

Gowsalya, M., Krushitha, K., & Valliyammai, C. (2014, December). "Predicting the risk of readmission of diabetic patients using MapReduce". In *2014 Sixth International Conference on Advanced Computing (ICoAC)* (pp. 297–301). Chennai, India, IEEE.

Gravili, G., Benvenuto, M., Avram, A., & Viola, C. (2018). "The influence of the digital divide on big data generation within supply chain management". *The International Journal of Logistics Management, 29*(2), pp. 592–628.

Jin, Q., Wu, B., Nishimura, S., & Ogihara, A. (2016, August). "Ubi-Liven: A human-centric safe and secure framework of ubiquitous living environments for the elderly". In *2016 International Conference on Advanced Cloud and Big Data (CBD)* (pp. 304–309). Chengdu, China, IEEE.

Kamble, S. S., Gunasekaran, A., Goswami, M., & Manda, J. (2018). "A systematic perspective on the applications of big data analytics in healthcare management". *International Journal of Healthcare Management, 12*(3), pp. 226–240.

Khanra, S., Dhir, A., Islam, A. N., & Mäntymäki, M. (2020). "Big data analytics in healthcare: a systematic literature review". *Enterprise Information Systems, 14*(7), 878–912.

Ma, X., Wang, Z., Zhou, S., Wen, H., & Zhang, Y. (2018, June). "Intelligent healthcare systems assisted by data analytics and mobile computing". In *2018 14th International Wireless Communications & Mobile Computing Conference (IWCMC)* (pp. 1317–1322). Limassol, Cyprus, IEEE.

Moutselos, K., Kyriazis, D., Diamantopoulou, V., & Maglogiannis, I. (2018, December). "Trustworthy data processing for health analytics tasks". In *2018 IEEE International Conference on Big Data (Big Data)* (pp. 3774–3779). Seattle, WA, USA, IEEE.

Navaz, A. N., Serhani, M. A., Al-Qirim, N., & Gergely, M. (2018). "Towards an efficient and Energy-Aware mobile big health data architecture". *Computer Methods and Programs in Biomedicine, 166*, 137–154.

Sabharwal, S., Gupta, S., & Thirunavukkarasu, K. (2016, April). "Insight of big data analytics in healthcare industry". In *2016 International Conference on Computing, Communication and Automation (ICCCA)* (pp. 95–100). Noida, India, IEEE.

Salomi, M., & Balamurugan, S. A. A. (2016). "Need, application and characteristics of big data analytics in healthcare – A survey". *Indian Journal of Science and Technology, 9*(16), 1–5.

Wu, J., Li, H., Liu, L., & Zheng, H. (2017). "Adoption of big data and analytics in mobile healthcare market: An economic perspective". *Electronic Commerce Research and Applications, 22*, pp. 24–41.

# *Section II*

## *Medical Imaging*

# 5 Diagnosis of Schizophrenia

## *A Study on Clinical and Computational Aspects*

*Indranath Chatterjee and Khushboo Mittal*

**CONTENTS**



DOI: 10.1201/9781003142751-7

## 5.1 INTRODUCTION

### 5.1.1 Schizophrenia

Schizophrenia is a severe, chronic mental disorder in which people live with delusions and hallucinations (Chatterjee et al., 2019). The person starts losing their real-world connections. Although it is not a common disorder, it is chronic and disabling. It generally presents in childhood and is experienced again and reinforced in a later period of life. It can happen because of

- genetics (heredity; Kallmann, 1938),
- brain chemistry and circuits,
- brain abnormality, and
- environment.

There are no laboratory tests to diagnose schizophrenia precisely. Whenever the following symptoms present in persons, they can be diagnosed:

- Delusions
- Hallucinations
- Disorganized speech
- Disorganized or catatonic behavior

Addressing schizophrenia involves developing a stress-free environment, a sense of hope, self-reliance, and a personalized awareness of current strengths and challenges (Lysaker & Buck, 2008).

### 5.1.2 Occurrence of Schizophrenia

Schizophrenia is a severe mental illness that affects a considerable amount of the population worldwide. The prevalence of the disorder is increasing. About 1% of the global population is affected by schizophrenia. It affects people regardless of race, ethnicity, geographical location, and so on. In 2000, the World Health Organization (WHO) found that the illness's prevalence and incidence are similar worldwide. According to a recent survey, it was found that around 1.2% of Americans are affected by schizophrenia. According to the data released for schizophrenia by WHO in 2004, it was found that the age-standardized DALY (disability-adjusted life-years) rate in India was 286.903 per 100,000 inhabitants (Chatterjee & Mittal, 2019).

Schizophrenia can occur in people throughout their lifetimes, but new instances of the illness are most likely to happen in early adulthood. It occurs primarily in the late adolescent years (the early 20s) to early adulthood (early 30s). People with schizophrenia start experiencing the primary symptoms between ages 16 to 25. It was found that men tend to develop the disorder earlier than women; that is, early diagnosis is more frequent in males than females. The onset of the illness is rare in the older population, as in childhood.

### 5.1.3 Reasons for Schizophrenia

It is challenging to determine the causes of schizophrenia. Some studies suggest schizophrenia can be caused due to some failure, sorrow, pain, genetic variants, social stress, and the like. Various factors, such as physical, genetic, psychological, and environmental, can cause a person to develop the illness. Some factors that may increase the risk for the development of the disease are discussed next.

#### 5.1.3.1 Genetics

Schizophrenia is sometimes found to occur in families. It is heritable, but a single gene is not responsible for the illness's circulation. A different combination of various genes is more likely to be vulnerable to the condition. Researchers have identified 108 genes associated with the risk of schizophrenia (Ripke et al., 2014). Most of the identified genes are dopamine receptors, glutamate transmission, immune system, and synaptic plasticity. It was also found that it is twin concordance; that is, in identical twins, if one has schizophrenia, the other has a 50% chance of developing the disorder, even if the children are raised separately. However, the genetic factor is not only the reason behind the illness.

#### 5.1.3.2 Development of the Brain

When people with schizophrenia are studied, differences in the brain structure suggest that the illness may also occur due to additional development of the brain. Nevertheless, developmental changes may be noticed in people without any mental illness.

#### 5.1.3.3 Neurotransmitters

Neurotransmitters are endogenous chemicals that help in the neurotransmission of information. It is found that drugs that change the levels of neurotransmitters often diminish some symptoms of schizophrenia. It shows that neurotransmitters may have some connections with the illness. Schizophrenia may be caused by an alteration in the level of dopamine and serotonin or sometimes may be caused due to an imbalance of these two neurotransmitters (Bansal & Chatterjee, 2021).

#### 5.1.3.4 Birth-Time Complications

Research suggests that experiencing birth-time complications may affect whether a person develops schizophrenia. Before birth, complications such as low birth weight, premature labor, and asphyxia during birth are more likely to affect people's brain development.

#### 5.1.3.5 Social and Personal Triggers

Triggers are the various factors that are mainly responsible for the cause of schizophrenia. The triggers include stress, such as from one's job and relationships and emotional, physical, and sexual abuse. These stressful situations do not necessarily cause schizophrenia but can trigger their development.

#### 5.1.3.6 Drug Intake

Studies show different drugs such as cannabis, cocaine, or LSD may affect the human brain to trigger mental illness development. However, medicines do not cause schizophrenia directly (Uzun et al., 2003).

## 5.2 CLINICAL DIAGNOSIS OF SCHIZOPHRENIA

### 5.2.1 Primary Symptoms

In recent days, schizophrenia is considered a neurodevelopmental disorder. Patients with this mental illness suffer from various symptoms (Charernboon, 2020). According to the *Diagnostic and Statistical Manual of Mental Disorders, Fifth Edition* (*DSM-5*) criteria (American Psychiatric Association, 2013), if two of the following five main symptoms persist, a patient is said to have schizophrenia. Figure 5.1 shows the most common forms of positive and negative symptoms of schizophrenia.

The main symptoms are as follows:

1. Delusions:
   Delusions are false beliefs out of reality. The patient thinks several fake thoughts about being harmed or harassed; they gain sudden exceptional fame; some catastrophe may happen. It is prevalent in schizophrenia.
2. Hallucinations:
   Hallucinations can be visual or auditory. During an auditory hallucination, the patient listens to individual, indistinct voices; noise; someone's cry; or calls for them that do not exist in actuality. Visual hallucinations cause the patient to see unwanted things or some known/unknown person who does not live. The person with schizophrenia has a deep conviction in these beliefs.
3. Disorganized thinking (speech):
   Disorganized thinking is also a common symptom of schizophrenia. It is inferred from disorganized speech—the patient who suffers from the illness experiences an impaired ability of effective communication. In rare cases, the patient uses a combination of meaningless words that may not be understood.
4. Disorganized or abnormal motor behavior:
   In some cases of schizophrenia, the patient shows an abnormality in their motor behavior in several ways, such as childlike silliness, resistance to any instructions, inappropriate postures, and unresponsiveness to queries.
5. Negative symptoms:
   Negative symptoms lead to an inability to function in daily life properly. The patients suffering from negative symptoms show a lack of interest in daily activities, withdraw themselves from society and social gatherings, show a lack of emotions, and even show negligence to personal hygiene (Ulas et al., 2007). Negative symptoms often are seen much before the occurrence of positive traits.

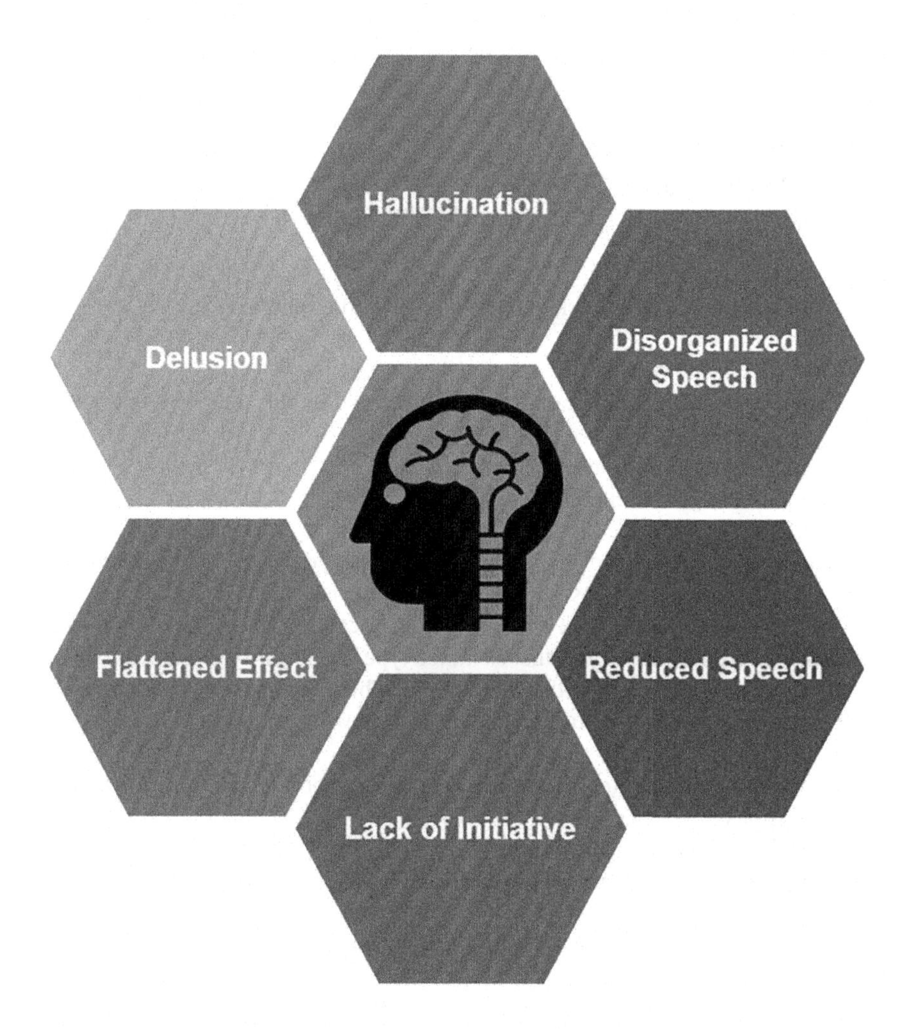


**FIGURE 5.1** Symptoms of Schizophrenia.

#### 5.2.1.1 Early Symptoms or Prodrome of Schizophrenia

Most patients with schizophrenia (about 80–90%) have a prodrome, characterized by some attenuated or subthreshold symptoms that appear along with delusions and hallucinations (Yung & McGorry, 1996). The primary prodromal symptoms include perplexity, diminished understanding, hearing a few indistinct noises, highly unusual beliefs, and guardedness, among others.

According to some studies, the prodromal phase can vary from person to person, but it lasts for almost one year in each patient. Most schizophrenic patients seek help and seem distressed while in the clinical high-risk (CHR) state.

Bleuler made the primary classification of schizophrenia. He recognized two significant classes of the symptoms, namely, positive and negative symptoms. In 1980, Crow suggested that schizophrenia can be classified into type 1 and type 2.

The characteristics of type 1 schizophrenia include the following:

1. Positive symptoms
2. Normal brain structure
3. Comparatively good response to treatment
4. Dopamine is assumed to be responsible for the neurochemical process governing the disease

The characteristics of type 2 schizophrenia include the following:

1. Negative symptoms
2. Some abnormalities in brain structure observed on the computed tomography scan
3. Relatively weak responses to the treatment
4. Impaired cognitive function

Type 1 schizophrenia with positive symptoms was named positive schizophrenia, and type 2 with negative symptoms was called negative schizophrenia. Andreasen (1987) suggested that some people have both positive and negative schizophrenic symptoms. The author added one more type, that is, mixed type, along with the previous two. The symptoms that mainly occur in the two types of schizophrenia are as follows:

Positive Schizophrenia—At least one of the following are present:

1. Hallucinations (auditory, visual, haptic, or olfactory; Andreasen, 1987)
2. Delusions
3. Positive formal thought disorder
4. The repeated occurrence of disorganized behavior

Negative Schizophrenia—At least two of the following are present to a marked degree:

1. Affective flattening
2. Anhedonia asociality (e.g., not able to feel intimacy or experience pleasure in social contacts)
3. Avolition—apathy (e.g., impersistence, anergia at work);
4. Attentional impairment.
5. Alogia (e.g., poverty of speech, poor content of speech; Andreasen, 1987)

Mixed Schizophrenia:

This category incorporates patients who show both positive and negative symptoms or did not fulfill the criteria for either one. Table 5.1 shows the list of important research papers stating the process of clinical diagnosis of schizophrenia.

### 5.2.2 Treatments of Schizophrenia

As schizophrenia was considered a brain disease or a set of diseases due to the transmission of genetic or prenatal or perinatal insult, the assumption was that

**TABLE 5.1**
**List of Important Research Papers Stating the Process of Clinical Diagnosis of Schizophrenia**

| Paper | Author | Year |
|---|---|---|
| "Diagnosis and Prognosis of Schizophrenia" | Langfeldt | 1960 |
| "A Checklist for the Diagnosis of Schizophrenia" | Astrachan et al. | 1972 |
| "Biological Homogeneity, Symptom Heterogeneity, and the Diagnosis of Schizophrenia" | Buchsbaum and Haier | 1978 |
| "Diagnosis of Schizophrenia: A Critical Review of Current Diagnostic Systems" | Fenton, Mosher, and Matthews | 1981 |
| "The Diagnosis of Schizophrenia" | Andreasen | 1987 |
| "The Diagnosis of Schizophrenia: A Review of Onset and Duration Issues" | Keith and Matthews | 1991 |
| "Symptom Assessment in Casenotes and the Clinical Diagnosis of Schizophrenia" | Lützhøft et al. | 1995 |
| "Diagnosis of Schizophrenia: A Review" | Pull | 1999 |
| "Toward Reformulating the Diagnosis of Schizophrenia" | Tsuang, Stone, and Faraone | 2000 |
| "Validation of a Blood-Based Laboratory Test to Aid in the Confirmation of a Diagnosis of Schizophrenia" | Schwarz et al. | 2010 |
| "Chronic Smoking and Cognition in Patients with Schizophrenia: A Meta-Analysis" | Coustals et al. | 2020 |

psychological interventions cannot be used to treat this disorder. Many schizophrenia patients are still untreated due to unawareness. Treatment is sometimes misguided due to the manifestation of similar symptoms in psychosis. However, psychosis is associated not only with schizophrenia but also with other disorders, such as dementia, Parkinson's disease, stroke, brain tumors, drug abuse, and alcohol usage. Schizophrenia is a psychosis primarily associated with delusion and hallucinations. Figure 5.2 shows the general stages of a standard clinical diagnosis process of schizophrenia. This figure also shows the symptoms that occur at each stage and its diagnosis methodology.

Schizophrenia patients have a premature death, around 14.5 years earlier than the healthy population's average mortal age. It is found that 40% of premature deaths are due to suicides.

#### 5.2.2.1 The Process of Treatment

The treatment goals of schizophrenia patients mainly aim to reduce the symptoms, prevent the relapse of the symptoms, and enhance the basic functionality, helping patients live healthy social lives.

Until the mid-1900s, no drugs for schizophrenia were available. The treatment was mostly like confining the patients in asylums, and potent doses of sedative drugs were administered to suppress the behavioral changes. The electric shock treatment, also known as electroconvulsive therapy (ECT), was widespread to control

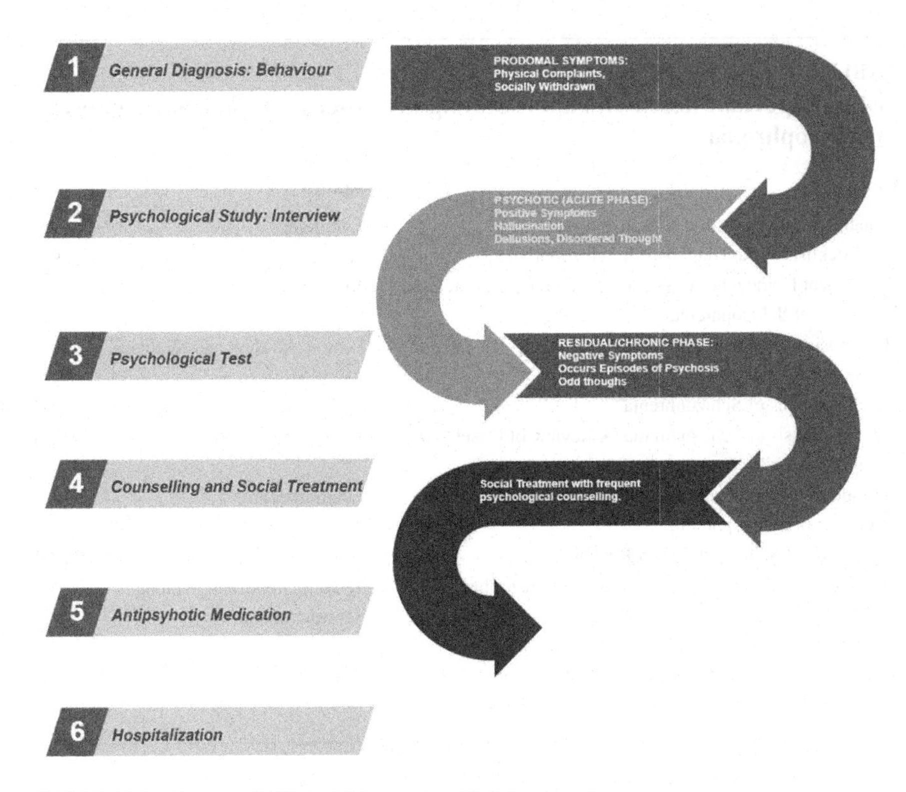


**FIGURE 5.2** Stages of Clinical Diagnosis of Schizophrenia.

depression and psychotic abnormal behaviors. The application of ECT is still controversial and is used today for acute psychotic administration.

The treatment of schizophrenia can be broadly categorized into two types, namely, nonpharmacological therapy and pharmacological therapy. With pharmacotherapy being necessary in treating schizophrenia, the residual symptoms still may remain active. Thus, non-pharmacotherapy, like psychotherapy, is essential (Üçok et al., 2004).

Psychotherapy may be again categorized into three types:

1. Individual level
   - Counseling
   - Personal therapy
2. Group level
   - Social therapy
3. Cognitive therapy
   - Cognitive behavioral therapy
   - Cognitive enhancement therapy

Other psychotherapies include social skills training, rehabilitation, family education, and joining self-help groups.

##### *5.2.2.1.1 Pharmacological Therapy*

Pharmacotherapy involves the administration of medicine. Medication is the most vital method for treating schizophrenia (Hursitoglu et al., 2020). Antipsychotic medications are the common drugs prescribed to schizophrenia patients. They can control and reduce the symptoms like hallucinations and delusions by affecting the brain's neurotransmitters. The dosage and selection of the drug are very crucial in antipsychotic medication. It is always targeted to manage the illness with the lowest possible dose. Antipsychotic medication usually takes 1 to 2 weeks to start working.

As the medicines for schizophrenia have serious side effects, patients sometimes become hesitant to take them. Thus, it is also vital to support and help the patients bring their willingness to cooperate with the treatment process. Recently, a new treatment option, an alternative to oral medication in the form of injection, became available. It is called long-acting treatment or long-acting injectable, which can be administered once every few weeks to once every few months, depending on the severity of long-term effects on the body.

There are primarily two types of antipsychotic medicines: first generation and second generation. The first-generation drugs show common and potential side effects in the brain, sometimes causing temporary or permanent motion disorder. First-generation antipsychotic drugs include the following (Leucht et al., 2009; Patel et al., 2014):

1. Chlorpromazine
2. Fluphenazine
3. Haloperidol
4. Perphenazine

These drugs are comparatively cheaper than the second generation. Thus, these drugs may be administered to patients when long-term treatment is needed.

Second-generation antipsychotics are generally recommended and prescribed due to their fewer side effects on the brain. Second-generation antipsychotics include aripiprazole, asenapine, brexpiprazole, clozapine, and olanzapine, among others (Leucht et al., 2009; Patel et al., 2014).

##### *5.2.2.1.2 Alternate Ways of Treatment*

*Hospitalization* When symptoms become very frequent for severe cases and patients start showing abnormal behavior, hospitalization may be prescribed to provide proper nutrition, adequate sleep, and safety.

*ECT* When adult schizophrenic patients stop responding to medication, ECT may be prescribed in rare cases. ECT may pose helpful for patients suffering from depression too. While many schizophrenia patients are still untreated due to social exclusion and stigma and negative stereotypes, lifelong treatment, including medication and psychotherapy, is required in schizophrenia, even after symptoms subside. The proper dosage of medications and psychosocial therapy can help in the management of the disorder.

## 5.3 COMPUTATIONAL DIAGNOSIS OF SCHIZOPHRENIA

### 5.3.1 fMRI

fMRI stands for functional magnetic resonance imaging. It is an imaging technique used to identify functional activations across different brain regions (Chatterjee et al., 2020b). The fMRI data are four-dimensional (4D) data, consisting of 3D images over time (Chatterjee et al., 2020c). Three-dimensional data is a sequence of 2D images (or slices) over the entire brain. Each slice consists of small units of brain volume, which are called a voxel. Therefore, a specific spot of the brain is represented by a voxel. So, by comparing the fMRI data of schizophrenia patients and healthy controls, we can diagnose the disease by detecting the active regions during a particular task in schizophrenia patients and healthy controls.

#### 5.3.1.1 Principle of an fMRI

When neurons get activated, they are supplied with raised cerebral blood flow and the oxygen supply, usually, more than they required, by the adjacent capillaries by the hemodynamic response process. They do not have any internal set aside of energy, neither in the form of oxygen nor glucose. Based on their distinctive magnetic susceptibilities, MRI detects the activations by measuring oxyhemoglobin and deoxyhemoglobin (Chatterjee et al., 2020a). This technique is called blood oxygen level–dependent (BOLD) imaging. According to recent research, BOLD signals intent by arteries partial pressure of both $O_2$ and $CO_2$. These BOLD signals are the main foundation of fMRI. We can construct maps of brain regions active during a particular task or react to detectable changes in the neural activity at a low frequency (0.01–0.01 Hz; Chatterjee et al., 2018).

#### 5.3.1.2 Why Do We Use fMRI

Because of its excellent spatial resolution, we use fMRI, from which we can detect the regions of the brain that are active during tasks. Through fMRI, researchers get technical expertise in small subcortical nuclei by visualizing transient activity. It is used to study functional connectivity, inter-region coupling, and networked computation in large-scale brain networks that support perceptual and cognitive processes. Therefore, it is appropriate to compare and fit simple structural models; it considers the structural and connectivity model selection procedures for large-scale areas.

fMRI is an indirect measure of synaptic and neuronal activity. It monitors the regional changes in blood oxygenation, which is happening in neural activity. However, some technical issues remain to be resolved, like incapability for localizing primary sensory and motor areas.

### 5.3.2 rs-fMRI

Our brain is functionally and metabolically active during the rs. rs-fMRI is widely used for mental, neurological, and neuropsychiatric disorders. Unlike task-based fMRI, rs-fMRI does not prescribe that subjects do any cognitive (mental, emotional) task

(Savio et al., 2014). The spontaneous neural activities can be related to low-frequency oscillations (0.01–0.1 Hz) of an rs-fMRI. It focuses on low-frequency fluctuations and spontaneous BOLD signals. Functional low connectivity is defined as the temporal dependency of neuronal activation patterns of anatomically separated brain regions.

rs-fMRI measures the correlation between random activation patterns of brain regions. Identifying correlation patterns in the spontaneous fluctuations of the BOLD signal has increased fMRI translation into clinical care. This technique has allowed identifying various resting-state networks or different brain areas that represent synchronous BOLD fluctuations at rest.

We can transform rs-fMRI data analysis in many ways, with the popularly known independent component analysis (ICA) and seed-based analysis techniques. Before that, we have to do preprocessing of data by reorganizing and eliminating confounding artifacts. Moreover, rs-fMRI data require corrections to slice timing, spatial filtering, motion compensation, and normalization in these preprocessing steps.

#### 5.3.2.1 Challenges of rs-fMRI

The removal of noise, that is, temporal filtering, must be done carefully so that there is no removal of appropriate or closely connected low-frequency rs signals. Noise levels from movement and physiological (like breathing) needs in rs-fMRI are very high. There are fewer than 5% of total signals that derive from neurological activity. rs-fMRI studies' main problems are the frequencies of cardiac and respiratory pulsations, which are close to the rs networks.

One weakness of rs-fMRI lies in the difference between the analysis of spontaneous fluctuations. This functional connectivity is calculated by measuring the correlation coefficient of temporal similarity of the BOLD time series in voxels. Voxels whose value of coefficient crossed the statistical threshold were considered functionally connected, giving frequent, spontaneous fluctuations of left and right motor cortices. Since the two time series are measured simultaneously, functional connectivity is affected by any non-neural-related activity process, thus giving a spurious result. These deceptive similarities increase the apparent functional connectivity, and if any differential confounds between regions are found, it reduces the connectivity metric. This will be problematic if we compare connectivity between behaviorally and physiologically different groups at "rest" by using the temporal similarity metric.

#### 5.3.2.2 Methods of Analysis of rs-fMRI Data

A large amount of rs-fMRI data can be easily compared with the help of graphs and maps. Various methods can be used to analyze rs-fMRI data and models, like the amplitude of low-frequency fluctuations (ALFF), fractional ALFF, regional homogeneity (ReHo), functional connectivity density, seed-based functional connectivity analysis or Region-of-Interest (ROI)-based functional connectivity, and ICA (Chatterjee et al., 2019).

The ALFF methods measure the strength of the BOLD signals within a low-frequency range between 0.01 Hz and 0.1 Hz. Both fractional-ALFF and ALFF reveal the densities of the active brain regions. Neither provides details about the functional connectivity within areas of the brain. These methods are known for their simplicity of analysis without any hypothesis. Both give long-term test–retest reliability and high temporal stability.

ReHo analysis is a voxel-based method with providing information about the various features of regional neurological activity. It finds the correlation between the BOLD signal of a given voxel and its nearby neighbors. It is calculated by the Kendall coefficient of the corresponding BOLD signals. A higher value of ReHo represents centrality and higher coherence of regional brain activity (Chyzhyk et al., 2015).

Seed-based functional connectivity is also called ROI-based functional connectivity. It finds the regions which are active at the same time when the seed region is active. It visualizes the correlation between the seed and the rest of the brain regions by the connectivity matrix, which shows all connections' strength. These regions may not be directly connected, yet the coupling of the activation of different brain regions shows their involvement in the same functional process and indicates that they are functionally connected.

ICA is used to separate spatially independent figures (Chatterjee et al., 2019) from their unswerving, mixed BOLD signals in the form of spatial maps, which are temporally correlated.

#### 5.3.2.3 Need rs-fMRI

Infants and neuropsychiatric disordered patients cannot perform the task in a task-based fMRI. rs-fMRI possesses no burden on them to perform any task. It focuses on spontaneous low-frequency BOLD fluctuations <0.1 Hz and investigates synchronous activations between spatially different regions to identify rs networks (RSNs) in the absence of tasks. It provides information about neural activities or active areas, free from demanding external stimuli. It examines the relationship between multiple RSNs and independently measured physiological and behavioral traits. The demonstration of RSNs has helped make fMRI an investigative tool for brain dynamics. The rs-fMRI can find a marker to distinguish an individual with major depressive disorder from a healthy control (Chatterjee et al., 2018).

### 5.3.3 Computer-Aided Diagnosis of Schizophrenia Using rs-fMRI

Computational approaches towards the diagnosis of schizophrenia are getting popular these days. However, besides the task-based fMRI, rs-fMRI is also gaining attention from computational neuroscientists. The process of computational diagnosis of schizophrenia involves giant steps to tackle the challenges. In the computer-aided process of schizophrenia diagnosis, we first apply feature selection and feature extraction methods that give the voxels or regions to discriminate between healthy and schizophrenia subjects. After this, we classify the selected voxels. Such regions/voxels are considered the biomarkers for the diseases for further studies. This will increase the classifier's predictive accuracy, as the features that give higher accuracy will have greater value as biomarkers.

First of all, we have to select a dataset that contains rs-fMRI of healthy and schizophrenia subjects. After collecting the dataset, we have to preprocess it. Standard preprocessing routines includes slice timing correction, motion correction, and denoising; smoothing and spatial normalization to the Montreal Neurological

Institute template; temporal filtering; linear trend removal; and, finally, the functional data of each subject's co-registered with the structural data of its corresponding image in addition to the registration with MNI152 template.

We apply feature selection methods due to high data dimensionality for selecting the most relevant voxels in the next step. Due to the registration error, noise in many voxels does not seem able to discriminate against healthy and patients. So we have to select such feature selection methods that select such voxels that achieve high discriminative power to construct the features space for classification. Some of the feature selection methods are as follows:

- Pearson correlation:
  Pearson's correlation test measures the relationship, the association, between the two continuous independent variables. It is based on the method of covariance. It gives the magnitude of the ties, association as well as the direction of the relationship.
- Statistical hypothesis testing (*t* test):
  The *t* test is a statistical hypothesis testing method. It is primarily used when the *t*-statistics follow a normal distribution. The *t* test is used for determining the significant difference between two groups of data samples, hypothesizing their population means the same for both groups. Other similar tests are the *F* test and the chi-squared test.
- Chi-squared test:
  It is a test applied to the categorical data to evaluate the correlation or association between them using their frequency distribution.

Alongside these filter-based approaches for feature selection, we may also use various wrapper methods for selecting the appropriate features out of the rs-fMRI data. We may use multiple computational approaches involving evolutionary approaches, such as genetic algorithms and particle swarm optimization, and other well-known methods, such as singular value decomposition and deep learning techniques. Figure 5.3 shows an overview of the entire process of the computational diagnosis of schizophrenia. It shows the different analysis levels, starting from fMRI data acquisition to computer-aided diagnosis and medical intervention.

Finally, we can verify the efficacy of the approaches used and identify the obtained features' different capacities using various classifiers (Gautam & Chatterjee, 2020).

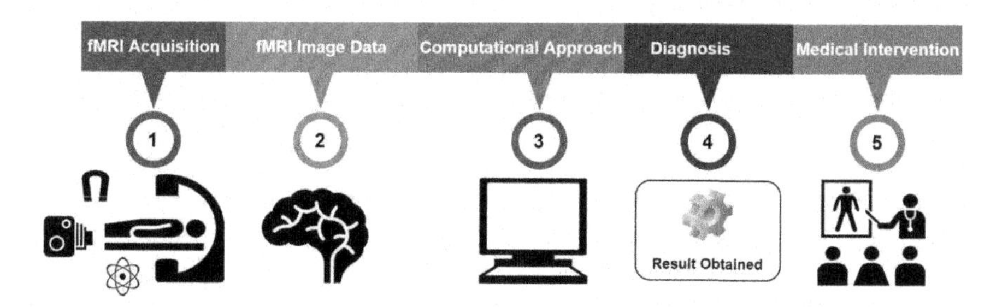


**FIGURE 5.3** Overview of the Entire Process of Computational Diagnosis of Schizophrenia.

**TABLE 5.2**
**List of Important Research Papers Stating the Process of Computational Diagnosis of Schizophrenia**

| Paper | Author | Year | Computational Approach |
|---|---|---|---|
| A computational approach to prefrontal cortex, cognitive control and schizophrenia: recent developments . . . | Cohen, Braver, and O' Reilly | 1996 | Review of first-generation neural networks |
| Computer-aided neurocognitive remediation as an enhancing strategy for schizophrenia rehabilitation | Cavallaro et al | 2009 | Analysis of variance (ANOVA) |
| Computer-aided diagnosis of schizophrenia based on local-activity measures of resting-state fMRI | Savio, Chyzhyk, and Grana | 2014 | ReHo; brain local activity measures, voxel saliency, and feature extraction parameters |
| Computer-aided diagnosis of schizophrenia based on local-activity measures of resting-state fMRI | Chyzhyk, Savio, and Grana | 2015 | Single-layer feed-forward networks (SLFNs), ELM |
| A computer-aided diagnosis system with EEG based on the P3b wave during an auditory oddball task . . . | Santos-Mayo, San-Jose-Revuelta, and Arribas | 2016 | Parameter optimization for various machine learning models |
| A computer-aided diagnosis system with EEG based on the P3b wave during an auditory odd-ball task | Valton et al. | 2017 | Machine learning classifers |
| Bi-objective approach for computer-aided diagnosis of schizophrenia patients using fMRI data | Chatterjee et al. | 2018 | Evolutionary algorithm, NSGA-II |
| Mean deviation based identification of activated voxels from time-series fMRI data of schizophrenia patients | Chatterjee | 2018 | Statistical approach, machine learning |
| Hybrid functional brain network with first-order and second-order information for computer-aided diagnosis . . . | Zhu et al. | 2019 | Triplet correlation, brain functional connectivity network (BFCN) analysis |
| Multi-modality low-rank learning fused first-order and second-order information for computer-aided . . . | Li et al. | 2019 | Multimodal BFCN feature selection, classification method |
| Diagnosis of schizophrenia from R-fMRI data using Ripplet transform and OLPP | Sartipi, Kalbkhani, and Shayesteh | 2020 | Orthogonal locality-preserving projection (OLPP), ICA, Ripple-II transform algorithm |

The widely used classifiers used in neuroimaging studies are Support Vector Machine (SVM), k-nearest neighbor (KNN), Random Forest, Extreme Learning Machine (ELM), naïve Bayes classifier, and others (Kumar & Chatterjee, 2016). To obtain high accuracy, often cross-validation techniques are also used. The most used cross-validation technique in neuroimaging studies is Leave-one-out cross-validation, popularly known as LOOCV. It is widely used due to the unavailability of a large number of samples. Researchers have also used k-gold and hold-out schemes as well. Table 5.2 shows the list of important research papers stating the process of computational diagnosis of schizophrenia.

#### 5.3.3.1 Advantages of rs-fMRI

Unlike task-based fMRI, rs-fMRI can be easily performed by infants, young children, and a person suffering from neurological and psychiatric diseases. rs functional connectivity determines the brain's tightly coupled functional networks using spontaneous synchronized fluctuations in the BOLD signal. These spontaneous fluctuations are postulated to coordinate, organize, and maintain operational brain systems (Erhardt et al., 2011). rs-fMRI places a minimal mental burden on participants and only takes a little time.

rs-fMRI offers a 3:1 improvement in signal-to-noise ratio compared to task-based fMRI. This can be demonstrated by a simple task in which patients were asked to press a button with their right hand during the scanning session. The patient pressed the button only once in the whole scanning section. At this time, the task-based BOLD modulation is maximal. So the task-based standard fMRI session discards more than 80% of the BOLD modulation as noise. In contrast, rs-fMRI focuses on this spontaneous activity as the signal rather than scrapping it as noise.

## 5.4 CONCLUSION

In this chapter, first, we elaborately discussed schizophrenia, its symptoms, its causes, and its treatments. This study presented the clinical diagnosis of the disorder in detail. Second, we talked about fMRI, followed by the working principles and introduction of rs-fMRI. Finally, we have described the methods of the computer-aided diagnosis of schizophrenia using rs-fMRI data. We also added a concise description of the data analysis of rs-fMRI and its advantages over task-based fMRI. This paper throws light on rs-fMRI's applicability for the diagnosis of schizophrenia and mentions the new way of computational diagnosis, beneficial over traditional clinical diagnosis. As a scope of future work, researchers, after gaining basic knowledge of schizophrenia and fMRI, may further study the various pathophysiological aspects of schizophrenia and may develop a computer-aided diagnosis tool using rs-fMRI.

**Disclosure Statement**

The authors declared no disclosure statement.

**Consent of Patients and Hospital**

This research does not involve any clinical study of patients. So the consent of patients and the hospital are not applicable here.

**Funding**

The authors declared no funding support for this research.

**Statement for Declaration of Interest**

The authors declared no conflict of interest.

## REFERENCES


American Psychiatric Association. (2013). *Diagnostic and Statistical Manual of Mental Disorders* (5th ed.). Washington, DC: Author.

Andreasen, N. C. (1987). The diagnosis of schizophrenia. *Schizophrenia Bulletin*, 13(1), 9–22. DOI: https://doi.org/10.1093/schbul/13.1.9

Astrachan, B. M., Harrow, M., Adler, D., Brauer, L., Schwartz, A., Schwartz, C., & Tucker, G. (1972). A checklist for the diagnosis of schizophrenia. *The British Journal of Psychiatry*, 121(564), 529–539. DOI: https://doi.org/10.1192/bjp.121.5.529

Bansal, V., & Chatterjee, I. (2021). Role of neurotransmitters in schizophrenia: A comprehensive study. *Kuwait Journal of Science*, 48(2). DOI: https://doi.org/10.48129/kjs.v48i2.9264

Buchsbaum, M. S., & Haier, R. J. (1978). Biological homogeneity, symptom heterogeneity, and the diagnosis of schizophrenia. *Schizophrenia Bulletin*, 4(4), 473–475. DOI: https://doi.org/10.1093/schbul/4.4.473

Cavallaro, R., Anselmetti, S., Poletti, S., Bechi, M., Ermoli, E., Cocchi, F., . . . & Smeraldi, E. (2009). Computer-aided neurocognitive remediation as an enhancing strategy for schizophrenia rehabilitation. *Psychiatry Research*, 169(3), 191–196. DOI: https://doi.org/10.1016/j.psychres.2008.06.027

Charernboon, T. (2020). Differentiating the cognitive impairment of clinically stable schizophrenia from mild cognitive impairment. *Psychiatry and Clinical Psychopharmacology*, 30(2), 122–127.

Chatterjee, I. (2018). Mean deviation based identification of activated voxels from time-series fMRI data of schizophrenia patients. *F1000Research*, 7(1615). DOI: https://doi.org/10.12688/f1000research.16405.2

Chatterjee, I., Agarwal, M., Rana, B., Lakhyani, N., & Kumar, N. (2018). Bi-objective approach for computer-aided diagnosis of schizophrenia patients using fMRI data. *Multimedia Tools and Applications*, 77(20), 26991–27015. DOI: https://doi.org/10.1007/s11042-018-5901-0

Chatterjee, I., Kumar, V., Rana, B., Agarwal, M., & Kumar, N. (2020a). Identification of changes in grey matter volume using an evolutionary approach: An MRI study of schizophrenia. *Multimedia Systems*, 26, 383–396. DOI: https://doi.org/10.1007/s00530-020-00649-6

Chatterjee, I., Kumar, V., Rana, B., Agarwal, M., & Kumar, N. (2020b). Impact of ageing on the brain regions of the schizophrenia patients: An fMRI study using an evolutionary approach. *Multimedia Tools and Applications*, 79(33), 24757–24779. DOI: https://doi.org/10.1007/s11042-020-09183-z

Chatterjee, I., Kumar, V., Sharma, S., Dhingra, D., Rana, B., Agarwal, M., & Kumar, N. (2019). Identification of brain regions associated with working memory deficit in schizophrenia. *F1000Research*, 8(124). DOI: https://doi.org/10.12688/f1000research.17731.1

Chatterjee, I., & Mittal, K. (2019). A concise study of schizophrenia and resting-state fMRI data analysis. *Qeios*, 2019, 414(599711.2). DOI: https://doi.org/10.32388/599711.2

Chatterjee, I., Rana, B., Agarwal, M., & Kumar, N. (2020c). Study of working memory impairment in schizophrenia patients. *F1000Research*, 9(787). DOI: https://doi.org/10.7490/f1000research.1118089.1.

Chyzhyk, D., Savio, A., & Graña, M. (2015). Computer aided diagnosis of schizophrenia on resting state fMRI data by ensembles of ELM. *Neural Networks*, *68*, 23–33.

Cohen, J. D., Braver, T. S., & O' Reilly, R. (1996). A computational approach to prefrontal cortex, cognitive control and schizophrenia: Recent developments and current challenges. *Philosophical Transactions of the Royal Society of London. Series B: Biological Sciences*, 351(1346), 1515–1527. DOI: https://doi.org/10.1098/rstb.1996.0138

Coustals, N., Martelli, C., Brunet-Lecomte, M., Petillion, A., Romeo, B., & Benyamina, A. (2020). Chronic smoking and cognition in patients with schizophrenia: A meta-analysis. *Schizophrenia Research*, 222, 113–121. DOI: https://doi.org/10.1016/j.schres.2020.03.071

Erhardt, E. B., Rachakonda, S., Bedrick, E. J., Allen, E. A., Adali, T., & Calhoun, V. D. (2011). Comparison of multi-subject ICA methods for analysis of fMRI data. *Human Brain Mapping*, 32(12), 2075–2095. DOI: https://doi.org/10.1002/hbm.21170

Fenton, W. S., Mosher, L. R., & Matthews, S. M. (1981). Diagnosis of schizophrenia: A critical review of current diagnostic systems. *Schizophrenia Bulletin*, 7(3), 452–476.

Gautam, A., & Chatterjee, I. (2020). Big data and cloud computing: A critical review. *International Journal of Operations Research and Information Systems (IJORIS)*, 11(3), 19–38. DOI: https://doi.org/10.4018/IJORIS.2020070102

Hursitoglu, O., Orhan, F. O., Kurutas, E. B., Doganer, A., Durmuş, H. T., & Bozkus, O. (2020). Evaluation serum levels of g protein-coupled estrogen receptor and its diagnostic value in patients with schizophrenia. *Psychiatry and Clinical Psychopharmacology*, 30(2), 115–121.

Kallmann, F. J. (1938). *The Genetics of Schizophrenia*. New York: J. J. Augustin.

Keith, S. J., & Matthews, S. M. (1991). The diagnosis of schizophrenia: A review of onset and duration issues. *Schizophrenia Bulletin*, 17(1), 51–68. DOI: https://doi.org/10.1093/schbul/17.1.51

Kumar, A., & Chatterjee, I. (2016). Data mining: An experimental approach with WEKA on UCI dataset. *International Journal of Computer Applications*, 138(13). DOI: https://doi.org/10.5120/ijca2016909050

Langfeldt, G. (1960). Diagnosis and prognosis of schizophrenia. *Proceedings of the Royal Society of Medicine*, 53(12), 1047–1052.

Leucht, S., Corves, C., Arbter, D., Engel, R. R., Li, C., & Davis, J. M. (2009). Second-generation versus first-generation antipsychotic drugs for schizophrenia: A meta-analysis. *The Lancet*, 373(9657), 31–41.

Li, H., Zhu, Q., Zhang, R., & Zhang, D. (2019, October). Multi-modality low-rank learning fused first-order and second-order information for computer-aided diagnosis of schizophrenia. In *International Conference on Intelligent Science and Big Data Engineering* (pp. 356–368). Cham: Springer. DOI: https://doi.org/10.1007/978-3-030-36204-1_30

Lützhøft, J. H., Skadhede, S., Fätkenheuer, B., Häfner, H., Löffler, W., Riecher-Rössler, A., & Maurer, K. (1995). Symptom assessment in casenotes and the clinical diagnosis of schizophrenia. *Psychopathology*, 28(3), 131–139.

Lysaker, P. H., & Buck, K. D. (2008). Is recovery from schizophrenia possible? An overview of concepts, evidence, and clinical implications. *Primary Psychiatry*, 15(6).

Patel, K. R., Cherian, J., Gohil, K., & Atkinson, D. (2014). Schizophrenia: Overview and treatment options. *Pharmacy and Therapeutics*, 39(9), 638.

Pull, C. B. (1999). *Diagnosis of Schizophrenia: A Review – Schizophrenia* (pp. 1–37). Chichester: Wiley. DOI: https://doi.org/10.1002/0470842334.ch1

Ripke, S., Neale, B. M., Corvin, A., Walters, J. T., Farh, K. H., Holmans, P. A., . . . & Pers, T. H. (2014). Biological insights from 108 schizophrenia-associated genetic loci. *Nature*, 511(7510), 421–427.

Santos-Mayo, L., San-José-Revuelta, L. M., & Arribas, J. I. (2016). A computer-aided diagnosis system with EEG based on the P3b wave during an auditory odd-ball task in schizophrenia. *IEEE Transactions on Biomedical Engineering*, 64(2), 395–407. DOI: https://doi.org/10.1109/tbme.2016.2558824

Sartipi, S., Kalbkhani, H., & Shayesteh, M. G. (2020). Diagnosis of schizophrenia from R-fMRI data using Ripplet transform and OLPP. *Multimedia Tools and Applications* (pp. 1–23)

Savio, A., Chyzhyk, D., & Graña, M. (2014, June). Computer aided diagnosis of schizophrenia based on local-activity measures of resting-state fMRI. In *International Conference on Hybrid Artificial Intelligence Systems* (pp. 1–12). Cham: Springer. DOI: https://doi.org/10.1007/978-3-319-07617-1_1

Schwarz, E., Izmailov, R., Spain, M., Barnes, A., Mapes, J. P., Guest, P. C., . . . & Steiner, J. (2010). Validation of a blood-based laboratory test to aid in the confirmation of a diagnosis of schizophrenia. *Biomarker Insights*, 5, BMI-S4877.

Tsuang, M. T., Stone, W. S., & Faraone, S. V. (2000). Toward reformulating the diagnosis of schizophrenia. *American Journal of Psychiatry*, 157(7), 1041–1050. DOI: https://doi.org/10.1176/appi.ajp.157.7.1041

Üçok, A., Polat, A., Sartorius, N., Erkoc, S., & Atakli, C. (2004). Attitudes of psychiatrists toward patients with schizophrenia. *Psychiatry and Clinical Neurosciences*, 58(1), 89–91.

Ulas, H., Alptekin, K., Akdede, B. B., Tumuklu, M., Akvardar, Y., Kitis, A., & Polat, S. (2007). Panic symptoms in schizophrenia: Comorbidity and clinical correlates. *Psychiatry and Clinical Neurosciences*, 61(6), 678–680.

Uzun, Ö., Cansever, A., Basoğlu, C., & Özşahin, A. (2003). Smoking and substance abuse in outpatients with schizophrenia: A 2-year follow-up study in Turkey. *Drug and Alcohol Dependence*, 70(2), 187–192.

Valton, V., Romaniuk, L., Steele, J. D., Lawrie, S., & Seriès, P. (2017). Comprehensive review: Computational modelling of schizophrenia. *Neuroscience & Biobehavioral Reviews*, 83, 631–646.

Yung, A. R., & McGorry, P. D. (1996). The prodromal phase of first-episode psychosis: Past and current conceptualizations. *Schizophrenia Bulletin*, 22(2), 353–370. DOI: https://doi.org/10.1093/schbul/22.2.353

Zhu, Q., Li, H., Huang, J., Xu, X., Guan, D., & Zhang, D. (2019). Hybrid functional brain network with first-order and second-order information for computer-aided diagnosis of schizophrenia. *Frontiers in Neuroscience*, 13, 603.

# 6 Deep Learning in Medical Imaging

*Arjun Sarkar*

## CONTENTS



## 6.1 INTRODUCTION

AI, especially with the emergence of deep learning algorithms (Lecun et al., 2015) such as convolutional neural networks (CNNs), has proved to be a resounding success in computer vision and has later been implemented in medical imaging. AI is now one of the most discussed and debated topics in medical imaging research. AI has already shown its dominance over deterministic and mathematical approaches in image acquisition, image classification, image processing, image reconstruction, image enhancement, and image generation. With the ability to harvest big data, AI finds its applications even in medical prognosis and medical data analysis.

In 2012, deep learning gained its fame in computer vision when it first beat traditional machine learning algorithms by a substantial margin on the famous ImageNet Large-Scale Visual Recognition Challenge (ILSVRC; Krizhevsky et al., 2017). CNN, a type of deep learning algorithm used in the challenge, now outperforms human performance with ease on ILSVRC and various other image classification and recognition tasks.

Hospitals and healthcare institutions generate vast amounts of useful data every day. It is no more feasible to process such bulk data using traditional analysis methods. That is precisely where machine learning in the form of deep learning comes

DOI: 10.1201/9781003142751-8

into the picture. Healthcare applications range from one-dimensional (1D) bio-signal (Ganapathy et al., 2018) analysis to medical prognosis applications, such as clinical decision-making and survival analysis (Katzman et al., 2018), drug discovery (Jiménez et al., 2018), analysis of electronic health records (Rajkomar et al., 2018; Shickel et al., 2018) and 2D (X-ray imaging) and 3D (computed tomography, CT) image analysis (Chartrand et al., 2017; Domingues et al., 2020; Rajpurkar et al., 2017).

This chapter focuses on using deep learning in medical imaging across all healthcare sectors, including radiology, ophthalmology, and digital pathology. AI can provide higher efficiency to clinicians and radiologists, allowing them to perform more value-added tasks. But with the increasing demand for AI in the medical field, there is ever more demand for the algorithms to be easy to be stable, ergonomic, and explainable. There is a widespread belief that AI will benefit the healthcare sector immensely, especially in diagnosis and medical imaging.

## 6.2 DEEP LEARNING AND CNNs

Machine learning is a subclass of AI, and deep learning is a further subcategory of machine learning. Deep learning makes use of neural network architectures to learn features and solve various problems. Deep neural networks contain a series of computational units called neurons, arranged in layers. Every neural network typically includes an input layer (first layer), which feeds in the data, followed by one or multiple hidden layers—each containing several neurons—before terminating in an output layer (final layer), which generates the network's outputs. Experts (clinicians and radiologists in a healthcare environment) expected predictions are predisposed to the network as specific labels or masks in supervised learning. The network identifies patterns in the labeled training data, measures the difference in its predicted outputs and the actual labels, and retrains itself by an algorithm known as back-propagation. Each neuron's values are tuned until the network produces good results from the training data. In the training phase, the neural network learns meaningful features. Once the network learns the necessary patterns, it can then make predictions on new, unseen data.

The neuron takes a bias $w_0$ and a weight vector $w = (w_1, w_2, \ldots, w_m)$ as parameters to model a decision. Inside each neuron in the hidden layers, two operations occur: The weights and biases are multiplied with their corresponding input values and summed together, followed by applying an activation function or a nonlinearity function to the weighted sum. Finally, it generates the output for the next neuron (Lundervold & Lundervold, 2019; Maier et al., 2019).

1. **Activation Functions:** The activation function introduces nonlinearity in the modelling capabilities of the network (Nwankpa et al., 2018). The following typical activation functions are used:
   a. Sigmoid function: Unlike a threshold function, sigmoid is a smooth function and is particularly useful in the output layer of binary classification problems.

$$\varnothing(x) = 1/(1+e^{(-x)}) \tag{6.1}$$

b. Rectified Linear Unit (ReLU) function: ReLU is the most famous and the most commonly used activation function in various neural networks. It passes the maximum of either 0 or the value.

$$\varnothing(x) = max(x,0) \tag{6.2}$$

c. Hyperbolic tangent (tanh) function: The tanh function is very similar to the sigmoid function, except it fits values between −1 and 1 rather than from 0 to 1, as in the sigmoid function.

$$\varnothing(x) = \left(1 - e^{(-2x)}\right) / \left(1 + e^{(-2x)}\right) \tag{6.3}$$

d. SoftMax function: This probability function is used when predicting multiple classes.

$$\varnothing(\vec{z})_i = e^{z_i} / \sum_{j-1}^{K} e^{z_i} \tag{6.4}$$

2. **Loss/Cost Functions:** In the optimization algorithms, the functions used to evaluate predicted output to the actual output are known as loss functions. Typically, with neural networks, we seek to minimize the error. Typical loss functions used across various problem cases are as follow:
   a. Regression problem: In a problem in which the system predicts a real value, the loss function used is mean squared error.
   b. Binary classification problem: In a problem in which the network classifies an input into one or two classes, binary cross-entropy, typically referred to as logarithmic loss, is used.
   c. Multiclass classification problem: In a problem in which the network classifies an input belonging to multiple classes, the loss functions used are categorical cross-entropy or sparse categorical cross-entropy.
   d. Segmentation problem: In a problem in which the network segments an image and compares the prediction to an output segmentation mask, the loss function generally used is dice loss or soft dice loss (Bertels et al., 2020; Sudre et al., 2017).
3. **Backpropagation and Optimization Algorithms:** After calculating the loss, the backpropagation algorithm enables information to propagate backward in the neural network to update the neurons' weights in the hidden layers. However, the hidden-layer neurons only receive a fraction of the total loss signal based on each neuron's contributions toward the final output. This process repeats across each layer of the neural network.
   Optimization algorithms help reduce the loss to provide the most accurate results possible. Optimization algorithms change the network parameters such as weights and learning rate to reduce the loss. The following are a few of the commonly used optimization algorithms:
   a. Gradient descent (GD): Gradient descent is the most fundamental optimization algorithm. It is a first-order iterative derivative aimed at locating the local minima of the loss function.

b. Stochastic GD (SGD): Stochastic gradient descent is an updated gradient descent. GD requires a lot of memory, as it loads the entire dataset all at once. SGD reduces the amount of computation immensely by randomly picking one data point from the whole dataset during each iteration.
c. Stochastic GD with momentum: A significant disadvantage of mini batch-SGD is that updated weights are very noisy. SGD with momentum helps overcome this drawback by denoising the gradients.
d. Adaptive gradient (AdaGrad): In SGD, the learning rate remains constant. AdaGrad updates the weights using adaptive learning. In AdaGrad, the learning rate decreases with the decrease in the number of iterations (Duchi et al., 2010).
e. AdaDelta: With AdaGrad, the learning rate becomes very small, which leads to very slow convergence. To avoid this problem, AdaDelta uses the exponential decaying average (for Computing Machinery et al., 2015).
f. RMSprop (root mean square prop): RMSprop adjusts the learning rate automatically and chooses a different learning rate for each parameter. When nearing the local/global minima RMSprop decreases the gradient steps' size accordingly, if the step sizes are too large (Tieleman & Hinton, 2012).
g. Adam: While AdaDelta stores exponential decaying averages of the square of gradients to modify the learning rate, the Adam optimization algorithm stores both the first and second order of the moment of the gradient. Adam is considered one of the best optimization algorithms (Kingma & Ba, 2015).

Figure 6.1 shows a simple network with multiple input values, which feeds forward into a single neuron, where an activation function is applied on the sum of all the weights to generate an output prediction. The generated output is compared with the actual value; the loss is calculated and is fed backward into the network via backpropagation to update the weights and biases for the next iteration. This is how the network trains. The training process continues till the loss function reaches its minimum possible value (error is minimized).

4. **CNNs:** The disadvantage of traditional artificial neural networks is that they do not scale well to large images. CNNs help overcome this problem by using various convolutional layers.

CNNs (or ConvNets) take in an image or time series as input, learn features from the input image, and can differentiate or segment objects in the image. The convolution layers are the core building blocks of CNN models. It implements the use of filters/kernels to identify the spatial and temporal features in the images. CNN layers can learn a large number of filters automatically without someone having to handcraft the filters. A CNN model has three significant layers: the convolutional layer, the pooling layer, and the fully connected layer (optional). Figure 6.2 shows a convolutional layer and the two commonly used pooling layers.

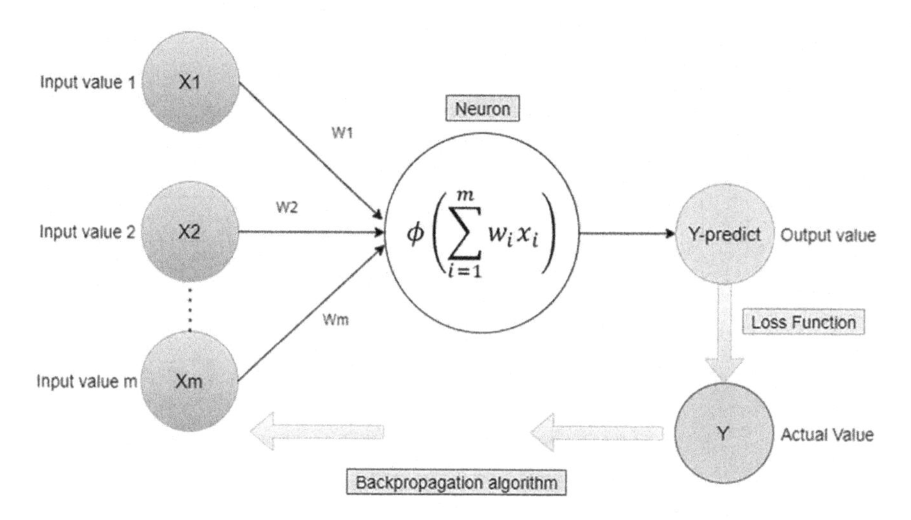


**FIGURE 6.1** The figure shows how a network learns. A simple network with multiple inputs (X1, X2, . . ., Xm) with weights wl, w2, . . ., wm, which feed forward into a neuron, where the weights and biases are multiplied and the weighted sum is passed through a nonlinear function to come up with an output (Y-predict). The predicted output is compared with the actual output Y, and the error (loss) is calculated. This is fed back into the network using backpropagation algorithm, and the weights are updated. This process continues till the loss function is minimized.

*Source:* Created by the author.

a. Convolutional layer: Convolution layers use multiple learnable filters that act as feature detectors and pattern recognition units (like vertical/horizontal edge detectors). This helps reduce the number of parameters across neurons. In a convolutional layer, the size of filters, the number of filters, the stride of the filters, and the input image's padding are manually adjustable.
b. Pooling layer: Similar to a convolutional layer, the pooling layer is also responsible for reducing the convolved feature's spatial size. It helps bring about a dimensionality reduction by reducing the computational power. Moreover, pooling allows the extraction of the dominant features from an image. Two common types of pooling used are max pooling and average pooling. Max pooling returns the maximum value, whereas average pooling returns the average of all the pixel values in an image covered by the filter.
c. Fully connected layer: These are similar to the hidden layers in an artificial neural network. They are optional layers in a CNN model.

As seen in Table 6.1, various CNN architectures have worked wonders in computer vision over the last decade. There are countless other network architectures, each with its unique design and built to perform better than the previous networks while being more efficient. Like all other computer vision fields, researchers in medical imaging have also been quick to take up deep learning. The implementation of

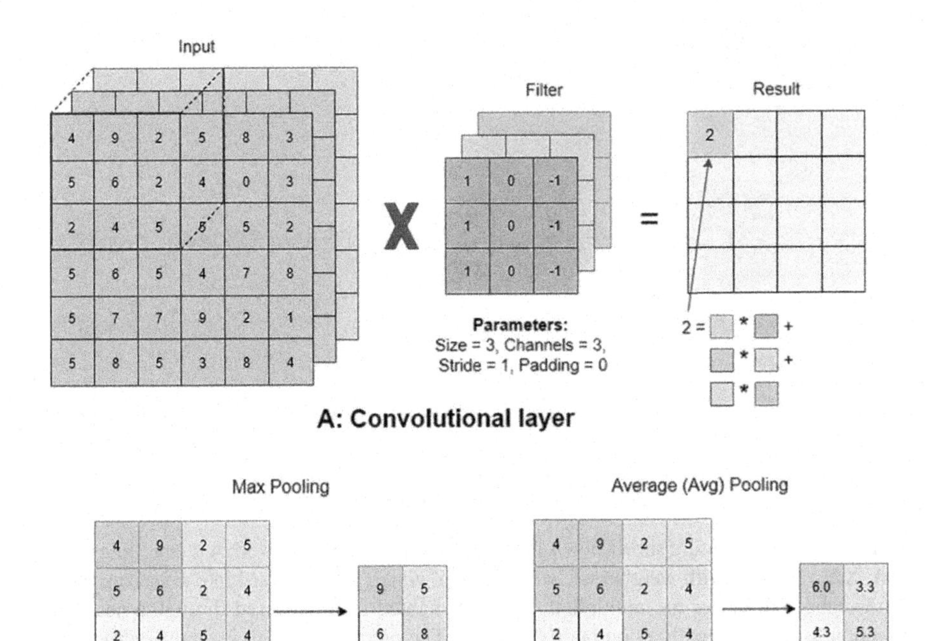


**FIGURE 6.2** (A) A Convolutional Layer with Three 3 × 3 Filters Passes over a Three-Channel Image to Generate an Output. The convolutional layer helps in edge detection and feature detection. (B) Max Pooling and Average Pooling—The Two Common Pooling Layers. Max pooling finds the maximum and average pooling finds the average of the pixels on the part of the image over which the filter is placed, respectively.

*Source:* Created by the author.

**TABLE 6.1**
**An Overview of Some Famous Deep Learning Architectures in Computer Vision**

| CNN Architecture | Description |
|---|---|
| AlexNet (Krizhevsky et al., 2012) | The CNN architecture that won the 2012 ILSVRC challenge. This architecture used ReLU activation function, dropout, regularization, and using multiple graphics processing units (GPUs) for computation. AlexNet helped launch deep learning into the spotlight. |
| ZFNet (Zeiler & Fergus, 2014) | ZFNet architecture won the ILSVRC challenge in 2013. ZFNet significantly improved the error rate as compared to AlexNet (Krizhevsky et al., 2012). |
| VGG (Simonyan & Zisserman, 2015) | VGG used much smaller filter kernels than the previous networks. It went up to 19 layers, compared to 7 layers for AlexNet (Krizhevsky et al., 2012), making it a much deeper network. |

| CNN Architecture | Description |
|---|---|
| ResNet (He et al., 2016) | ResNet added skip connections to the network architecture, reducing the vanishing gradient problem and allowing the network architecture to go much deeper. A 152-layer ResNet won the ILSVRC challenge in 2015. The deepest ResNet has 1,001 layers. |
| GoogleLeNet (Inception) (Szegedy et al., 2015) | Inception added stacks of different filter kernels in inception blocks and concatenated the outputs to extract various features simultaneously. It won the 2014 ILSVRC challenge. |
| DenseNet (Huang et al., 2017) | DenseNet uses ResNet's (He et al., 2016) idea but instead adds one layer to some later layer; it concatenates all the layers. It helps reduce the number of parameters significantly. |
| Xception (Chollet, 2017) | The Xception network makes use of separable convolutional layers to significantly reduce the dimensionality of the filters, hence reducing computation cost. |
| ResNext (Xie et al., 2017) | ResNext architecture uses the idea of both ResNet (He et al., 2016) and GoogleLeNet (Szegedy et al., 2015) . That is, it adds inception modules between skip connections. |
| SENets (Hu et al., 2020) | SENet is a Squeeze-and-Excitation (SE) network. It contains several SE blocks that allow the network to model the channel and spatial information separately. The network won the 2017 ILSVRC competition. |
| NASNet (Zoph et al., 2018) | Built using AutoML, Google Brain's reinforcement learning approach allowed a neural network to design the NASNet CNN model (Bello et al., 2017). NASNet beat all the previous state-of-the-art models on ILSVRC data. |
| MobileNet (Howard et al., 2017) | MobileNet makes use of depth-wise, separable convolutions, making it very efficient and significantly reducing computation and model size. The network aims to be useful in mobile vision applications. |
| YOLO (Redmon et al., 2016) | YOLO is a neural network that performs object detection and image classification simultaneously. Its outputs bounds boxes on the objects in an image and gives class probabilities. |
| GAN (Goodfellow et al., 2014) | Generative Adversarial Network (GAN) uses two neural networks, each trying to outmatch each other. GAN helps create artificial images. |
| Siamese Net (van der Spoel et al., 2015) | Siamese Nets are useful for performing one-shot learning, that is, training from a single example. |

deep learning is seen widely in many medical imaging modalities. Deep learning has proved to be extremely useful in various medical imaging applications, such as image classification, object detection, image segmentation, image registration, image generation, and image transformation.

## 6.3 DEEP LEARNING IN MEDICAL IMAGING

### 6.3.1 Image Classification

One of the critical tasks in medical imaging is the classification of images based on various disease types and the presence and absence of certain features. CNN architectures perform exceptionally well in classification tasks. Various applications include the diagnosis of tuberculosis (Lakhani & Sundaram, 2017), skin cancers (Esteva et al., 2017), pneumonia (Stephen et al., 2019), and diabetic retinopathy (Gulshan et al., 2016). Deep learning architectures require a substantial amount of data to attain a human-level or better performance and avoid overfitting. With the rise of AI in medical imaging, more and more hospitals and health institutes are willing to share data with deep learning researchers, thus enabling better CNN classifiers. While most CNN architectures perform well with a large dataset of images, enough data might not be at the disposal of AI researchers in some instances. Medical images also have much more complicated patterns of various sizes and disease types, unlike basic image classification tasks; hence, it is difficult for CNN models to learn these complicated patterns directly. Even though labeled datasets are more readily available now than a few years ago, there is often a class imbalance problem. Like in the classification of COVID-19 images from pneumonia and normal (no infection) classes from chest X-rays, while many images belong to pneumonia and normal class, there are very few images in the COVID-19 class (Wang et al., 2020). Images need to be upsampled in the minority classes using image transformation methods such as rotation, flipping, scaling, and translation to overcome this problem (Mikołajczyk & Grochowski, 2018). Curriculum learning is a technique often applied when easier-to-classify images are added to the neural network by radiologists and gradually the task difficulty is increased (Bengio et al., 2009). It conveys the impression that learning the overall concept on a few easy examples and only later fine-tuning the concept with other complicated examples give the continuously learning network a recognizable improvement over a network that needs to grasp the entire concept at once. Various studies implement the idea of using active learning, which implements incremental learning, allowing the algorithm to self-learn over time in experts' presence (Santosh, 2020). The studies aim to create a model that iteratively learns and adapts to new data without forgetting what it has previously learned. In radiology, often for small datasets, it is essential to implement algorithms that work well even with significantly fewer training images. A study showed that using radiomic features and supervised learning or multilayer perceptron after feature selection can obtain better accuracy (Yun et al., 2019). Most deep learning algorithms have already proved how they learn much more abstract features that are very useful for medical image analysis than hand-designed feature extraction algorithms.

With the implementation of picture archiving and communication systems across hospital systems and more doctors and radiologists keen to take up AI as an aid, sizable medical image databases may not be hard to obtain in the next decade. We expect the datasets to be available more publicly for research purposes after considering the technical and privacy issues.

### 6.3.2 Object Detection

Object detection is the location and identification of particular objects of interest in any given image. In medical imaging, object detection is mostly performed by adding a bounding box to the interest object. In a region proposal-based algorithm (Girshick et al., 2014) (Girshick, 2015), diverse varieties of regions of interest (ROIs) are extracted by using selective search from input images. When the network is trained, it decides on the bounding boxes, decides if multiple objects are present in each block of ROI, and further analyzes and segregates the objects based on the ROI. The region proposed network is said to boost the speed of the object detection process (Ren et al., 2017). Often regression models are also used for object detection (Redmon et al., 2016). They are one-stage models and provide better in terms of both speed and accuracy (W. Liu et al., 2016).

Attention-based models are slowly gaining momentum in medical image detection. Sliding window approaches are often used for histological pattern analysis in microscopy images. However, this method requires a laborious annotation process. Grid-based attention models achieved higher accuracy than sliding window methods in detecting cancerous and precancerous esophagus tissue (Tomita et al., 2019). Liver lesion detection using a Single Shot Multibox Detector (SSD) from CT volumes shows that grouped convolutions performed better than SSD's naive application, which suffered from a generalization gap (Lee et al., 2018). Deep learning frameworks can now identify and localize tumors as small as 100 × 100 pixels from gigapixel microscopy images that are about 100,000 × 100,000 pixels. While human pathologists obtained a sensitivity of 73.2% on the Camelyon16 dataset, deep learning models could get a sensitivity of 92.4% (Y. Liu et al., 2017). RetinaNet uses focal loss and to obtain state-of-the-art results using one-stage detectors (Lin et al., 2020). Aubreville et al. (2017) use guided spatial transformers for cell classification and detection. The transformer network allows refinement in detection from the histological images before classification.

### 6.3.3 Image Segmentation

Image segmentation refers to identifying pixels of organs and deformities to outline the structure as precisely as possible. Image segmentation is considered the holy grail of medical image analysis and has benefited immensely due to the development of various deep learning algorithms.

Classical image segmentation methods used edge detection filters, mathematical algorithms, and thousands of lines of code. Several methods, such as dependent thresholding and close-contour methods, were used (Aslam et al., 2015). The introduction of CNNs has improved image segmentation significantly and has even surpassed human-level performance in various applications. Cascaded deep learning CNN containing two subnetworks helped obtain a dice similarity coefficient of 0.89 on the BRATS 2015 dataset (brain tumor segmentation; Cui et al., 2018). The network consists of a tumor localization network, which uses transfer learning to process the MRI images, characterize the tumor region from the MRI slice, and an intratumor classification network helped label the tumor region further into numerous subregions. End-to-end incremental deep CNN using ensemble learning produced a dice score of 0.88 on the BRATS-2017 dataset (naceur et al., 2018). The efficient design of the deep learning models enabled brain segmentation results in approximately 20.87 seconds.

Two-and-a-half-dimensional CNN models have also been used for segmentation tasks. The advantage of using 2.5D CNN is that it has much more spatial information from neighboring pixels than a 2D CNN but has significantly less computational cost than a 3D CNN (Moeskops et al., 2016). Three orthogonal 2D patches are extracted from the images, using 2D filter kernels, in the XY, YZ, and XZ planes. The advantage of using 2.5D CNN was evident in a study on knee cartilage segmentation, which achieved a dice coefficient of 0.82 while significantly reducing computation cost (Prasoon et al., 2013). Three-dimensional CNNs were necessary, as 2.5D CNNs were only limited to 2D kernels. A 3D CNN is very similar to a 2D CNN except it uses 3D convolutional and subsampling layers rather than its 2D counterparts.

One of the most famous biomedical image segmentation deep learning models is the U-Net (Ronneberger et al., 2015). U-Net employs downsampling and upsampling the images while passing information to the upsampling layers via skip connections. Based on U-Net, various other segmentation models have been developed (Gordienko et al., 2019; Zeng et al., 2017). DCAN (H. Chen et al., 2016), another segmentation network, uses an auxiliary classifier on top of the U-Net. DCAN obtained approximately 2% higher results on gland segmentation than the original U-Net model. V-Net (Milletari et al., 2016) is a 3D U-Net-based model that can learn to predict the whole 3D volume segmentation at once.

Other than CNN models, various recurrent neural network (RNN) models have also aided in medical image segmentation. Contextual Long Short-Term Memory (CLSTM; Cai et al., 2017) has been shown to have improved dice scores more than U-Net models when applied in the output layer of a deep CNN. It helps attain precise segmentation by capturing information across the adjacent image slides. Similarly, bidirectional CLSTM (J. Chen et al., 2016), when applied to a U-Net structure (Ronneberger et al., 2015), has produced better results by capturing sequential data in two directions rather than in a single direction.

### 6.3.4 Image Registration

While medical image classification and segmentation using deep learning have quickly gained momentum, image registration is a branch, which is significantly less explored. Image registration refers to analyzing images procured from different viewpoints, dissimilar times, or unassociated imaging modalities and merging them into a single coordinate system.

Unimodal registration was first performed by Wu et al. (2013, 2016), who used deep learning to obtain application-specific similarity metrics for 3D brain MRI images, using a convolutional stacked autoencoder (CAE). A combination of CNN-based descriptors and handcrafted Markov Random Filed-based descriptor helped in lung CT registration (Blendowski & Heinrich, 2019). Deep learning plays a crucial role in multimodal registration, as changed manually crafted methods have had very little success in multimodal registration. Stacked denoising autoencoders help access the alignment of CT and MRI images by learning a similarity metric that evaluates the standard of the alignment (Cheng et al., 2018). A five-layer neural network has outperformed mutual information-optimization-based registration in 3D US/MR (ultrasound/magnetic resonance imaging) abdominal scans (Sedghi et al., 2019). RNN models such as Long Short-Term Memory (LSTM) spatial co-transformer is used by Wright et al. (2018) to register MR and US volumes group-wise iteratively. Other than CNN- and RNN-based methods, reinforcement learning has also been used often to perform registration. Typically, rigid transformation models are used for reinforcement learning–based registrations. Reinforcement learning–based registration was first used to carry out rigid registration on cardiac and abdominal 3D CT images and cone-beam CT images (Liao et al., 2017). Q-learning-based methods helped determine the depth of projected images in rigid registration of MR/CT chest images (Ma et al., 2017). This method used a single agent to train the model. Multiagent-based reinforcement learning was used by Miao et al. (2018) for the rigid registration of X-ray and CT images of the spine.

### 6.3.5 Image Restoration

Image restoration refers to the process of removing artifacts from an image and denoising it. It has been employed in medical imaging studies for many years now, and various mathematical methods are applied, such as Bayesian Markov random field models (Baselice et al., 2017), independent component analysis (Salimi-Khorshidi et al., 2014), or higher order singular value decomposition (X. Zhang et al., 2015). Deep learning methods have brought about significant improvements in this application as well. Bermudez et al. (2018) implemented autoencoders with skip connections to denoise brain MRI. The autoencoder performed better than one of the best denoising software—the FSL SUSAN. Jifara et al. (2019) proposed a network for medical image denoising using residual learning. The method learns the noise from noisy images, unlike other methods where the deep learning model learns the noise from latent clean images.

Ran et al. (2019) proposed a network that uses residual encoder–decoder Wasserstein generative adversarial network (RED-WGAN) to denoise 3D MRI images. Residual autoencoders, along with deconvolution operations, were introduced in the network. The performance of RED-WGAN surpassed many state-of-the-art techniques of image denoising in both simulated and clinical data. Arterial spin labeling (ASL) perfusion imaging is a type of MRI acquisition method that also suffers from a low signal-to-noise ratio (Petcharunpaisan, 2010). When used to learn spatiotemporal properties of ASL signal, deep convolution joint filter architecture can achieve remarkable performance in both denoising and artifact removal (Owen et al., 2018). It helped improve the signal-to-noise ratio in ASL signals by up to 50%. Incorporating multi-contrast images with nonlinear, spatial variant multilateral filtering, and finally tuning the final denoising level with deep neural network models helped significantly accelerate ASL image acquisition and improve image quality.

Often due to magnetic field inhomogeneities, patient movement, and inappropriate water or lipid suppression in MR imaging, there is spurious noise. Tiled CNN models tuned by Bayesian optimization methods helped detect such artifacts (Gurbani et al., 2018). For detecting and correcting spurious echo ghost signals, a combination of CNN and stacked autoencoders was used, and the authors achieved an accuracy of almost 100% (Kyathanahally et al., 2018). Due to the long acquisition time, MR images often have motion artifacts. A CNN-based classifier trained to interpret motion artifacts on a per-patch basis helped identify and locate the motion artifacts in a test dataset (Küstner et al., 2018). The CNN architecture achieved around 97%/100% accuracy in the head and 75%/100% accuracy in the abdomen, respectively, based on a per-patch/per-volunteer basis.

For accurate quantification of radiotracer distribution in positron emission tomography (PET) imaging, attenuation correction is an important step. Previously, to determine the attenuation coefficient in PET/MRI, segmentation and the atlas-based algorithm have been used (Zhu et al., 2019). Maximum-likelihood reconstruction of activity and attenuation (MLAA) is a new technique to generate images that can produce the attenuation coefficients directly from the emission data without CT or MRI (Heußer et al., 2017). One problem with MLAA is that it generates crosstalk artifacts. Three different CNN architectures were tested to eliminate crosstalk artifacts and generate less noisy images (Hwang et al., 2018). The architectures used were CAE, U-Net (Ronneberger et al., 2015), and hybrid CAE. Various other studies using various deep learning models have been implemented to overcome most problems occurring due to crosstalk artifacts (Han, 2017; Leynes et al., 2017; Torrado-Carvajal et al., 2019). Angiography uses live video to view the body's blood vessels in real time. However, these videos have various random noise, which affects the quality of the video. Deep learning networks have helped reduce Gaussian noise, speckle noise, salt-and-pepper noise from these angiography videos (Sadda & Qarni, 2018).

### 6.3.6 Image Super-Resolution

Image super-resolution is the reconstruction of high-resolution images from low-resolution images. For image super-resolution, mathematical models such as tricubic interpolation (TCI) (Lekien & Marsden, 2005), Fourier interpolation (FI), and image sparse-coding super-resolution (ScSR) methods are used in the clinical environment (Ropele et al., 2010; Shilling et al., 2009). Super-resolution techniques can help improve the acquired image's resolution, signal-to-noise ratio and help reduce image acquisition time (Plenge et al., 2012). In MR imaging, super-resolution can help produce 7T-like MRI images from 3T MRI scanners (Bahrami et al., 2017). Deep learning architectures have already surpassed the accuracies produced by these mathematical models.

One of the most famous deep learning networks used to achieve super-resolution in MR images is the DeepResolve network (Chaudhari et al., 2018). DeepResolve helped generate high-resolution thin slice MR images (knee images) using low-resolution thick slice MR inputs. The network predicted the residuals from the input images and learned a residual map by finding the high-resolution and low-resolution images' differences. Once the network's residuals were generated, they were added with the new low-resolution images to obtain high-resolution images. DeepResolve (Chaudhari et al., 2018) performed much better in all parameters, such as contrast, sharpness, signal-to-noise ratio, and artifacts when compared with TCI (Lekien & Marsden, 2005).

### 6.3.7 Image Synthesis

One of the biggest problems while training deep learning models on medical image datasets is a class imbalance. Class imbalance refers to an unequal number of images in various classes. While some unequal data is considered normal, when there is a difference of tens of thousands of images between the various classes in the training set, the loss function tends to favor the class with more images than the class with fewer images. Various data augmentation techniques are used for upsampling or downsampling images of certain classes to overcome this problem. While this technique can somewhat improve the deep learning model's accuracy, better methods are being sought to overcome class imbalance and challenges in obtaining high-quality images. Several studies implement GANs (Goodfellow et al., 2014) to generate realistic X-ray, CT, or MRI images to overcome these problems.

GANs (Goodfellow et al., 2014) combine two different neural networks, a generator model trained to generate new images and a discriminator model that tries to classify if the generator's images are real or fake. The two models undergo simultaneous training until the generator model starts producing images, which the discriminator model can no longer verify as fake.

GANs (Goodfellow et al., 2014) helped generate high-quality liver lesion ROIs from a limited CT image dataset (Frid-Adar et al., 2018). It helped improve the CNN network's performance from 78.6% sensitivity and 88.4% specificity to 85.7% sensitivity and 92.4% specificity, respectively. GAN-based histopathological color-stain normalization framework and the preservation of detailed structural information outperformed all other color normalization methods, thus giving better classification and segmentation accuracies (Mahapatra et al., 2020). MedGAN (Armanious et al., 2020) is a GAN-based architecture that merges the adversarial framework with non-adversarial losses. Additionally, a new generator architecture, CasNet (Schlegl et al., 2017), enhances the sharpness of the generated medical images by using encoder–decoder pairs. MedGAN helped in PET–CT translation, MRI motion artifact correction, and PET image denoising. Direct application of MedGAN is possible on several medical tasks without any alternation of the hyperparameters (Armanious et al., 2020). Various other GAN-based networks (Goodfellow et al., 2014) are implemented across all the imaging modalities such as X-ray, CT, MRI, PET, histopathology, and retinal images.

GANs (Goodfellow et al., 2014) are also capable of anomaly detection in medical images. AnoGAN (Schlegl et al., 2017) aided in the anomaly detection from optical coherence tomography (OCT) images of the retina by correctly identifying retinal fluid or hyperreflective foci. An updated AnoGAN, f-AnoGAN (fast AnoGAN; Schlegl et al., 2019), outperforms traditional models and yields high accuracy. The network even passed a visual Turing test with two retina experts who agreed that the retina images were indistinguishable from real typical retinal OCT images.

## 6.4 DISCUSSION

Deep learning algorithms can detect, identify, and locate diseased or infected medical scans, reduce the burden on medical doctors, reduce the time of diagnosis, and improve the healthcare system's overall quality. Still, there remains a black box about understanding how the neural network is predicting its results. For example, while diagnosing pneumonia from X-ray images, the network should predict its results from the presence of pneumonia in the lungs and not some artifact from any other part of the X-ray image. Many companies are now working on making AI explainable. Grad-CAM (Selvaraju et al., 2017; B. Zhou et al., 2016), LRP (layer-wise relevance propagation; Bach et al., 2015), DeepLIFT (Shrikumar et al., 2016), PRM (Peak Response Mapping; Y. Zhou et al., 2018), LIME (Local Interpretable Model-Agnostic Explanations; Ribeiro et al., 2016), and DeconvNet (Zeiler & Fergus, 2014) are some of the interpretability methods used for understanding the predictions or learning of deep learning algorithms. Various reinforcement learning or unsupervised learning are also well suited to understand the black-box nature of deep learning models.

With many new machine learning and data science competitions focusing on medical imaging, researchers are coming up with various deep learning models that often push the state of the art to new levels. Many such models are made open-source, and developers work to tune or add new layers and change hyperparameters to design new networks to achieve even better accuracy metrics. Table 6.2 lists some popular biomedical imaging–based deep learning architectures and toolkits available open source for medical image analysis, classification, and segmentation.

While AI in medical imaging faces many challenges, one particular challenge is building an intuitive tool that can easily integrate with clinical workflow. All deep

**TABLE 6.2**
**Overview of Some Famous Open-Source Deep Learning Architectures and Toolkits Used in Medical Imaging**

| | |
|---|---|
| Nifty-Net (Gibson et al., 2018) | Nifty-Net is a TensorFlow (Abadi et al., 2016)–based open-source CNN for medical image analysis and image-guided therapy. It supports 2D, 2.5D, 3D, and 4D inputs. |
| DeepMedic (Kamnitsas et al., 2017) | It is an 11-layer multiscale 3D CNN for brain lesion segmentation. |
| DLTK (Deep Learning Toolkit) (Pawlowski et al., 2017) | A deep learning toolkit, it was built using TensorFlow (Abadi et al., 2016) for biomedical image analysis. |
| U-Net (Ronneberger et al., 2015) | U-Net is one of the most famous biomedical image segmentation models. It uses downsampling and upsampling and skips connections to segment images. |
| V-Net (Milletari et al., 2016) | V-Net is a modified version of U-Net (Ronneberger et al., 2015) for volumetric image segmentation. |
| Seg-Net (Badrinarayanan et al., 2017) | A semantic pixel-wise segmentation network, with an encoder network similar to VGG16 [23] and a decoder network that maps low-resolution encoder feature maps into high-resolution feature maps for classification. |
| HighRes3dNet (Larroza et al., 2019) | This is a high-resolution compact CNN for volumetric image segmentation. |
| GANCS (Mardani et al., 2017) | GANCS uses state-of-the-art compressed sensing (CS) framework that uses GANs to convert undersampled MRI into high-quality MRI. |

learning algorithms are code in some programming language, and radiologists and doctors need not learn to code to implement the AI models in their clinical environment. Hence it is necessary to build easy-to-use graphic user interfaces (GUIs). Various medical companies and startups have stepped up to the task, and there are many U.S. Food and Drug Administration (FDA) cleared AI solutions already in use in the clinical environment. The following are a few such FDA-cleared solutions (American College of Radiology Data Science Institute, 2019; Benjamens et al., 2020):

a. Accipiolx: Accipiolx is a tool developed by MaxQ AI Ltd. to assess non-contrast head CT images to detect intracranial hemorrhage.
b. ADAS 3D: ADAS 3D is an imaging platform by Adas 3D Medical (Galgo Medical) to visualize and analyze MRI and CT images of the heart.
c. Aidoc BriefCase: It is a radiological computer-aided software designed by Aidoc Medical for analysis of the cervical spine, intracranial hemorrhage, pulmonary embolism, large vessel occlusion, intra-abdominal free gas pathologies.
d. AI-Rad Companion: Developed by Siemens Medical Solutions, AI-Rad Companion is one of the most famous AI tools in the healthcare sector. It has functionalities across various domains and helps in the qualitative analysis and segmentation of brain structures from MRI, analysis of CT Digital Imaging and Communications in Medicine images in cardiology and the musculoskeletal system, post-processing image analysis of prostate MRI, and in the evaluation and assessment of diseases in the lungs.
e. Arterys Cardio and Oncology DL: It is a web-accessible medical imaging analysis software for viewing, manipulating, 3D visualization, and quantifying images from various imaging modalities.
f. ClearView cCAD: It is a software application by Clearview Diagnostics Inc. for classifying the shape and orientation of breast ultrasound images.
g. Critical Care Suite: Critical Care Suite is software designed by GE Medical Systems for computer-aided analysis of chest X-ray images.
h. Densitas Densityai and DM-Density: This software application by Densitas uses full-field digital mammography modality to assess breast tissue composition.
i. FerriSmart Analysis System: Resonance Health Analysis Service developed this software for the measurement of R2 and iron concentration in the liver from MRI.
j. FractureDetect (FX): Developed by Imagen Technologies, FX is a computer-assisted detection tool that helps detect fractures from X-ray images.
k. HealthCCS, HealthCXR, HealthICH, HealthMammo, and HealthVCF: These are all AI software developed by Zebra Medical Vision, used to evaluate calcified plaques in coronary arteries from CT images, to evaluate chest X-rays for pleural effusion, to aid in the clinical assessment of non-contrast head CT images for detection of intracranial hemorrhage, to screen mammogram images for suspicious findings, and to analyze chest and abdominal CT images for the presence of vertebral compression.

l. Icobrain and Icobrain-ctp: Developed by Icometrix NV, this software helps in the automatic labeling, visualization, analysis, and volumetric quantification of brain MR and CT images.
m. Quantib Brain and Quantib ND: Quantib designed this software for segmentation and volumetric measurement of grey matter, white matter, and cerebrospinal fluid from brain MR images.
n. StoneChecker: It is a post-processing software by Imaging Biometrics used to diagnose kidney stones from abdominal CT images.
o. SubtleMR and SubtlePET: These image processing software programs by Subtle Medical are used for image enhancement, transfer, storage, and noise reduction of MRI and PET images.

While these are some of the AI algorithms that have achieved success in clinical environments, many studies have failed. Google AI team worked on the detection of diabetic retinopathy with hospitals from India and the United States. They achieved high accuracies of over 90% with their deep learning model on the trained dataset but failed to achieve similar results when deployed across various clinics. Their deep learning model trained on high-quality images, but most images were of low quality due to inadequate lighting conditions in clinical settings.

Deep learning algorithms require a massive amount of data to train the models. Big data in medical imaging will be a big boost to more deep learning implementation in the field, as many studies have shown how the accuracy of deep learning models improves logarithmically with the amount of data used to train the model (Sun et al., 2017).

### 6.4.1 Case Study: Deep Learning Algorithms Beat Human Performance

a. **Breast cancer:** According to a study in 2020, researchers from Google DeepMind AI could outperform radiologists in breast cancer detection (McKinney et al., 2020). Google trained the algorithm on thousands of breast cancer images. The model reduced the false-positive rate by 5.7% and the false-negative rate by 9.4% when used to analyze 18,000 U.S. patients against the initial analysis by radiologists. In the United Kingdom, the model reduced false positives by 1.2% and false negatives by 2.7%.
b. **Brain Tumor:** In a competition in 2018, a deep learning algorithm from the Artificial Intelligence Research Centre for Neurological Disorders at the Beijing Tiantan Hospital, called the BioMind AI system, beat doctors in the detection of brain tumor and hematoma expansion (Yamei, 2018). Out of the 225 brain tumor cases, the AI could make 87% accurate guesses in 15 minutes, while 15 senior doctors could only obtain an accuracy of 66%. Also, it reached an accuracy of 83% in detecting brain hematoma expansion, while doctors could get an accuracy of 63% only.
c. **COVID-19:** A 2020 study on the detection of COVID-19 from chest CT scan images showed how a deep learning algorithm performs similarly to senior radiologists and enhanced the performance of junior radiologists (K. Zhang et al., 2020). The four senior radiologists had 12 to 25 years of

clinical experience, whereas the four junior radiologists had 5 to 15 years of clinical experience. The AI system performed much better than junior radiologists and gave a comparable performance as the senior radiologists. The AI yielded a weighted error of 9.29%, while the radiologists yielded a mean error of 13.55%. With the help of the AI system, the performance of junior radiologists improved significantly and was similar to that of senior radiologists.

## 6.5 CONCLUSION

Deep learning is advancing rapidly, and with the meteoric development of state-of-the-art algorithms, it is gradually getting better than humans at visual and auditory recognition tasks. These networks, especially from computer vision, are adopted or fine-tuned to work on medical images. With the rise of big data in the medical field and hospitals and research institutions willing to cooperate with AI firms, the adoption of deep learning in various medical imaging domains is imminent. According to a Signify Research report [123], AI in medical imaging will undergo robust growth and top 2 billion USD by 2023. AI will help not just in improving accuracy but will also boost productivity and curate to a more personalized treatment plan. Even though safety and understanding of networks, privacy concerns, integration into the hospital workflow are significant challenges that need addressing, the ever-increasing interest of AI giants, such as Google, Microsoft, Tencent, Alibaba, IBM, and Nvidia, and various start-ups in the medical imaging field suggests that we will soon overcome these barriers.

## REFERENCES

Abadi, M., Barham, P., Chen, J., Chen, Z., Davis, A., Dean, J., Devin, M., Ghemawat, S., Irving, G., Isard, M., Kudlur, M., Levenberg, J., Monga, R., Moore, S., Murray, D. G., Steiner, B., Tucker, P., Vasudevan, V., Warden, P., . . . Zheng, X. (2016). TensorFlow: A system for large-scale machine learning. In *Proceedings of the 12th USENIX Symposium on Operating Systems Design and Implementation, OSDI 2016*. https://www.tensorflow.org/

American College of Radiology Data Science Institute. (2019). *FDA Cleared AI Algorithms*. American College of Radiology.

Armanious, K., Jiang, C., Fischer, M., Küstner, T., Hepp, T., Nikolaou, K., Gatidis, S., & Yang, B. (2020). MedGAN: Medical image translation using GANs. *Computerized Medical Imaging and Graphics*, 79, 101684. https://doi.org/10.1016/j.compmedimag.2019.101684

Aslam, A., Khan, E., & Beg, M. M. S. (2015). Improved edge detection algorithm for brain tumor segmentation. *Procedia Computer Science*, 58, 430–437. https://doi.org/10.1016/j.procs.2015.08.057

Aubreville, M., Krappmann, M., Bertram, C., Klopfleisch, R., & Maier, A. (2017). A guided spatial transformer network for histology cell differentiation. In *VCBM 2017 – Eurographics Workshop on Visual Computing for Biology and Medicine*. https://doi.org/10.2312/vcbm.20171233

Bach, S., Binder, A., Montavon, G., Klauschen, F., Müller, K. R., & Samek, W. (2015). On pixel-wise explanations for non-linear classifier decisions by layer-wise relevance propagation. *PLoS ONE*, 10(7), e0130140. https://doi.org/10.1371/journal.pone.0130140

Badrinarayanan, V., Kendall, A., & Cipolla, R. (2017). SegNet: A deep convolutional encoder-decoder architecture for image segmentation. *IEEE Transactions on Pattern Analysis and Machine Intelligence,* 39(12), 2481–2495. https://doi.org/10.1109/TPAMI.2016.2644615

Bahrami, K., Shi, F., Rekik, I., Gao, Y., & Shen, D. (2017). 7T-guided super-resolution of 3T MRI. *Medical Physics,* 44(5), 1661–1667. https://doi.org/10.1002/mp.12132

Baselice, F., Ferraioli, G., Pascazio, V., & Sorriso, A. (2017). Bayesian MRI denoising in complex domain. *Magnetic Resonance Imaging,* 38, 112–122. https://doi.org/10.1016/j.mri.2016.12.024

Bello, I., Zoph, B., Vasudevan, V., & Le, Q. V. (2017). Neural optimizer search with Reinforcement learning. In *Proceedings of the 34th International Conference on Machine Learning*, Sydney, Australia, PMLR 70, 2017, ML Research Press. arXiv preprint arXiv:1707.07012

Bengio, Y., Louradour, J., Collobert, R., & Weston, J. (2009). Curriculum learning. In *Proceedings of the 26th Annual International Conference on Machine Learning (ICML '09),* 41–48. Association for Computing Machinery, New York, NY, USA. https://doi.org/10.1145/1553374.1553380

Benjamens, S., Dhunnoo, P., & Meskó, B. (2020). The state of artificial intelligence-based FDA-approved medical devices and algorithms: An online database. *npj Digital Medicine*, 3, 118. https://doi.org/10.1038/s41746-020-00324-0

Bermudez, C., Plassard, A. J., Davis, L. T., Newton, A. T., Resnick, S. M., & Landman, B. A. (2018). Learning implicit brain MRI manifolds with deep learning. *Proceedings of the SPIE International Society for Optical Engineering*, 1–5, *arXiv.* https://doi.org/10.1117/12.2293515

Bertels, J., Robben, D., Vandermeulen, D., & Suetens, P. (2020). Optimization with soft dice can lead to a volumetric bias. In Crimi, A. & Bakas, S. (eds.), *Brainlesion: Glioma, Multiple Sclerosis, Stroke and Traumatic Brain Injuries. BrainLes 2019. Lecture Notes in Computer Science (Including Subseries Lecture Notes in Artificial Intelligence and Lecture Notes in Bioinformatics),* vol. 11992, 1–7. Springer, Cham. https://doi.org/10.1007/978-3-030-46640-4_9

Blendowski, M., & Heinrich, M. P. (2019). Combining MRF-based deformable registration and deep binary 3D-CNN descriptors for large lung motion estimation in COPD patients. *International Journal of Computer Assisted Radiology and Surgery,* 14, 43–52. https://doi.org/10.1007/s11548-018-1888-2

Cai, J., Lu, L., Xie, Y., Xing, F., & Yang, L. (2017). Improving deep pancreas segmentation in CT and MRI images via recurrent neural contextual learning and direct loss function. *arXiv.*

Chartrand, G., Cheng, P. M., Vorontsov, E., Drozdzal, M., Turcotte, S., Pal, C. J., Kadoury, S., & Tang, A. (2017). Deep learning: A primer for radiologists. *Radiographics,* 37, 2113–2131. https://doi.org/10.1148/rg.2017170077

Chaudhari, A. S., Fang, Z., Kogan, F., Wood, J., Stevens, K. J., Gibbons, E. K., Lee, J. H., Gold, G. E., & Hargreaves, B. A. (2018). Super-resolution musculoskeletal MRI using deep learning. *Magnetic Resonance in Medicine,* 80(5), 2139–2154. https://doi.org/10.1002/mrm.27178

Chen, H., Qi, X., Yu, L., & Heng, P. A. (2016). DCAN: Deep contour-aware networks for accurate gland segmentation. In *2016 IEEE Conference on Computer Vision and Pattern Recognition (CVPR),* 2487–2496. IEEE. https://doi.org/10.1109/CVPR.2016.273

Chen, J., Yang, L., Zhang, Y., Alber, M., & Chen, D. Z. (2016). Combining fully convolutional and recurrent neural networks for 3D biomedical image segmentation. *Advances in Neural Information Processing Systems*, 29, 3044–3052.

Cheng, X., Zhang, L., & Zheng, Y. (2018). Deep similarity learning for multimodal medical images. *Computer Methods in Biomechanics and Biomedical Engineering: Imaging and Visualization,* 6(3), 248–252. https://doi.org/10.1080/21681163.2015.1135299

Chollet, F. (2017). Xception: Deep learning with depthwise separable convolutions. In *2017 IEEE Conference on Computer Vision and Pattern Recognition (CVPR),* Honolulu, HI, USA, 1800–1807. IEEE. https://doi.org/10.1109/CVPR.2017.195

Cui, S., Mao, L., Jiang, J., Liu, C., & Xiong, S. (2018). Automatic semantic segmentation of brain gliomas from MRI images using a deep cascaded neural network. *Journal of Healthcare Engineering,* 2018, 14 pages. https://doi.org/10.1155/2018/4940593

Domingues, I., Pereira, G., Martins, P., Duarte, H., Santos, J., & Abreu, P. H. (2020). Using deep learning techniques in medical imaging: A systematic review of applications on CT and PET. *Artificial Intelligence Review,* 53, 4093–4160. https://doi.org/10.1007/s10462-019-09788-3

Duchi, J., Hazan, E., & Singer, Y. (2010). Adaptive subgradient methods for online learning and stochastic optimization. *Journal of Machine Learning Research,* 12(2011), 2121–2159.

Esteva, A., Kuprel, B., Novoa, R. A., Ko, J., Swetter, S. M., Blau, H. M., & Thrun, S. (2017). Dermatologist-level classification of skin cancer with deep neural networks. *Nature,* 542, 115–118. https://doi.org/10.1038/nature21056

Frid-Adar, M., Diamant, I., Klang, E., Amitai, M., Goldberger, J., & Greenspan, H. (2018). GAN-based synthetic medical image augmentation for increased CNN performance in liver lesion classification. *Neurocomputing,* 321, 321–331. arxiv:1803.01229v1. https://doi.org/10.1016/j.neucom.2018.09.013

Ganapathy, N., Swaminathan, R., & Deserno, T. M. (2018). Deep learning on 1-D biosignals: A taxonomy-based survey. *Yearbook of Medical Informatics,* 2018, 98–109. https://doi.org/10.1055/s-0038-1667083

Gibson, E., Li, W., Sudre, C., Fidon, L., Shakir, D. I., Wang, G., Eaton-Rosen, Z., Gray, R., Doel, T., Hu, Y., Whyntie, T., Nachev, P., Modat, M., Barratt, D. C., Ourselin, S., Cardoso, M. J., & Vercauteren, T. (2018). NiftyNet: A deep-learning platform for medical imaging. *Computer Methods and Programs in Biomedicine,* 158, 113–122. https://doi.org/10.1016/j.cmpb.2018.01.025

Girshick, R. (2015). Fast R-CNN. In *Proceedings of the IEEE International Conference on Computer Vision, (ICCV) (ICCV '15),* 1440–1448. IEEE Computer Society, New York, USA. https://doi.org/10.1109/ICCV.2015.169

Girshick, R., Donahue, J., Darrell, T., & Malik, J. (2014). Rich feature hierarchies for accurate object detection and semantic segmentation. In *2014 IEEE Conference on Computer Vision and Pattern Recognition*, 580–587. IEEE. https://doi.org/10.1109/CVPR.2014.81

Goodfellow, I. J., Pouget-Abadie, J., Mirza, M., Xu, B., Warde-Farley, D., Ozair, S., Courville, A., & Bengio, Y. (2014). Generative adversarial nets. In *NIPS'14: Proceedings of the 27th International Conference on Neural Information Processing Systems – Volume 2 (NIPS'14),* 2672–2680. MIT Press, Cambridge, MA, USA. https://doi.org/10.3156/jsoft.29.5_177_2

Gordienko, Y., Gang, P., Hui, J., Zeng, W., Kochura, Y., Alienin, O., Rokovyi, O., & Stirenko, S. (2019). Deep learning with lung segmentation and bone shadow exclusion techniques for chest X-ray analysis of lung cancer. In Z. Hu, S. Petoukhov, I. Dychka, & M. He (eds.) *Advances in Computer Science for Engineering and Education. ICCSEEA 2018. Advances in Intelligent Systems and Computing,* vol. 754. Springer, Cham. https://doi.org/10.1007/978-3-319-91008-6_63

Gulshan, V., Peng, L., Coram, M., Stumpe, M. C., Wu, D., Narayanaswamy, A., Venugopalan, S., Widner, K., Madams, T., Cuadros, J., Kim, R., Raman, R., Nelson, P. C., Mega, J.

L., & Webster, D. R. (2016). Development and validation of a deep learning algorithm for detection of diabetic retinopathy in retinal fundus photographs. *JAMA – Journal of the American Medical Association,* 316(22), 2402–2410. https://doi.org/10.1001/jama.2016.17216

Gurbani, S. S., Schreibmann, E., Maudsley, A. A., Cordova, J. S., Soher, B. J., Poptani, H., Verma, G., Barker, P. B., Shim, H., & Cooper, L. A. D. (2018). A convolutional neural network to filter artifacts in spectroscopic MRI. *Magnetic Resonance in Medicine,* 80(5), 1765–1775. https://doi.org/10.1002/mrm.27166

Han, X. (2017). MR-based synthetic CT generation using a deep convolutional neural network method. *Medical Physics,* 44(4), 1408–1419. https://doi.org/10.1002/mp.12155

He, K., Zhang, X., Ren, S., & Sun, J. (2016). Deep residual learning for image recognition. In *2016 IEEE Conference on Computer Vision and Pattern Recognition (CVPR),* 770–778. IEEE. https://doi.org/10.1109/CVPR.2016.90

Heußer, T., Rank, C. M., Berker, Y., Freitag, M. T., & Kachelrieß, M. (2017). MLAA-based attenuation correction of flexible hardware components in hybrid PET/MR imaging. *EJNMMI Physics 4.* Article number: 12. https://doi.org/10.1186/s40658-017-0177-4

Howard, A. G., Zhu, M., Chen, B., Kalenichenko, D., Wang, W., Weyand, T., Andreetto, M., & Adam, H. (2017). MobileNets: Efficient convolutional neural networks for mobile vision applications. *arXiv.*

Hu, J., Shen, L., Albanie, S., Sun, G., & Wu, E. (2020). Squeeze-and-excitation networks. *IEEE Transactions on Pattern Analysis and Machine Intelligence.* In *2018 IEEE/CVF Conference on Computer Vision and Pattern Recognition*, 7132–7141. https://10.1109/CVPR.2018.00745

Huang, G., Liu, Z., Van Der Maaten, L., & Weinberger, K. Q. (2017). Densely connected convolutional networks. In *2017 IEEE Conference on Computer Vision and Pattern Recognition (CVPR),* 2261–2269. IEEE. https://doi.org/10.1109/CVPR.2017.243

Hwang, D., Kim, K. Y., Kang, S. K., Seo, S., Paeng, J. C., Lee, D. S., & Lee, J. S. (2018). Improving the accuracy of simultaneously reconstructed activity and attenuation maps using deep learning. *Journal of Nuclear Medicine,* 59, 1624–1629. https://doi.org/10.2967/jnumed.117.202317

Jifara, W., Jiang, F., Rho, S., Cheng, M., & Liu, S. (2019). Medical image denoising using convolutional neural network: A residual learning approach. *Journal of Supercomputing,* 75, 704–718. https://doi.org/10.1007/s11227-017-2080-0

Jiménez, J., Škalič, M., Martínez-Rosell, G., & De Fabritiis, G. (2018). KDEEP: Protein-ligand absolute binding affinity prediction via 3D-convolutional neural networks. *Journal of Chemical Information and Modelling,* 58(2), 287–296. https://doi.org/10.1021/acs.jcim.7b00650

Kamnitsas, K., Ledig, C., Newcombe, V. F. J., Simpson, J. P., Kane, A. D., Menon, D. K., Rueckert, D., & Glocker, B. (2017). Efficient multi-scale 3D CNN with fully connected CRF for accurate brain lesion segmentation. *Medical Image Analysis,* 36, 61–78. https://doi.org/10.1016/j.media.2016.10.004

Katzman, J. L., Shaham, U., Cloninger, A., Bates, J., Jiang, T., & Kluger, Y. (2018). DeepSurv: Personalized treatment recommender system using a Cox proportional hazards deep neural network. *BMC Medical Research Methodology 18.* Article number: 24. https://doi.org/10.1186/s12874-018-0482-1

Kingma, D. P., & Ba, J. L. (2015). Adam: A method for stochastic optimization. In *3rd International Conference on Learning Representations, ICLR 2015 – Conference Track Proceedings*, CoRR, abs/1412.6980. https://arxiv.org/abs/1412.6980

Krizhevsky, A., Sutskever, I., & Hinton, G. E. (2012). 2012 AlexNet. In *NIPS'12: Proceedings of the 25th International Conference on Neural Information Processing Systems – Volume 1, December 2012*, 1097–1105.

Krizhevsky, A., Sutskever, I., & Hinton, G. E. (2017). ImageNet classification with deep convolutional neural networks. *Communications of the ACM*, 60(6), 84–90. https://doi.org/10.1145/3065386

Küstner, T., Liebgott, A., Mauch, L., Martirosian, P., Bamberg, F., Nikolaou, K., Yang, B., Schick, F., & Gatidis, S. (2018). Automated reference-free detection of motion artifacts in magnetic resonance images. *Magnetic Resonance Materials in Physics, Biology and Medicine,* 31, 243–256. https://doi.org/10.1007/s10334-017-0650-z

Kyathanahally, S. P., Döring, A., & Kreis, R. (2018). Deep learning approaches for detection and removal of ghosting artifacts in MR spectroscopy. *Magnetic Resonance in Medicine*, 80(3), 851–863. https://doi.org/10.1002/mrm.27096

Lakhani, P., & Sundaram, B. (2017). Deep learning at chest radiography: Automated classification of pulmonary tuberculosis by using convolutional neural networks. *Radiology*, 284, 574–582. https://doi.org/10.1148/radiol.2017162326

Larroza, A., Moliner, L., Alvarez-Gomez, J. M., Oliver, S., Espinos-Morato, H., Vergara-DIaz, M., & Rodriguez-Alvarez, M. J. (2019). Deep learning for MRI-based CT synthesis: A comparison of MRI sequences and neural network architectures. In *2019 IEEE Nuclear Science Symposium and Medical Imaging Conference (NSS/MIC 2019),* 1–4. https://doi.org/10.1109/NSS/MIC42101.2019.9060051

Lecun, Y., Bengio, Y., & Hinton, G. (2015). Deep learning. *Nature*, 521, 436–444. https://doi.org/10.1038/nature14539

Lee, S. G., Bae, J. S., Kim, H., Kim, J. H., & Yoon, S. (2018). Liver lesion detection from weakly-labeled multi-phase CT volumes with a grouped single shot multibox detector. In *Lecture Notes in Computer Science (Including Subseries Lecture Notes in Artificial Intelligence and Lecture Notes in Bioinformatics),* 693–701. https://doi.org/10.1007/978-3-030-00934-2_77

Lekien, F., & Marsden, J. (2005). Tricubic interpolation in three dimensions. *International Journal for Numerical Methods in Engineering,* 63, 455–471. https://doi.org/10.1002/nme.1296

Leynes, A. P., Yang, J., Wiesinger, F., Kaushik, S. S., Shanbhag, D. D., Seo, Y., Hope, T. A., & Larson, P. E. Z. (2017). Direct PseudoCT generation for pelvis PET/MRI attenuation correction using deep convolutional neural networks with multi-parametric MRI: Zero echo-time and dixon deep pseudoCT (ZeDD-CT). *Journal of Nuclear Medicine*, 117, 198051. https://doi.org/10.2967/jnumed.117.198051

Liao, R., Miao, S., De Tournemire, P., Grbic, S., Kamen, A., Mansi, T., & Comaniciu, D. (2017). An artificial agent for robust image registration. In *31st AAAI Conference on Artificial Intelligence, AAAI 2017,* 4168–4175. Arxiv.

Lin, T. Y., Goyal, P., Girshick, R., He, K., & Dollar, P. (2020). Focal loss for dense object detection. *IEEE Transactions on Pattern Analysis and Machine Intelligence,* 42(2), 318–327. https://doi.org/10.1109/TPAMI.2018.2858826

Liu, W., Anguelov, D., Erhan, D., Szegedy, C., Reed, S., Fu, C. Y., & Berg, A. C. (2016). SSD: Single shot multibox detector. In *Lecture Notes in Computer Science (Including Subseries Lecture Notes in Artificial Intelligence and Lecture Notes in Bioinformatics),* vol. 9905, 21–37. Springer, Cham. https://doi.org/10.1007/978-3-319-46448-0_2

Liu, Y., Gadepalli, K., Norouzi, M., Dahl, G. E., Kohlberger, T., Boyko, A., Venugopalan, S., Timofeev, A., Nelson, P. Q., Corrado, G. S., Hipp, J. D., Peng, L., & Stumpe, M. C. (2017). Detecting cancer metastases on gigapixel pathology images. *arXiv.*

Lundervold, A. S., & Lundervold, A. (2019). An overview of deep learning in medical imaging focusing on MRI. *Zeitschrift fur Medizinische Physik,* 29(2), 102–127. https://doi.org/10.1016/j.zemedi.2018.11.002

Ma, K., Wang, J., Singh, V., Tamersoy, B., Chang, Y. J., Wimmer, A., & Chen, T. (2017). Multimodal image registration with deep context reinforcement learning. In *Lecture*

*Notes in Computer Science (Including Subseries Lecture Notes in Artificial Intelligence and Lecture Notes in Bioinformatics),* vol. 10433, 240–248. Springer, Cham. https://doi.org/10.1007/978-3-319-66182-7_28

Mahapatra, D., Bozorgtabar, B., Thiran, J. P., & Shao, L. (2020). Structure preserving stain normalization of histopathology images using self supervised semantic guidance. In *Lecture Notes in Computer Science (Including Subseries Lecture Notes in Artificial Intelligence and Lecture Notes in Bioinformatics),* vol. 12265, 309–319. Springer, Cham. https://doi.org/10.1007/978-3-030-59722-1_30

Maier, A., Syben, C., Lasser, T., & Riess, C. (2019). A gentle introduction to deep learning in medical image processing. *Zeitschrift fur Medizinische Physik,* 29(2), 86–101. https://doi.org/10.1016/j.zemedi.2018.12.003

Mardani, M., Gong, E., Cheng, J. Y., Vasanawala, S., Zaharchuk, G., Alley, M., Thakur, N., Han, S., Dally, W., Pauly, J. M., & Xing, L. (2017). Deep generative adversarial networks for compressed sensing (GANCS) automates MRI. *arXiv.*

McKinney, S. M., Sieniek, M., Godbole, V., Godwin, J., Antropova, N., Ashrafian, H., Back, T., Chesus, M., Corrado, G. C., Darzi, A., Etemadi, M., Garcia-Vicente, F., Gilbert, F. J., Halling-Brown, M., Hassabis, D., Jansen, S., Karthikesalingam, A., Kelly, C. J., King, D., . . . Shetty, S. (2020). International evaluation of an AI system for breast cancer screening. *Nature.* https://doi.org/10.1038/s41586-019-1799-6

Miao, S., Piat, S., Fischer, P., Tuysuzoglu, A., Mewes, P., Mansi, T., & Liao, R. (2018). Dilated FCN for multi-agent 2D/3D medical image registration. In *32nd AAAI Conference on Artificial Intelligence, AAAI 2018.*

Mikołajczyk, A., & Grochowski, M. (2018). Data augmentation for improving deep learning in image classification problem. In *2018 International Interdisciplinary PhD Workshop, IIPhDW 2018*, 117–122. https://doi.org/10.1109/IIPHDW.2018.8388338

Milletari, F., Navab, N., & Ahmadi, S. A. (2016). V-Net: Fully convolutional neural networks for volumetric medical image segmentation. In *Proceedings – 2016 4th International Conference on 3D Vision, 3DV 2016.* https://doi.org/10.1109/3DV.2016.79, arxiv.

Moeskops, P., Wolterink, J. M., van der Velden, B. H. M., Gilhuijs, K. G. A., Leiner, T., Viergever, M. A., & Išgum, I. (2016). Deep learning for multi-task medical image segmentation in multiple modalities. In *Lecture Notes in Computer Science (Including Subseries Lecture Notes in Artificial Intelligence and Lecture Notes in Bioinformatics),* vol. 9901, 478–486. Springer, Cham. https://doi.org/10.1007/978-3-319-46723-8_55

Naceur, M. B., Saouli, R., Akil, M., & Kachouri, R. (2018). Fully automatic brain tumor segmentation using end-to-end incremental deep neural networks in MRI images. *Computer Methods and Programs in Biomedicine,* 39–49. https://doi.org/10.1016/j.cmpb.2018.09.007

Nwankpa, C. E., Ijomah, W., Gachagan, A., & Marshall, S. (2018). Activation functions: Comparison of trends in practice and research for deep learning. *arXiv.*

Owen, D., Melbourne, A., Eaton-Rosen, Z., Thomas, D. L., Marlow, N., Rohrer, J., & Ourselin, S. (2018). Deep convolutional filtering for spatio-temporal denoising and artifact removal in arterial spin labelling MRI. In *Lecture Notes in Computer Science (Including Subseries Lecture Notes in Artificial Intelligence and Lecture Notes in Bioinformatics),* vol. 11070, 21–29. Springer, Cham. https://doi.org/10.1007/978-3-030-00928-1_3

Pawlowski, N., Ira, S., Matthew, K., Lee, C. H., Kainz, B., Rueckert, D., Glocker, B., & Rajchl, M. (2017). DLTK: State of the art reference implementations for deep learning on medical images. *arXiv.*

Petcharunpaisan, S. (2010). Arterial spin labeling in neuroimaging. *World Journal of Radiology,* 2(10), 384–398. https://doi.org/10.4329/wjr.v2.i10.384

Plenge, E., Poot, D. H. J., Bernsen, M., Kotek, G., Houston, G., Wielopolski, P., Van Der Weerd, L., Niessen, W. J., & Meijering, E. (2012). Super-resolution methods in MRI: Can they improve the trade-off between resolution, signal-to-noise ratio, and acquisition time? *Magnetic Resonance in Medicine,* 68(6), 1983–1993. https://doi.org/10.1002/mrm.24187

Prasoon, A., Petersen, K., Igel, C., Lauze, F., Dam, E., & Nielsen, M. (2013). Deep feature learning for knee cartilage segmentation using a triplanar convolutional neural network. In K. Mori, I. Sakuma, Y. Sato, C. Barillot, & N. Navab (eds.), *Medical Image Computing and Computer-Assisted Intervention – MICCAI 2013. MICCAI 2013. Lecture Notes in Computer Science (Including Subseries Lecture Notes in Artificial Intelligence and Lecture Notes in Bioinformatics),* vol. 8150, 246–253. Springer, Berlin, Heidelberg. https://doi.org/10.1007/978-3-642-40763-5_31

Rajkomar, A., Oren, E., Chen, K., Dai, A. M., Hajaj, N., Hardt, M., Liu, P. J., Liu, X., Marcus, J., Sun, M., Sundberg, P., Yee, H., Zhang, K., Zhang, Y., Flores, G., Duggan, G. E., Irvine, J., Le, Q., Litsch, K., . . . Dean, J. (2018). Scalable and accurate deep learning with electronic health records. *arXiv.* https://doi.org/10.1038/s41746-018-0029-1

Rajpurkar, P., Irvin, J., Zhu, K., Yang, B., Mehta, H., Duan, T., Ding, D., Bagul, A., Ball, R. L., Langlotz, C., Shpanskaya, K., Lungren, M. P., & Ng, A. Y. (2017). CheXNet: Radiologist-level pneumonia detection on chest X-rays with deep learning. *arXiv.*

Ran, M., Hu, J., Chen, Y., Chen, H., Sun, H., Zhou, J., & Zhang, Y. (2019). Denoising of 3D magnetic resonance images using a residual encoder–decoder Wasserstein generative adversarial network. *Medical Image Analysis,* vol. 55, 165–180. Elsevier. https://doi.org/10.1016/j.media.2019.05.001

Redmon, J., Divvala, S., Girshick, R., & Farhadi, A. (2016). You only look once: Unified, real-time object detection. In *2016 IEEE Conference on Computer Vision and Pattern Recognition (CVPR),* 779–788. IEEE. https://doi.org/10.1109/CVPR.2016.91

Ren, S., He, K., Girshick, R., & Sun, J. (2017). Faster R-CNN: Towards real-time object detection with region proposal networks. In *NIPS'15: Proceedings of the 28th International Conference on Neural Information Processing Systems,* vol. 1, 91–99. https://doi.org/10.1109/TPAMI.2016.2577031

Ribeiro, M. T., Singh, S., & Guestrin, C. (2016). "Why should I trust you?" Explaining the predictions of any classifier. In *KDD '16: Proceedings of the ACM SIGKDD International Conference on Knowledge Discovery and Data Mining,* 1135–1144. Association for Computing Machinery, New York, NY, USA. https://doi.org/10.1145/2939672.2939778

Ronneberger, O., Fischer, P., & Brox, T. (2015). U-net: Convolutional networks for biomedical image segmentation. In O. Ronneberger, P. Fischer, & T. Brox (eds.), *Medical Image Computing and Computer-Assisted Intervention (MICCAI). Lecture Notes in Computer Science (Including Subseries Lecture Notes in Artificial Intelligence and Lecture Notes in Bioinformatics),* vol. 9351, 234–241. Springer, LNCS. https://doi.org/10.1007/978-3-319-24574-4_28

Ropele, S., Ebner, F., Fazekas, F., & Reishofer, G. (2010). Super-resolution MRI using microscopic spatial modulation of magnetization. *Magnetic Resonance in Medicine,* 64(6), 1671–1675. https://doi.org/10.1002/mrm.22616

Sadda, P., & Qarni, T. (2018). Real-time medical video denoising with deep learning: Application to angiography. *International Journal of Applied Information Systems,* 12(13), 22–28. https://doi.org/10.5120/ijais2018451755.

Salimi-Khorshidi, G., Douaud, G., Beckmann, C. F., Glasser, M. F., Griffanti, L., & Smith, S. M. (2014). Automatic denoising of functional MRI data: Combining independent component analysis and hierarchical fusion of classifiers. *NeuroImage,* 90, 449–468. https://doi.org/10.1016/j.neuroimage.2013.11.046

Santosh, K. C. (2020). AI-driven tools for coronavirus outbreak: Need of active learning and cross-population train/test models on multitudinal/multimodal data. *Journal of Medical Systems,* 44, 93. https://doi.org/10.1007/s10916-020-01562-1

Schlegl, T., Seeböck, P., Waldstein, S. M., Langs, G., & Schmidt-Erfurth, U. (2019). f-AnoGAN: Fast unsupervised anomaly detection with generative adversarial networks. *Medical Image Analysis,* 54, 30–44. https://doi.org/10.1016/j.media.2019.01.010

Schlegl, T., Seeböck, P., Waldstein, S. M., Schmidt-Erfurth, U., & Langs, G. (2017). Unsupervised anomaly detection with generative adversarial networks to guide marker discovery. In M. Niethammer et al. (eds.), *Information Processing in Medical Imaging. IPMI 2017. Lecture Notes in Computer Science (Including Subseries Lecture Notes in Artificial Intelligence and Lecture Notes in Bioinformatics),* vol. 10265, 146–157. Springer, Cham. https://doi.org/10.1007/978-3-319-59050-9_12

Sedghi, A., Luo, J., Mehrtash, A., Pieper, S., Tempany, C., Kapur, T., Mousavi, P., & Wells, W. (2019). Semi-supervised image registration using deep learning. In *Proceedings Volume 10951, Medical Imaging 2019: Image-Guided Procedures, Robotic Interventions, and Modelling*, 109511G. https://doi.org/10.1117/12.2513020

Selvaraju, R. R., Cogswell, M., Das, A., Vedantam, R., Parikh, D., & Batra, D. (2017). Grad-CAM. Visual explanations from deep networks via gradient-based localization. *International Journal of Computer Vision,* 128, 336–359. https://doi.org/10.1007/s11263-019-01228-7

Shickel, B., Tighe, P. J., Bihorac, A., & Rashidi, P. (2018). Deep EHR: A survey of recent advances in deep learning techniques for Electronic Health Record (EHR) analysis. *IEEE Journal of Biomedical and Health Informatics,* 22(5), 1589–1604. https://doi.org/10.1109/JBHI.2017.2767063

Shilling, R. Z., Robbie, T. Q., Bailloeul, T., Mewes, K., Mersereau, R. M., & Brummer, M. E. (2009). A super-resolution framework for 3-D high-resolution and high-contrast imaging using 2-D multislice MRI. In *IEEE Transactions on Medical Imaging,* 28(5), 633–644. https://doi.org/10.1109/TMI.2008.2007348

Shrikumar, A., Greenside, P., Shcherbina, A., & Kundaje, A. (2016). Not just a black box: Interpretable deep learning by propagating activation differences. *ArXiv.*

Simonyan, K., & Zisserman, A. (2015). Very deep convolutional networks for large-scale image recognition. In *3rd International Conference on Learning Representations, ICLR 2015,* San Diego, CA, USA, May 7–9, 2015, Conference Track Proceedings. https://arxiv.org/abs/1409.1556.

Stephen, O., Sain, M., Maduh, U. J., & Jeong, D. U. (2019). An efficient deep learning approach to pneumonia classification in healthcare. *Journal of Healthcare Engineering,* 2019, Article ID 4180949. https://doi.org/10.1155/2019/4180949

Sudre, C. H., Li, W., Vercauteren, T., Ourselin, S., & Jorge Cardoso, M. (2017). Generalised dice overlap as a deep learning loss function for highly unbalanced segmentations. In *Deep learning in medical image analysis and multimodal learning for clinical decision support: Third International Workshop, DLMIA 2017, and 7th International Workshop, ML-CDS 2017, held in conjunction with MICCAI 2017 Quebec City, QC. Lecture Notes in Computer Science (Including Subseries Lecture Notes in Artificial Intelligence and Lecture Notes in Bioinformatics),* 240–248. Springer, Cham. https://doi.org/10.1007/978-3-319-67558-9_28.

Sun, C., Shrivastava, A., Singh, S., & Gupta, A. (2017). Revisiting unreasonable effectiveness of data in deep learning era. In *2017 IEEE International Conference on Computer Vision (ICCV),* 843–852, IEEE. https://doi.org/10.1109/ICCV.2017.97

Szegedy, C., Liu, W., Jia, Y., Sermanet, P., Reed, S., Anguelov, D., Erhan, D., Vanhoucke, V., & Rabinovich, A. (2015). Going deeper with convolutions. In *2015 IEEE Conference on*

*Computer Vision and Pattern Recognition (CVPR),* 1–9. IEEE. https://doi.org/10.1109/CVPR.2015.7298594.

Tieleman, T., & Hinton, G. (2012). Lecture 6.5-rmsprop: Divide the gradient by a running average of its recent magnitude. In *COURSERA: Neural Networks for Machine Learning*. COURSERA.

Tomita, N., Abdollahi, B., Wei, J., Ren, B., Suriawinata, A., & Hassanpour, S. (2019). Attention-based deep neural networks for detection of cancerous and precancerous esophagus tissue on histopathological slides. *JAMA Network Open,* 2(11), e1914645. https://doi.org/10.1001/jamanetworkopen.2019.14645

Torrado-Carvajal, A., Vera-Olmos, J., Izquierdo-Garcia, D., Catalano, O. A., Morales, M. A., Margolin, J., Soricelli, A., Salvatore, M., Malpica, N., & Catana, C. (2019). Dixon-vibe deep learning (divide) pseudo-CT synthesis for pelvis PET/MR attenuation correction. *Journal of Nuclear Medicine,* 60, 429–435. https://doi.org/10.2967/jnumed.118.209288

van der Spoel, E., Rozing, M. P., Houwing-Duistermaat, J. J., Eline Slagboom, P., Beekman, M., de Craen, A. J. M., Westendorp, R. G. J., & van Heemst, D. (2015). Siamese neural networks for one-shot image recognition. In *Proceedings of the 32nd International Conference on Machine Learning, Lille, France, 2015*, vol. 37. JMLR: W&CP.

Wang, L., Lin, Z. Q., & Wong, A. (2020). COVID-Net: A tailored deep convolutional neural network design for detection of COVID-19 cases from chest X-ray images. *Scientific Reports,* 10, 19549. https://doi.org/10.1038/s41598-020-76550-z

Wright, R., Khanal, B., Gomez, A., Skelton, E., Matthew, J., Hajnal, J. V., Rueckert, D., & Schnabel, J. A. (2018). LSTM spatial co-transformer networks for registration of 3D fetal US and MR brain images. In A. Melbourne et al. (eds.), *Data Driven Treatment Response Assessment and Preterm, Perinatal, and Paediatric Image Analysis. PIPPI 2018, DATRA 2018. Lecture Notes in Computer Science (Including Subseries Lecture Notes in Artificial Intelligence and Lecture Notes in Bioinformatics),* vol. 11076, 149–159. Springer, Cham. https://doi.org/10.1007/978-3-030-00807-9_15

Wu, G., Kim, M., Wang, Q., Gao, Y., Liao, S., & Shen, D. (2013). Unsupervised deep feature learning for deformable registration of MR brain images. In K. Mori, I. Sakuma, Y. Sato, C. Barillot, & N. Navab (eds.), *Medical Image Computing and Computer-Assisted Intervention – MICCAI 2013. MICCAI 2013. Lecture Notes in Computer Science (Including Subseries Lecture Notes in Artificial Intelligence and Lecture Notes in Bioinformatics)*, vol. 8150, 649–656. Springer, Berlin, Heidelberg. https://doi.org/10.1007/978-3-642-40763-5_80

Wu, G., Kim, M., Wang, Q., Munsell, B. C., & Shen, D. (2016). Scalable high-performance image registration framework by unsupervised deep feature representations learning. *IEEE Transactions on Biomedical Engineering,* 63(7), 1505–1516. https://doi.org/10.1109/TBME.2015.2496253

Xie, S., Girshick, R., Dollár, P., Tu, Z., & He, K. (2017). Aggregated residual transformations for deep neural networks. In *2017 IEEE Conference on Computer Vision and Pattern Recognition (CVPR),* 5987–5995. IEEE. https://doi.org/10.1109/CVPR.2017.634

Yun, J., Park, J. E., Lee, H., Ham, S., Kim, N., & Kim, H. S. (2019). Radiomic features and multilayer perceptron network classifier: A robust MRI classification strategy for distinguishing glioblastoma from primary central nervous system lymphoma. *Scientific Reports,* 9, 5746. https://doi.org/10.1038/s41598-019-42276-w

Zeiler, M. D. (2012). ADADELTA: An adaptive learning rate method. *ArXiv*, abs/1212.5701.

Zeiler, M. D., & Fergus, R. (2014). Visualizing and understanding convolutional networks. In D. Fleet, T. Pajdla, B. Schiele, & T. Tuytelaars (eds.), *Computer Vision – ECCV 2014. ECCV 2014. Lecture Notes in Computer Science (Including Subseries Lecture Notes in Artificial Intelligence and Lecture Notes in Bioinformatics),* vol. 8689, 818–833. Springer, Cham. https://doi.org/10.1007/978-3-319-10590-1_53

Zeng, G., Yang, X., Li, J., Yu, L., Heng, P. A., & Zheng, G. (2017). 3D U-net with multi-level deep supervision: Fully automatic segmentation of proximal femur in 3D MR images. In Q. Wang, Y. Shi, H. I. Suk, & K. Suzuki (eds.), *Machine Learning in Medical Imaging. MLMI 2017. Lecture Notes in Computer Science (Including Subseries Lecture Notes in Artificial Intelligence and Lecture Notes in Bioinformatics),* vol. 10541, 274–282. Springer, Cham. https://doi.org/10.1007/978-3-319-67389-9_32

Zhang, K., Liu, X., Shen, J., Li, Z., Sang, Y., Wu, X., Zha, Y., Liang, W., Wang, C., Wang, K., Ye, L., Gao, M., Zhou, Z., Li, L., Wang, J., Yang, Z., Cai, H., Xu, J., Yang, L., . . . Wang, G. (2020). Clinically applicable AI system for accurate diagnosis, quantitative measurements, and prognosis of COVID-19 pneumonia using computed tomography. *Cell,* 181(6), 1423–1433. https://doi.org/10.1016/j.cell.2020.04.045

Zhang, X., Xu, Z., Jia, N., Yang, W., Feng, Q., Chen, W., & Feng, Y. (2015). Denoising of 3D magnetic resonance images by using higher-order singular value decomposition. *Medical Image Analysis,* 19(1), 75–86. https://doi.org/10.1016/j.media.2014.08.004

Zhou, B., Khosla, A., Lapedriza, A., Oliva, A., & Torralba, A. (2016). Learning deep features for discriminative localization. In *2016 IEEE Conference on Computer Vision and Pattern Recognition (CVPR),* 2921–2929. IEEE. https://doi.org/10.1109/CVPR.2016.319

Zhou, Y., Zhu, Y., Ye, Q., Qiu, Q., & Jiao, J. (2018). Weakly supervised instance segmentation using class peak response. In *2018 IEEE/CVF Conference on Computer Vision and Pattern Recognition,* 3791–3800. IEEE. https://doi.org/10.1109/CVPR.2018.00399

Zhu, G., Jiang, B., Tong, L., Xie, Y., Zaharchuk, G., & Wintermark, M. (2019). Applications of deep learning to neuro-imaging techniques. *Frontiers in Neurology,* 10, 869. https://doi.org/10.3389/fneur.2019.00869

Zoph, B., Vasudevan, V., Shlens, J., & Le, Q. V. (2018). Learning transferable architectures for scalable image recognition. In *2018 IEEE/CVF Conference on Computer Vision and Pattern Recognition,* 8697–8710. IEEE. https://doi.org/10.1109/CVPR.2018.00907

# 7 Integrated Neuroinformatics
## *Analytics and Application*

*Rajan Patel, Rahul Vaghela, Madhuri Chopade, Prakash Patel, and Dulari Bhatt*

**CONTENTS**



## 7.1 INTRODUCTION: INTEGRATED NEUROSCIENCE AND NEUROINFORMATICS

In terms of brainstorming, it is good to talk to Butler and Bellou about the following: The activity in the brain is similar to playing a musical instrument at a performance, learning to play, and creating a score while playing. To do this, the brain must practice nonstop reasoning functions, for example, acquiring hearing data performed on the instrument, reading without watching a game, remembering from working memory for part of a concert, deciding to choose the next step, and observing the lines used to find the next program. In other words, brain information processing involves viewing, transmitting, connecting, storing, and building information. All

DOI: 10.1201/9781003142751-9

**TABLE 7.1**
**Big Data Technologies in Neuroscience**

| |
|---|
| Brain as an information-processing organ (15 cm) |
| Physical and logical systems (5 cm) |
| Maps (1 centimeter) |
| Networks (1 millimeter) |
| Neuron (100 micrometer) |
| Synapses (1 micrometer) |
| Molecules (1 angstroms) |

*Source:* Boubela et al. (2016).

these processes are powerful and are associated with constant changes in the basic state of the brain or in a stable environment (Rabinovich et al. 2012)

Fortunately, there is no law on data storage in the human organ. Since the brain is an uneven material system able to produce evidence as a consequence of chronological instability, the value of such statistics can be determined by the entropy of Kolmogorov–Sinai, and the quality of the facts depends on how adequate it is for reasoning purposes, that is, how far it is between the information produced and relevant data in a particular logic.

This is a very puzzling and confrontational problem. With cognitive detail, unlike traditional communication systems, the definition of information, such as quality and semantics, is important compared to the amount of information. Here's an understanding of the cognitive process of the brain as mentioned in Table 7.1. We have to deal with context-based information and terms-based data. This means that understanding depends on the ongoing work of understanding and behavior. Information that travels through the ocean flows and converges. Such a flow produces a closed loop. A response is needed for successful behavior in the transformation of the external ecosphere and the creation of new info based on symbols from the internal ecosphere. Our brains are constantly trying to find meaning in organizations, relationships, data, and patterns. We try to do that from the majority of the information offered. The brain tries to connect new information with our past understandings and information stored in our minds. When it catches an arrangement that makes sense to us, it puts this arrangement on our map. When it comes to the information kept in our minds, we learn. When we are able to make that linking, we feel free from anxiety, confusion, or pressure associated with data, facts, and statistics. Yes, when we can't connect with past experiences, we feel confused, and we feel frustrated.

## 7.2 REPRESENTATIONS

Neurons or nerve cells, aka electrical cells, can detect process and transmit information through electrical and chemical signals. It is said that the average human organ, like the brain, consists of about 100 billion neuro cells (Popovych et al. 2019). The processing of 100 billion neuro cell data is clearly no small feat. Emerging methods such as big data are best suited to deal with neuronal data. It is very difficult to store such data on a standard hard drive because it requires a very large space. Therefore, scientists are now looking for opportunities and challenges in big data to deal with neuronal data. The body's response to sodium, potassium, chloride, and calcium ions in neuronal cells forms a gradient of voltage energy throughout its membrane. Emotional tension is produced when there is a change in strength between layers (Bowden et al. 2018). Linking the multifaceted structures and tasks of the nervous system requires harmonization among dissimilar purviews of understanding, the unification through numerous levels of findings, and the blending of apparently unlike technical tactics from molecules to behavior (Vaghela et al. 2021). The difficulty of neuroinformatics is to provide a cohesive computational information framework to qualify, assist, and raise such an enterprise. In practice, major advances in our understanding of the brain involve the development and application of suitable automated tools to handle, represent, transform, analyze, and synthesize digital neuroscience data.

Transferring datasets from major projects is an important component of scientific progress. However, progress in combining genes and neuronal behavior data has been slower than expected. Advanced bioinformatics tools for mathematical analysis and data in a neural toolkit (Abdurakhmonov 2018). The main reason is the name mismatch in the aeronautics preference, with information related to genes and speech expressions related to neurons. Here we show you simple ways to interact with different sources of information using different brain systems. Abdurakhmonov translates words as neuro for several objects using a name known in the linked section. The integrated company uses its individual tools and terms to make a databank for study. The aims of this chapter are to explain the following:

1. The dissimilarities between bioinformatics and neuroscience
2. The dissimilarities between logical diagnosis and data from mathematical analysis
3. The main functions of their implementation, brain map markers selected by different atlases for quantitative neurological testing

Our brain is made up of 86 billion neurons. Each neuron is associated with 1 to 200,000 neurons, and up to 100 billion nerve fibers flow through the brain (Popovych et al. 2019). This is more difficult due to differences in the brain between people. What brain structures and parts of the brain are connected by different personalities, and what areas work together to perform certain functions?

Scientists from 14 European research institutes are working on the subject in different ways. Based on differences in the distribution and size of neurons in the donor's brain, the Forcheng Centre in Germany developed a map of the human brain (Jubrain Atlas). Interactions between neurons such as polarized optical imaging and diffuse tensor imaging can detect electrical nerves and connections that are well known to those skilled in the arts. In collaboration with Jalich and EDA in Paris, groups in the Netherlands, France, Belgium, the United Kingdom, and Germany are using it to identify areas of the brain involved in visual or auditory and auditory functions and to reveal brain structure (Popovych et al. 2019).

## 7.3 VIVID BIG DATA PRACTICES TO PROCESS AND ANALYZE NEURONAL DATA

Recent advances in neuroscience such as sensory-sensory technology have increased sensitivity to sensory information, the degree of sensory design, and differences in neurological information. This player acting a key role in the collection of mathematical data is called "big data neuroscience," which has different time scales and indicators, allowing the neuroscience field in research to create different experiments. With the development of other data-based displays, serious cardiac problems can be identified. It helps with brain damage and long-term brain damage. There is a design and testing tool for a neurological system. But the inability of neuroscience to transfer large amounts of data has made it difficult for researchers to continue working due to the lack of control over the available data. Therefore, in this chapter, we specifically examined three large data centers: mapping, pulverization, and pig detection to determine the potential for experimental neuroscience. The three NeuroPigPen methods described in the analysis showed that high-tech data are the best solution to the problem (Rabinovich et al. 2012).

### 7.3.1 Role of Big Data in Neuroscience

Big data is a reliable way to gather information. The complexity of troubleshooting is limited. The big data features are confidentiality, storage, retrieval, analysis, retrieval, sharing, identification, investigation, broadcasting, transmission, and storage. Many large data formats such as Hadoop can be used to store data. Figure 7.1 shows the architecture of big data analytics (Hasson et al. 2008).

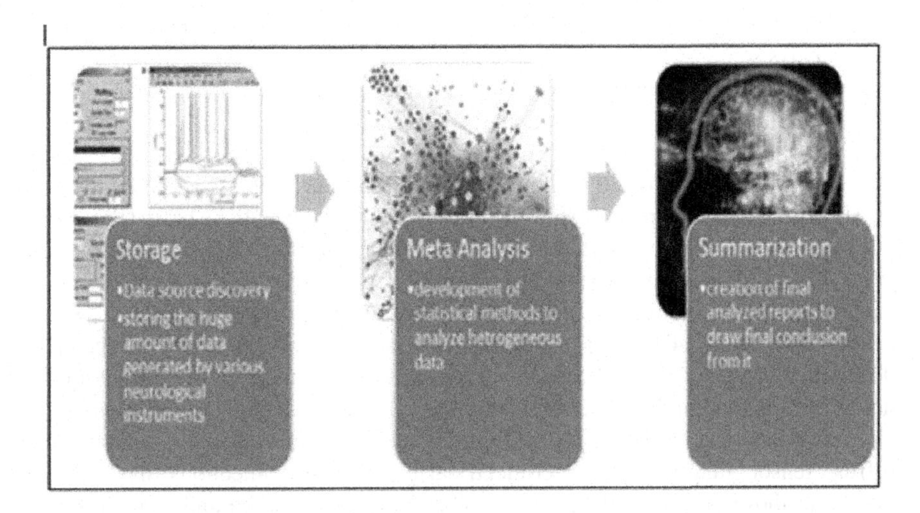


**FIGURE 7.1** Architecture of Big Data Analytics.

*Source:* Hasson et al. (2008).

## 7.3.2 Various Techniques of Big Data

### 7.3.2.1 Map Reduce and Hadoop for Neuron Data

Electroencephalogram data are an example of electronic data recorded by an electronic device in the brain. These electrodes show a vital part in the detection of brain damage and are also useful in diagnosing neurological disorders. Electrophysiological data seconds are also used in the study of brain integration. In neurological disorders such as epilepsy, anthrax testing is difficult because stereoencephalogram data show scaled and localized brain activity.

On-site analysis of physical and electrical data the electroencephalogram (EEG) is one of the most important tests or techniques for detecting pathogens (Wang et al. 2012).

### 7.3.2.2 Swipe the Apache Data

Unplanned analysis of large data file technology takes place in the computer space, but it cannot be used without data and research problems in neural thinking that require the use of a systematic computational tool, especially for data in the functional magnetic resonance imaging (fMRI) field. It shows the main sizes used for image analysis and explains how they use pumps.

## 7.3.3 How to Use Large Amounts of Data in Neural Processing

Experienced Interrupt Mode (EEMD) integration is excellent and imaginatively efficient. In a complete information system, the Hadoop host and MapReduce computer are considered as a modular mode. Related age ranges and variable measurement rates in EEMD show few solutions. The Hadoop clusters developed on the future grid test site laid the foundation for future promotion and equality building. EEMD approvals for design and implementation will be implemented through experimental testing. The future goal is to place larger EGEG files with more time support on the Hadoop Distributed File System (HDFS) and bypass further comparisons with EEMD on larger Hadoop clusters (Landhuis 2017).

### Use Apache Spark and a Graphics Processing Unit (GPU) to Fix fMRI Data Processing

Although big data diagnostics design has been launched in the fMRI space, it has little to do with neural data testing. Big data technology also allows for easy writing space. The ambiguity of the neuroscience data model with large data technology makes it difficult for neuroscience to detect (Boubela et al. 2016).

Scala is the only language we are interested in developing the Spark model in fMRI data analysis. Languages are available in forums such as Python and R, however, and Application Programming Interfaces (APIs) are available in one of these languages using the Spark format. However, it does not give full access to the technical toolkits found in the Scala API (Boubela et al. 2016).

Authors Roland N., Claudius, Huff, and Moser (2013) in the context of large data are said to have approached the examination of large fMRI facts using tools such as

Apache Spark and a GPU. Knowledge analysis was provided directly by the Spark MLib and Graph X libraries in the FMRI community (Roland et al. 2013).

Competitiveness in reading speed and early development capabilities is essential when developing effective search software and the same tools are available at different levels in different programming languages. Spark is an easy way to move numbers to collect or share information. A computer hidden from a developer simplifies the program compared to other devices that require obvious errors that can be easily accessed by integrating open data as a process in the form of a sharing system (Landhuis 2017).

## 7.4 DATA MODELS

Scientific data models should contain diverse stages of neuronal scale that is opening at the cellular level, ascending to the level of the casing and synapses moving to the dendritic branches and branch axons and ending at the local and systemic levels. Each level includes additional details, such as at the local level and data on proteins and ions unseen in a low representation. Each scale has different data models here we select a representative approach for each model. The systems shown in the following for each data model, especially in neuronal-based representation systems, are not intended to accurately define material schemes or features. They are provided here to illustrate typical modelling methods.

### 7.4.1 Organizing Data

The data should be sorted once the sample is received. These can be organized into flat files, table formats, structured files (such as XML), object-based, or layered-based schemes as seen in Figure 7.2 (Tsur 2018). The data in the archive file are stored in an unstructured format, so managing it requires reading it from memory.

The data can be organized into tables in which each value is guided by type and frequency and measured according to a scale. The Extensible Markup (XML) language is an identical data structure in which the data are organized as a scheme with each subsequent level increasing the previous size. XML accelerates the industry because it is simpler and more flexible enabling declarative clarity rather than coding. It allows to automatically convert model features to different formats. One of the main methods of data processing is the representation of material-based processed

**FIGURE 7.2** Schematics of Data-Structuring Paradigms.

*Source:* Tsur (2018).

data in which businesses are defined by a set of structures and linked by symbols. Object-based presentation enables the incorporation of internal data related to the diversity of a primary data source. Another approach is a layered approach (LOA) in which the associated (or layer) related language determines its model. The reason behind the LOA is that the computer model is not a "collection of flat equations" but a structure from which the basic concepts of biology originate.

### 7.4.2 Data Life Cycle

Researchers collected participant's participatory data in the form of clinical trials, intervention studies and brain thinking, motor and cognitive measures, and biological samples for genetic analysis. Personal health information (PHI) should be dealt with in accordance with the Personal Health Information Protection Act (PHIPA), 2004. The PHIPA, from a governance and contractual perspective, is prescribed by the regulations in ISO 27001 information management. To maximize the distribution of data and the ability to analyze Brain-CODE while enabling secure PHI integration, the process was upgraded to enable effective fragmentation of sensitive data while incorporating granular access controls to ensure data were only available to authorized Brain-CODE users.

Neuroscientific information comes from a variety of sources from government-sponsored consortiums in laboratories to individual laboratories around the world. We can see the schematics of data-structuring paradigms in Figure 7.3 (Hasson et al. 2008).

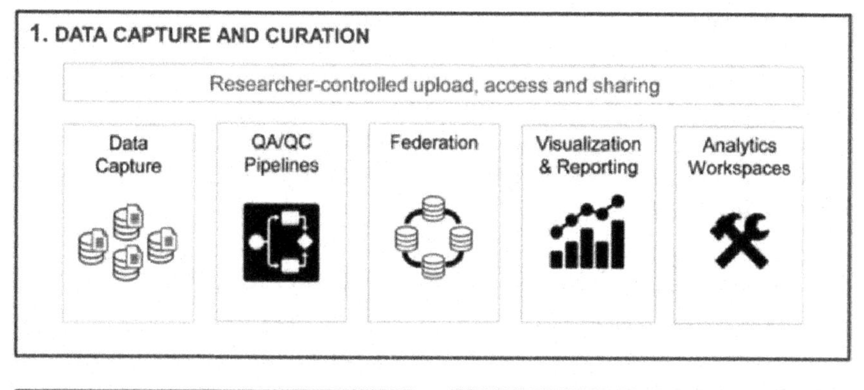


**FIGURE 7.3** Schematics of Data-Structuring Paradigms.

*Source:* Hasson et al. (2008).

## 7.5 A GIANT INITIATIVE IN SCIENCE

Today one of the most intensive experiments aimed at integrating human neurology in Human Brain Project (HBP; Bowden et al. 2018) is the EU-funded HBP project. Neuroinformatics is headquartered in HBP and is managed by Colbab, a cloud web affiliate program developed in neuroinformatics platform discussed earlier. Distributed as an SaaS, HBP provides powerful high performance analytics and computing platform (HPAC) that enables partners to store and distribute virtual machines and provide advanced data analysis. It must be linked to complex data types and ontologies in order to protect metadata storage and provide information recovery systems that the human brain can create and live with at various levels of information methods an Anatomy and Physiology. Again COLLAB data must be linked to external maps, data, and micrographs. Lead HBP is the Human Genome Program (HGP), a program that radically changes research methods in molecular training. Biology, from the point of view we see, has new features as a successor to HGP from genome-based drugs to genomic modelling. This has paved the way for new ways of building and maintaining data scientists such as the creation of public libraries where this approach can be equally disseminated. The goal of the HBP approach is to perform the same neuroscience.

With the support of HGP and HBP, the White House is launching a new scientific experiment called "Brain" in the United States: "Refreshing Our Understanding of the Human Brain" (Ascoli et al. 2009) and "Other Ways to Empower Every Laboratory and . . . Open Access to Information". The goal of National Institute of Health (human connectome project) funding is to reflect the workings and functioning of the human brain. The program collects large amounts of data from hundreds of patients with advanced 3D fMRI, EEG, and magnetoencephalography equipment. Complete genome sequencing and complete all lessons. Other government-sponsored neuroscience projects include the "Brain Canada" (Bowden et al. 2018) and the "China Brain Project" (Ascoli et al. 2009). However, keep in mind that the methods of generating generalizations and performance data, sharing information, and giving credit are crucial to the success of the project.

## 7.6 MODELS FOR COMPUTATIONAL NEUROSCIENCE

Linking neuroscience data to a simulated environment has deep roots in modelling and database neurons. In the working model of Alan et al. from the1970s (Tsur 2018), creating a widely used model of neuronal dynamics with the simulation scale of neural networks has increased. As computer resources become plentiful, neuronal mimicking begins to take place by increasing the number of laboratories creating a database of where available and built-in models are created.

The neuronal dynamics model encompasses all scale extraction, whereby each extraction level involves an increase in the number of data. The different computational models can be seen in Figure 7.4 (Tsur 2018).

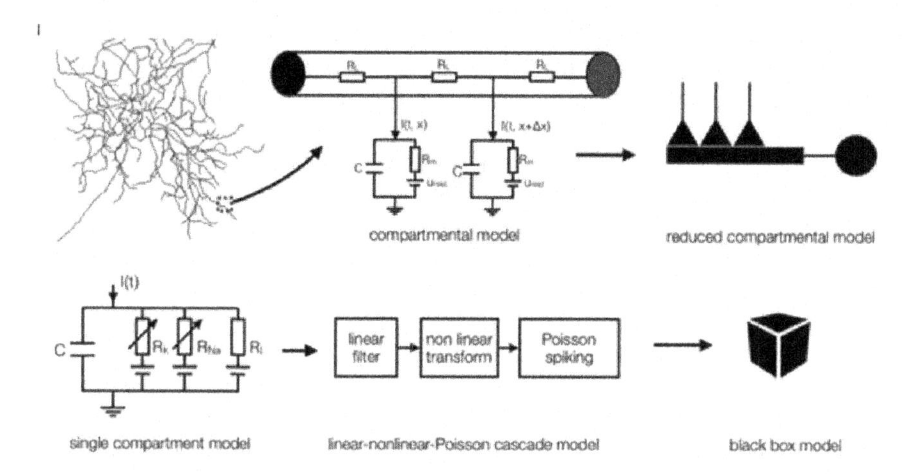


**FIGURE 7.4** Schematic of Computational Model.

*Source:* Tsur (2018).

### 7.6.1 Data and Computational Models

The term *neuroinformatics* describes scientific studies of the movement and cognitive development of the nervous system. Neuroinformatics is dedicated to building neurological information and basic knowledge, as well as modern computers and diagnostic sharing tools, integrating and analyzing experimental data, and working theory on neurotransmitters from NNCF data. The International Neuroinformatics Coordinating Facility is concerned with the scientific knowledge of information neuroscience critical data analysis, metadata oncology, diagnostic tools, and number system statistics. Important data include experiments and experimental conditions on genomes, molecules, cell systems, networks, systems, all types of systems, and general and differentiated preparations (Rabinovich et al. 2012).

The peripheral nervous system has two parts: the somatic system and the autonomic system. Let us first look at the nervous system of the body. The body's system consists of nerves connected by spontaneous muscles and all the nerves that receive command from brain. What do we mean by this state of mind? What does it feel like to hold an axe? Remember that the ion is the first neuron. Now a few hands or sticks correspond to what we call sensors. Let us look at an example of when we see a somatic system working. Say you want to shake hands with a friend of yours. You take your friend's hand and slide your hand over theirs. Therefore, this process uses the somatic nervous system to send commands from the brain and spinal cord. Often when you shake hands with a friend, you feel like you are holding hands with your thoughts. It's on your skin, on your hands. So we use two types of electrical nerves. The first is called the fiber and fits each axon that transmits sensory information from its tip. In this case, the hands lead to the central nervous system, that is, to the brain and spinal cord. To hold a friend's hand, commands from the brain are sent to the muscles of the hand. The healthy artery is then used. These are strong muscle fibers or ions that carry information from the orphaned central organs and from the brain back to the shoulders and legs. The second part of the cardiovascular system is the autonomic system. The autonomic system consists of nerves that connect to your internal organs such as the heart, blood vessels, smooth muscles, and glands. Under the cognitive component, automation works best and performs many important functions such as heart rate, digestion, and respiration. Moreover, it contributes to other important activities, such as combat or air reaction. Another important part of the nervous system is the central nervous system. The central nervous system is composed of the spinal cord and the brain.

In recent years, robust models have played an important role in improving our understanding of brain function and mental instability. Well-designed and well-expressed mathematics can be used to visualize the structure of the brain and to identify the strength of the nerves based on information from the nervous system. A comprehensive analysis of modular parameters and experimental concepts using reliable methods and mathematical physics can help identify and predict brain status. On the other hand, the in-depth search and pattern of the parameter set make it impossible to examine and analyze the data. Instead of the mode-based approach can differentiate between energy availability and invisible visual intelligence, consistent stability, deceleration time, noise trust, and many more. These statistics allow us to compare and contrast. Due to its impossibility and high kinetic power, brain structure and health and the risk of disease can be used as further indicators of a living brain's condition and behavior.

Modelling the human brain is a major problem because the data are dynamic and diverse, reflecting spatial differences and depicting human differences. Unlike advances in understanding the nervous system of nonhuman genes, most neuro-numerical images can be limited by information from physicians and without reproductive capacity. A major turning point in the history of neuroscience is based on the discovery of a specific biophysical regime. However, we are more interested in the organization of the system, so we do not create easily visible patterns.

## ACKNOWLEDGMENT

We would like to express our gratitude to the Gandhinagar Institute of Technology for support and Dr. H. N. Shah (Director, GIT) for constant guidance and encouragement.

## REFERENCES

Abdurakhmonov, I. Y. (Ed.). (2018). *Bioinformatics in the Era of Post Genomics and Big Data*. BoD – Books on Demand.

Ascoli, G. A., & Halavi, M. (2009). Neuroinformatics. *Encyclopedia of Neuroscience*, 477–484. https://doi.org/10.1016/b978-008045046-9.00872-x.

Boubela, R., Kalcher, K., Huf, W., Claudia, K., Filzmoser, P., & Moser, E. (2013). Beyond noise: Using temporal ICA to extract meaningful information from high-frequency fMRI signal fluctuations during rest. *Frontiers in Human Neuroscience,* 168.

Boubela, R. N., Kalcher, K., Huf, W., Našel, C., & Moser, E. (2016). Big data approaches for the analysis of large-scale fMRI data using apache spark and GPU processing: A demonstration on resting-state fMRI data from the human connectome project. *Frontiers in Neuroscience*, 9, 492.

Bowden, D. M., Dubach, M. F., & Dong, E. (2018). Informatics for interoperability of molecular-genetic and neurobehavioral databases. In *Molecular-Genetic and Statistical Techniques for Behavioral and Neural Research* (pp. 31–50). New York: Academic Press.

Hasson, U., Skipper, J. I., Wilde, M. J., Nusbaum, H. C., & Small, S. L. (2008). Improving the analysis, storage and sharing of neuroimaging data using relational databases and distributed computing. *NeuroImage*, 39(2), 693–706.

Landhuis, E. (2017). Neuroscience: Big brain, big data. *Nature*, 541(7638), 559–561.

Popovych, O. V., Manos, T., Hoffstaedter, F., & Eickhoff, S. B. (2019). What can computational models contribute to neuroimaging data analytics? *Frontiers in Systems Neuroscience*, 12, 68.

Rabinovich, M. I., Afraimovich, V. S., Bick, C., & Varona, P. (2012). Information flow dynamics in the brain. *Physics of Life Reviews*, 9(1), 51–73.

Tsur, E. E. (2018). Data models in neuroinformatics. In *Bioinformatics in the Era of Post Genomics and Big Data* (p. 133). London: IntechOpen.

Vaghela, R., & Solanki, K. (2021). Fuzzy logic based light control systems for heterogeneous traffic and prospectus in Ahmedabad (India). In *Data Science and Intelligent Applications* (pp. 239–246). Singapore: Springer.

Wang, L., Chen, D., Ranjan, R., Khan, S. U., Kołodziej, J., & Wang, J. (2012, December). Parallel processing of massive EEG data with MapReduce. In *2012 IEEE 18th International Conference on Parallel and Distributed Systems* (pp. 164–171).

# 8 A Fast and Reliable Detection System for Lung Cancer Using Image-Processing Techniques

*Anindya Banerjee and Himadri Sekhar Dutta*

## CONTENTS



## 8.1 INTRODUCTION

Cancer is the abnormal and unpredictable growth of cells into a tumor. Lung cancer is considered to have the highest mortality rate than other types of cancers (Chaudhary et al., 2012). The mortality rate is increasing each year in the United States. The primary causes of lung cancer include cigarette smoking and genetics for both men and women. This can be cured if it is detected early. According to the statistics from 2019, the American Cancer Society approximated 228,150 maiden lung cancer cases, and out of those, 62.53% resulted in death (American Cancer Society. Facts & Figures, 2019). With an age-standardized rate per 100,000, Hungary had been recorded the highest incidence of lung cancer at 56.7, and Slovakia has been recorded as the lowest, that is, 31.2 in 2018, as per data by the Global Cancer Observatory and retrieved from the International Agency for Research on Cancer (IARC; lung cancer statistics). Based on the tumor cells' size and rate of growth, there are four

DOI: 10.1201/9781003142751-10

phases: Level 1, Level 2, Level 3, and Level 4. These stages are solely dependent on tumor size, the quality of spreading, and many such conditions. One should also keep in mind that the nodule detected in computed tomography (CT) scan images does not have to be cancerous cells; it might be because of different diseases such as pneumonia, tuberculosis, and others. For the diagnosis of the suggested procedure, a CT image is preferable and is used as it is prone to less noise than X-ray images. We have used MATLAB R2017a version software throughout the experiment. This chapter aims to present a clear and concise idea of various types of lung cancer and some causes of lung cancer. A proposed method is discussed briefly. The output of each stage, including image acquisition, image enhancement, image segmentation, histogram equalization and so on, is discussed and explained. Finally, the chapter concludes with a discussion of the results of the proposed framework and the direction of future work.

## 8.2 RELATED WORKS

Mokhled S. Al-Tarawneh (2012) proposed an image improvement technique to detect lung cancer efficiently and quickly. Mask labeling with greater accuracy and pixel percentage were the primary detected features of this study. Auto enhancement, fast Fourier transfer (FFT) filters, and Gabor filters are used at the image enhancement stage in this study, with average final values of 38.025, 27.51, and 80.735, respectively. For feature extraction, a binarization approach and masking approach are used extensively in this study. Shakeel et al. (2019) developed a lung cancer detection system for CT images utilizing improved clustering technique and implementing deep learning that had a 98.42% accuracy observed with a classification error of 0.038. Vas et al. (2017) proposed a novel method using mathematical morphological operations dedicated to segment affected lung images. Alam et al. (2018) proposed an approach using a multiclass support vector machine (SVM) and achieved a superior percentage of accuracy for predicting as well as detecting lung cancer. Kumar et al. (2011) designed a dual-stage computer-aided design (CAD) system that is able to detect CT scan histological images of a cancerous lung or a normal lung automatically.

## 8.3 LUNG CANCER AND ITS TYPES

Lung cancer is a malignant tumor categorized by its uncontrollable rate of growth of lung tissues. It can also be entitled lung carcinoma. It can also be called lung carcinoma. Tumors are basically of two types, namely, benign and malignant, based on the growth rate of tissues. Every type of cancer is referred to as a malignant tumor. The growth rates of malignant tumors are unpredictable; they tend to expand to other parts of the body as well, while the growth rate of benign tumors never gets spread to different portions of the human body and the growth rate is slow and noncancerous. The initial symptoms are breathing problems, constant and long-lasting tiredness, and uninterrupted cough, among others. Various types of lung disease are chronic obstructive pulmonary disease, lung cancer, and acute respiratory distress syndrome

**TABLE 8.1**
**Comparison of Different Lung Cancer Detection and Classification Techniques**

| Methods | References | Images | Results |
|---|---|---|---|
| Image enhancement, image segmentation, feature extraction stage | Mokhled S. Al-Tarawneh | CT | Image enhancement: 80.735 (Gabor filter) Image segmentation: 85.165 (Watershed filter) |
| Improved profuse clustering technique and deep learning with instantaneously trained neural networks | Shakeel et al. | CT | Accuracy: 98.42% Classification error: 0.038 |
| Mathematical morphological operations | Vas et al. | CT | Accuracy: 92% |
| Watershed transform: gray-level co-occurrence matrix and SVM classifier | Alam et al. | CT | Detection: 97%, Prediction: 87% |
| Bio-orthogonal transform and region growing | Kumar et al. | CT | Detection: 86% |

(Chander et al., 2017). According to previous studies, higher odds ratios of lung cancer have been detected in men compared to women, but current studies show a minimal difference between men and women (Khuder, 2001).

Following the growth rate and size of the tumor, lung cancer could be split into two parts:

A. **Small cell lung cancer (SCLC):** SCLC contains nearly 13% of all recently diagnosed lung cancer (Govindan et al., 2006). While chemotherapy has proved to have good results for non–small cell lung cancer (NSCLC), but for SCLC, it is only beneficial for the short term (Slotman et al., 2007). SCLC can be detected in the early stage or the critical stage. In the early stage, it can be cured by using chemotherapy and using radiation at the actual location, but it is very much challenging for the survival of a patient (Kalemkerian et al., 2017).
B. **NSCLC:** Diagnosis and detection of NSCLC are usually evaluated after two sets of chemotherapy. NSCLC cells also have a relatively slower growth rate after being undergoing chemotherapy (Mac Manus et al., 2003, Geus-Oei et al., 2007). However, several former studies and experiments have reported that early detection of NSCLC using perfusion CT images, PET images and magnetic resonance imaging (MRI) exhibits significant results for detecting NSCLC in the early stages after chemotherapy (Yabuuchi et al., 2011).

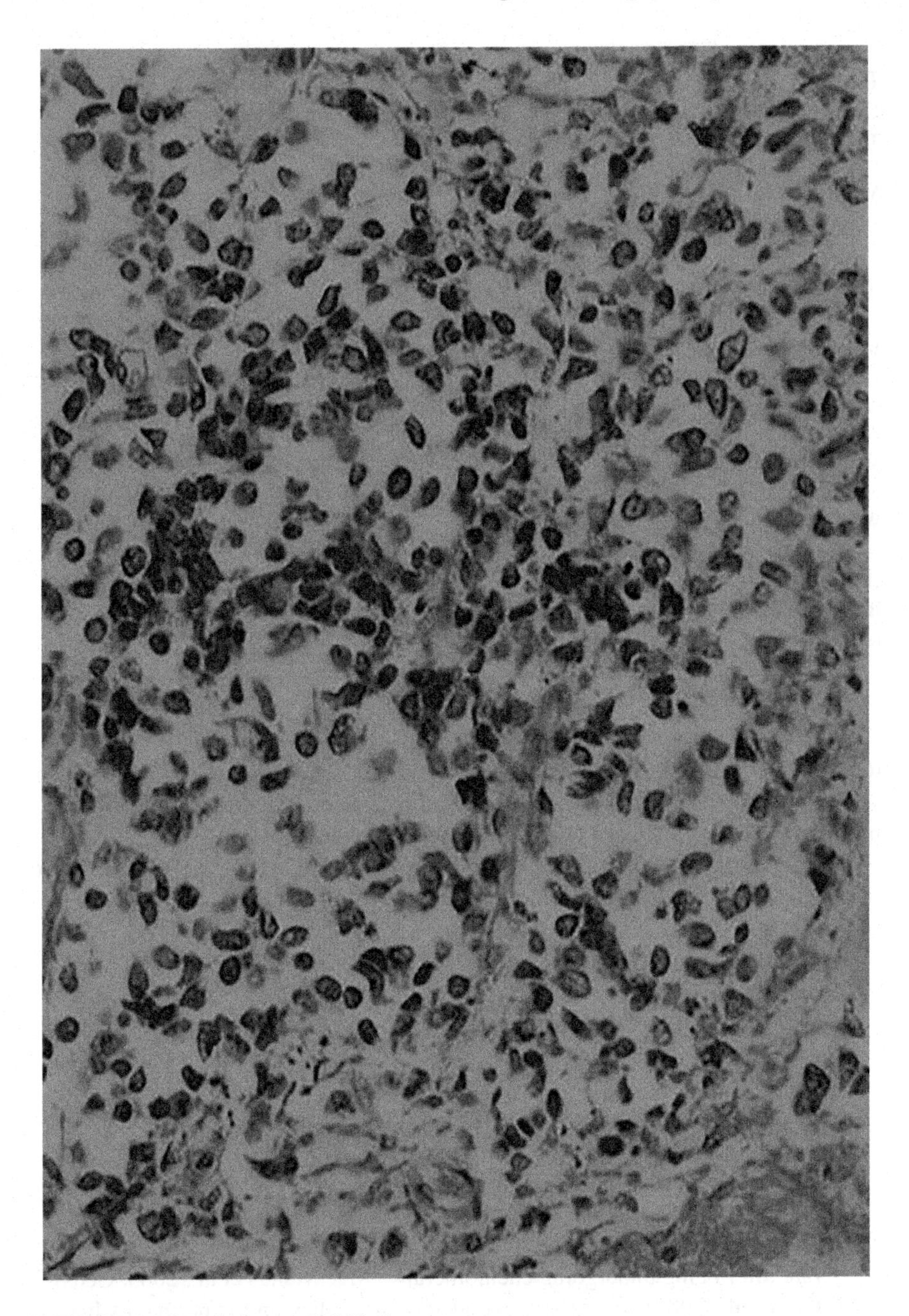

**FIGURE 8.1** Sample Image of SCLC.

*Source:* Adapted and modified from www.scitechnol.com.

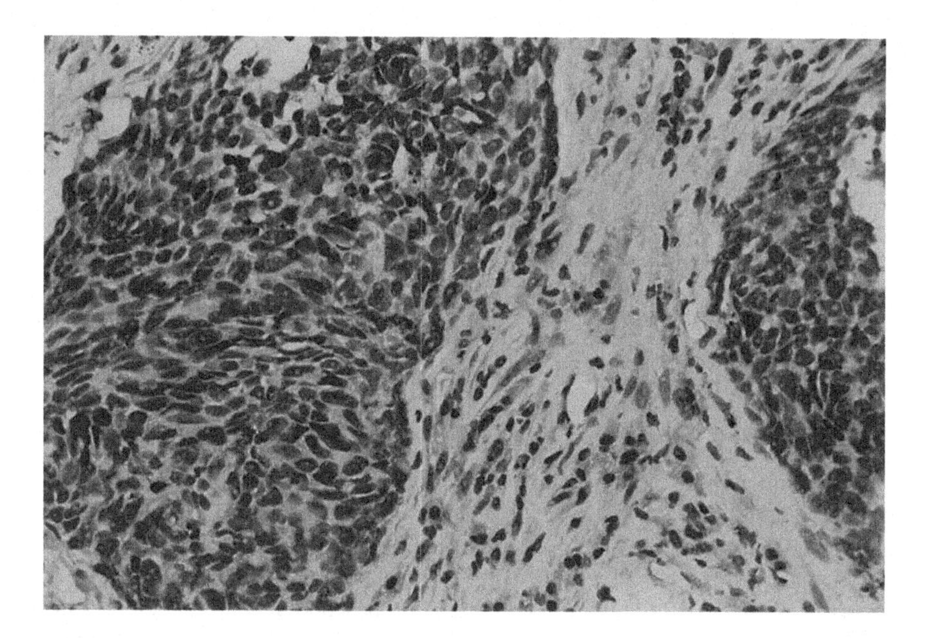

**FIGURE 8.2** Sample Image of NSCLC.

*Source:* Adapted and modified from www.emedicinehealth.com.

## 8.4 CAUSES OF LUNG CANCER

The causes of lung cancer can be a variety of reasons. Risk factors such as cigarette smoking, air pollution, odorless radon gas, and others increase the possibility of causing lung cancer. According to American Lung Association, smoking has been the primary reason for causing lung cancer. Nearly 10% to 15% of people affected by lung cancer had never indulged in smoking in their lifetime (Thun et al., 2008). The causes can be passive smoking, secondhand smoking, and exposure to harmful air pollutants, among others.

1. Tobacco smoking—Tobacco contains harmful chemical ingredients like nicotine, benzene, lead, and so on. In 2000, deaths caused by lung cancer due to tobacco smoking were about 90% to men and 70% to women (Peto et al., 2006).
2. Genetics—Genetics plays a huge role in inheriting and developing the possibility of lung cancer. Of overall lung cancer instances, 8% caused due to several familial variables (Yang et al., 2013).
3. Radon Gas—Radon gas is formed by the disruption of radioactive agents such as radium. It is often found in rock and soil in different parts of the world. In the United States, it's considered the second primary cause of lung cancer after tobacco smoking (Choi et al., 2014).

## 8.5 PROPOSED METHODOLOGY

The proposed set of algorithms focuses on four primary techniques, which are image enhancement, adaptive histogram equalization (AHE), image segmentation, histogram equalization. All the methods have been described in detail with experimental output results of each step. The histogram equalization of input images is compared with the histogram equalization curve of ideal cancerous cell images in the end stage. If both histogram curves are analogous, then a conclusion could be drawn as the image to be cancerous and vice versa. Figure 8.3 represents the flowchart of proposed algorithms for the quick detection of lung cancer. Selected experimental results are illustrated in Figures 8.14 and 8.15.

### 8.5.1 IMAGE ACQUISITION

Many CT images have been collected from open access online available database (lung cancer detection cases with CT images, Github). For the sake of efficacy and accuracy, CT scan images were chosen over X-ray images. CT scan images have significantly less noise than X-ray images. Our database consists of both normal and abnormal samples. All the images used in the proposed system are in JPEG format.

### 8.5.2 IMAGE ENHANCEMENT

It is the fundamental building block of image preprocessing. The objective is to facilitate the standard or quality of the picture. Image enhancement techniques also help to increase the brightness of the image on dark background. Unfortunately, there is no reliable theory for comparing the quality of a painting by human perception (Gujral, 2015), but by comparing the images to MATLAB software, we could quickly determine the inequalities. Image enhancement algorithms gained tremendous attention in biomedical engineering in recent years (Khandelwal, 2013). Here in this stage, we have also used a wiener filter to remove the noise that might be already present in the input image and the blurriness of the image. There are two types of image enhancement techniques:

1. **Frequency domain image enhancement techniques:** This technique's fundamental idea is to manipulate several types of transforms in an image. Transforms include discrete wavelength transform, discrete cosine transform, and Fourier transform, among others. Frequency domain image enhancement techniques are less complex and can be easily implemented in real-world problems (Gujral, 2015).
2. **Spatial domain image enhancement techniques:** These techniques are easy to understand and operate directly on image pixels.

Figures 8.6 and 8.7 are the experimental output results of a normal lung image and a cancerous lung image after applying the wiener filter.

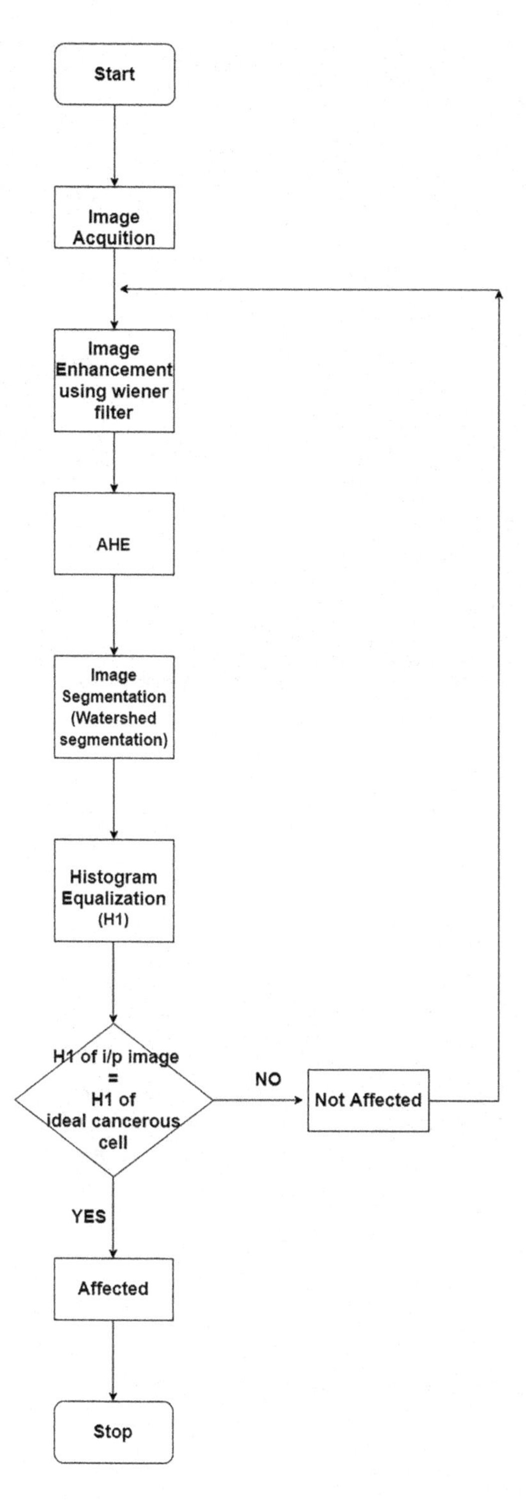


**FIGURE 8.3** Flowchart of Proposed CAD-Based Intelligent System.

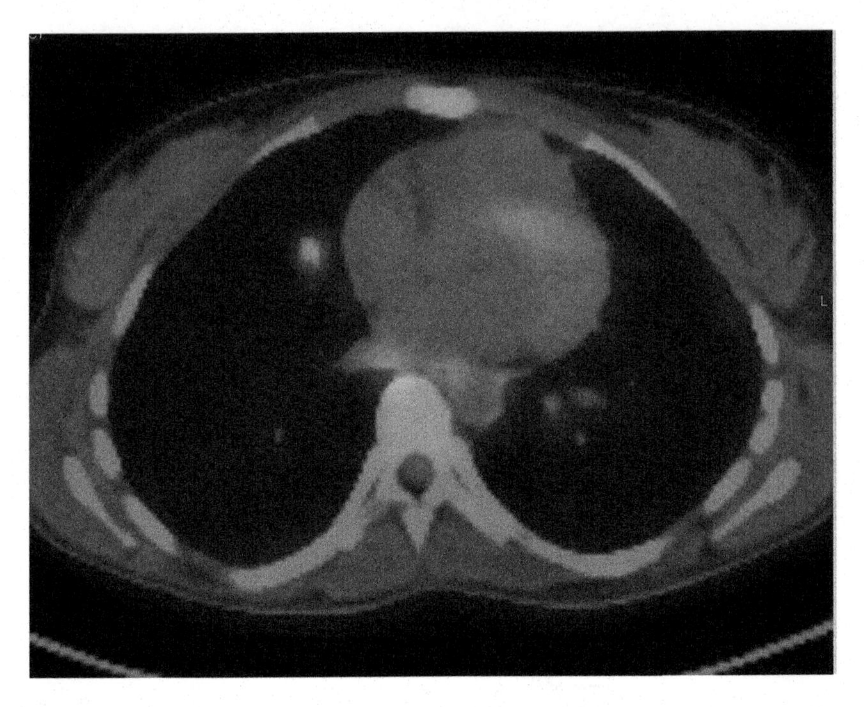

**FIGURE 8.4** Normal Cancerous Lung Cell Image.

*Source:* Adapted and modified from www.ctisus.com.

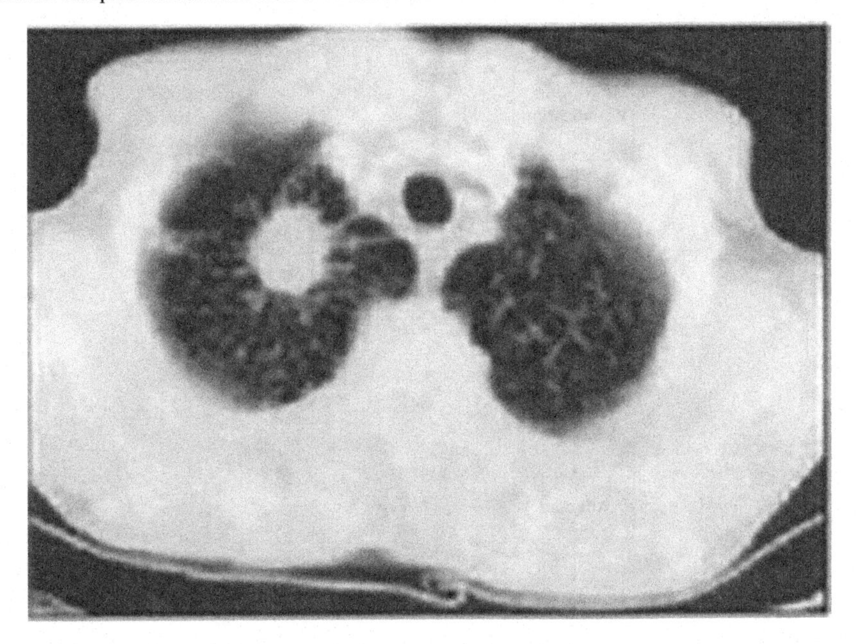

**FIGURE 8.5** Affected Cancerous Lung Cell Image.

*Source:* Adapted from www.doctorlib.info/.

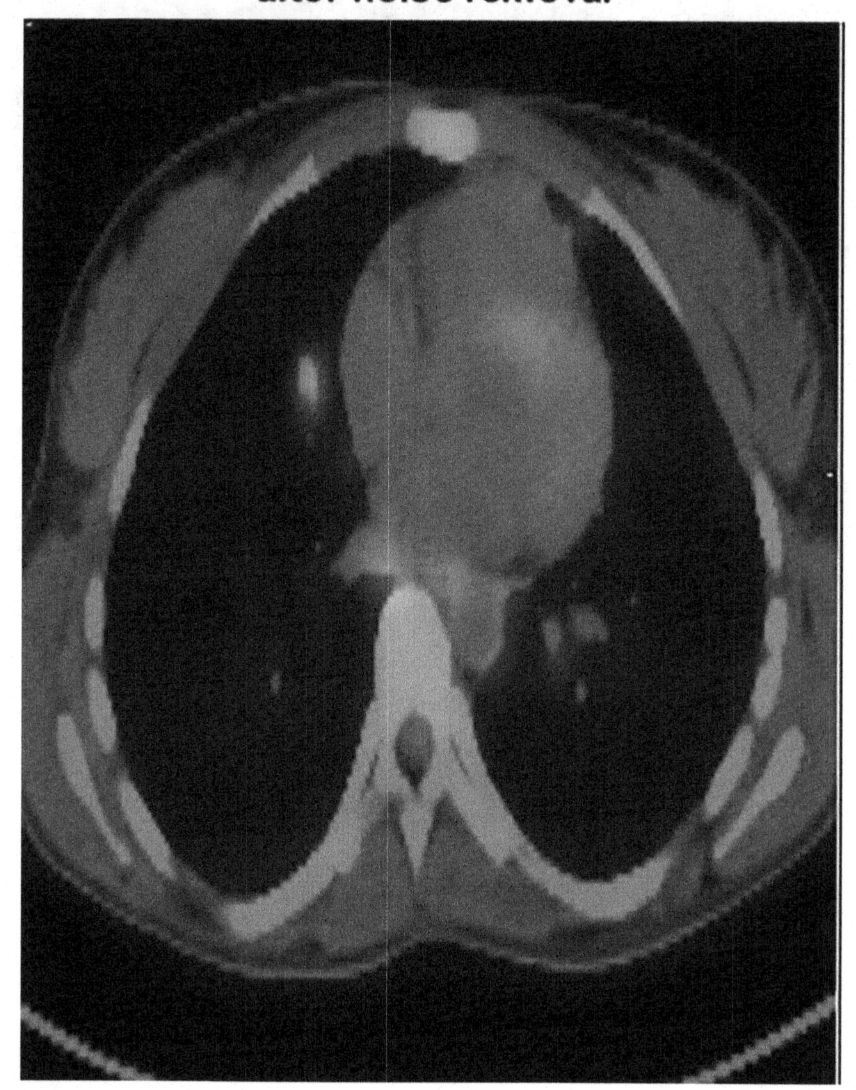


**FIGURE 8.6** After Applying a Wiener Filter to a Normal Lung Image.

Generally, noise reduction filters used in image processing could be categorized into two groups:

a. **Time-domain filtering**
b. **Frequency domain filtering**

Time-domain filters and frequency domain filters can be also be subdivided into the categories shown in Figure 8.8.

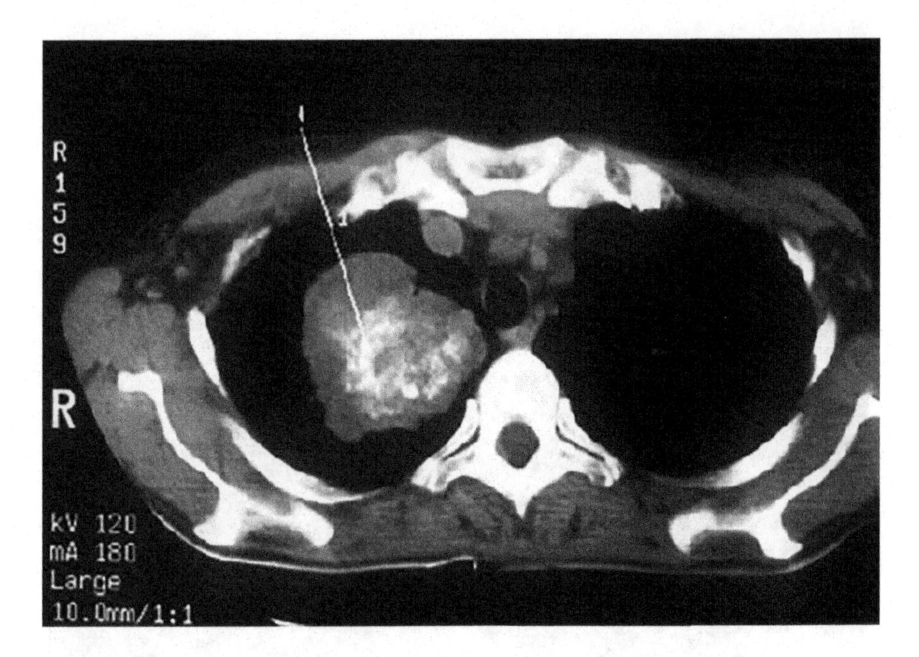


FIGURE 8.7 After Applying a Wiener Filter to the Cancerous Lung Image.

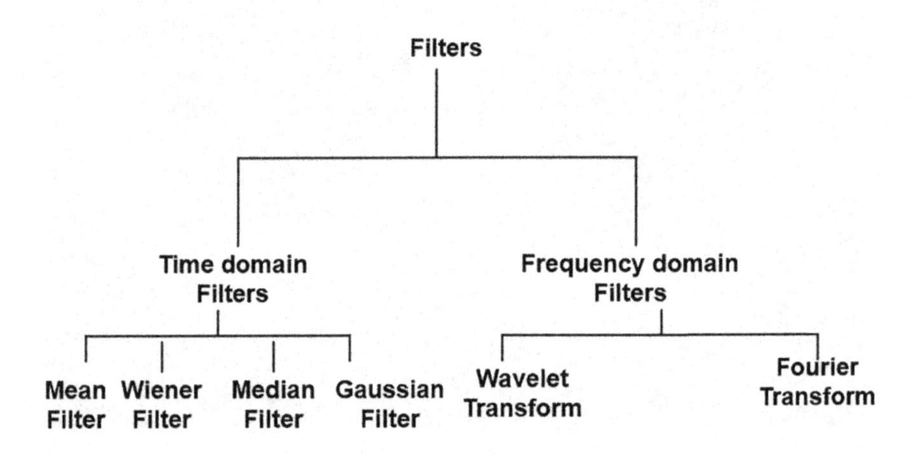


FIGURE 8.8 Different Categories of Filters.

### 8.5.3 AHE

AHE sketches the intensity values of input for a uniform distribution of the pixel values. For example, let variable g is the gray level of an image. T and s are considered as transformation function and transformed value (Hossain et al., 2007) respectively. Thus,

$$s = T(g) = \int_0^g p_g(w)\,dw. \tag{8.1}$$

$p_g$ is the probability density function (pdf) of g; $p_s$ can be easily achieved by the following formula:

$$p_s = p_g(g)\left|\frac{dg}{ds}\right|, \tag{8.2}$$

where $p_s$ is the probability density function of $s$.

Every pixel value is marked by its corresponding intensity level by its neighboring pixel value. The resulting pixels are assigned a new value (Wang et al., 2006). AHE is widely utilized to enhance the contrast of images simply by altering the image pixel values. In Figures 8.9 and 8.10, AHE computes histograms in each part of both normal and lung cancer cell images. This method is quite favorable for enhancing the edges of images.

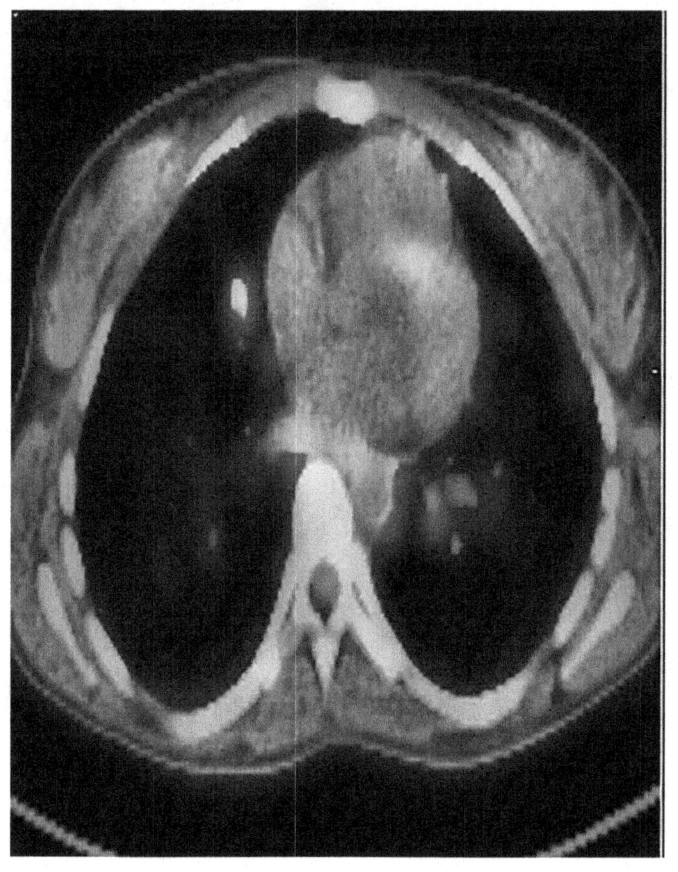


**FIGURE 8.9** AHE of a Normal Lung Image.

**AHE**

**after noise removal**

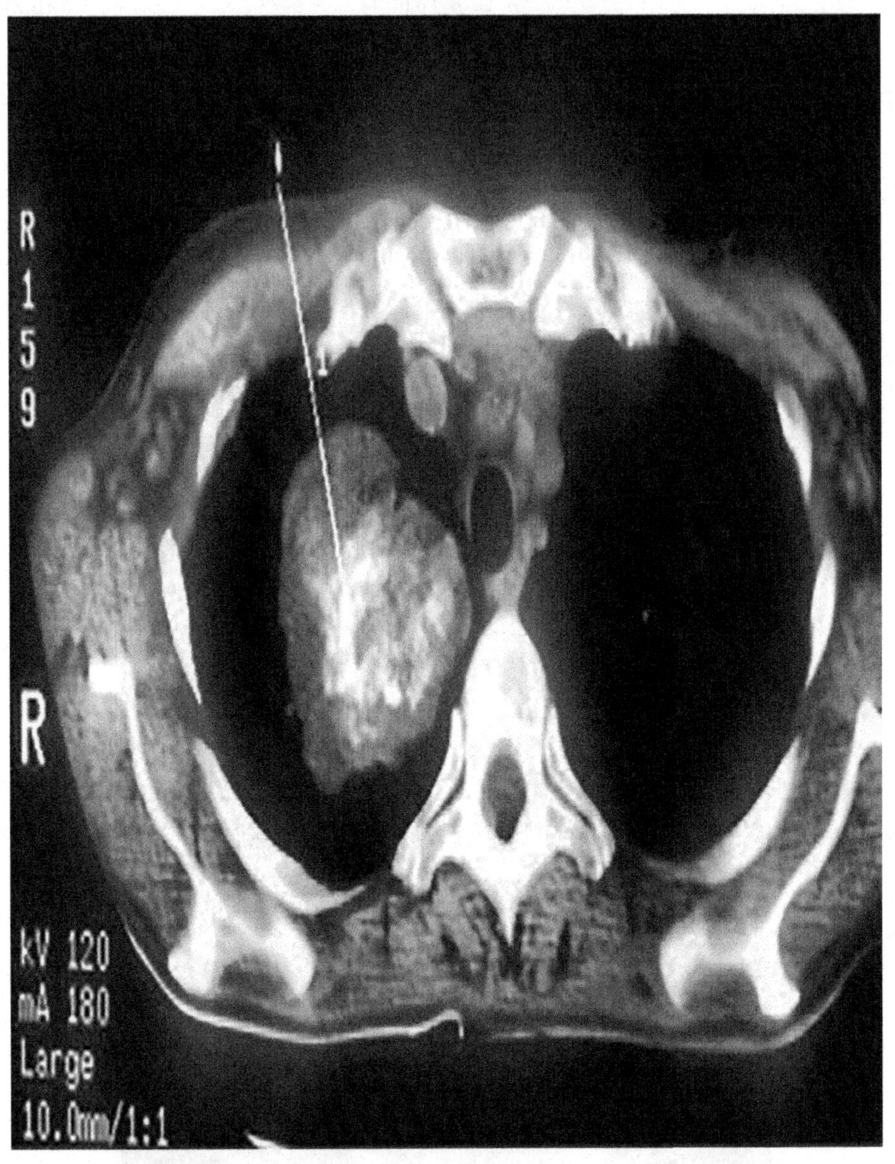


**FIGURE 8.10** AHE of an Affected Lung Cell Image.

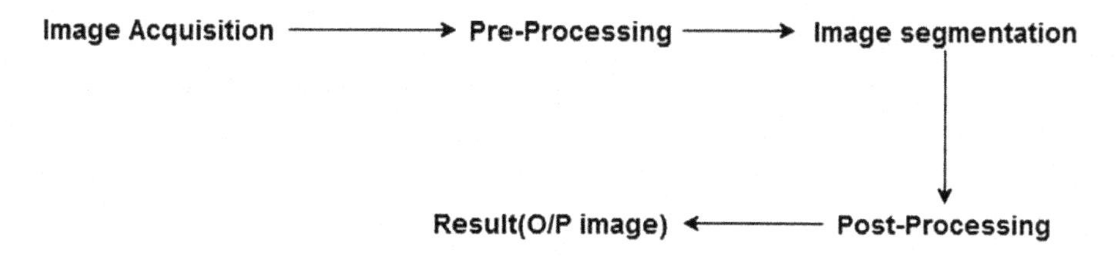


**FIGURE 8.11** Image Segmentation Methods.

### 8.5.4 Image Segmentation

An image is partitioned into a number of segments during image segmentation (Chouhan et al., 2018). Image segmentation applications, such as an autonomous vehicle, boundary measurement, and diabetic retinopathy (Minaee et al., 2020), play a crucial role as a preprocessing stage in nearly every pattern recognition and computer vision applications (Mesejo et al., 2016, Jiao et al., 2010, Jothi et al., 2017). There is no rigid rule or level for partitioning an image into its subparts. It purely depends on the problem (Yadav et al., 2019). Image segmentation methods are analyzed in Figure 8.11.

Image segmentation techniques could be split into two different categories:

1. Region-based image segmentation
2. Edge-detection segmentation

1. **Region-based image segmentation:** In region-based image segmentation (Senthilkumaran et al., 2009, Hore et al., 2016), the image is segmented into its identical regions (Krishna et al., 2019). Segmentation is performed with the help of gray-level values of image pixels (Yadav et al., 2019). These gray-level values of the foreground and background of the image are significantly different if the contrasts are identical. Watershed segmentation has been used in our experiment.

**Watershed segmentation**: watershed algorithms are applicable on grayscale images. In this algorithm, a typical line segments the region of interest of the target image and is also suitable for applying to marker location, which is considered local minima (Logesh Kumar et al., 2016, Al Tarawneh, 2012). However, it is observed that in some natural images, it causes undue segmentation (Amoda et al., 2013). The experimental results are represented in Figures 8.12 and 8.13.

2. **Edge-detection segmentation:** Edge-detection techniques are beneficial to find discontinuities in a grayscale image. An edge is considered the border between two regions with distinct grayscale values (Al-Amri et al., 2010). This algorithm is constructive for extracting relevant information from an image and filtering out the irrelevant parts of the image. Edge detectors such as Canny edge detectors and Sobel detectors are widely used in image processing, computer vision, and pattern recognition applications.

### 8.5.5 Histogram Equalization

Let a 2D image Z (j,k) comprising P distinct gray levels, characterized by $\{Z_0, Z_1, \ldots, Z_i, \ldots, Z_{p1}\}$. The probability density function or PDF of $Z_i$ can be expressed as follows:

$$\mathrm{PDF}(Z_i) = {}^{x_i}\!/_{x}\,. \tag{8.3}$$

While i = 0; 1; . . .; P − 1, $x_i$ = number of pixels. having value of i, x can be defined as the total sample number. It is to be noted that $\mathrm{PDF}(Z_i)$ is associated with a histogram of the input image.

That signifies pixel numbers having a certain intensity $Z_i$. A plot of $x_i$ vs. $Z_i$ has been referred to as image histogram Z (j,k) (Sim et al., 2007). Figures 8.14 and 8.15 shows the resultant experimental curves of both normal and cancerous lung cell images.

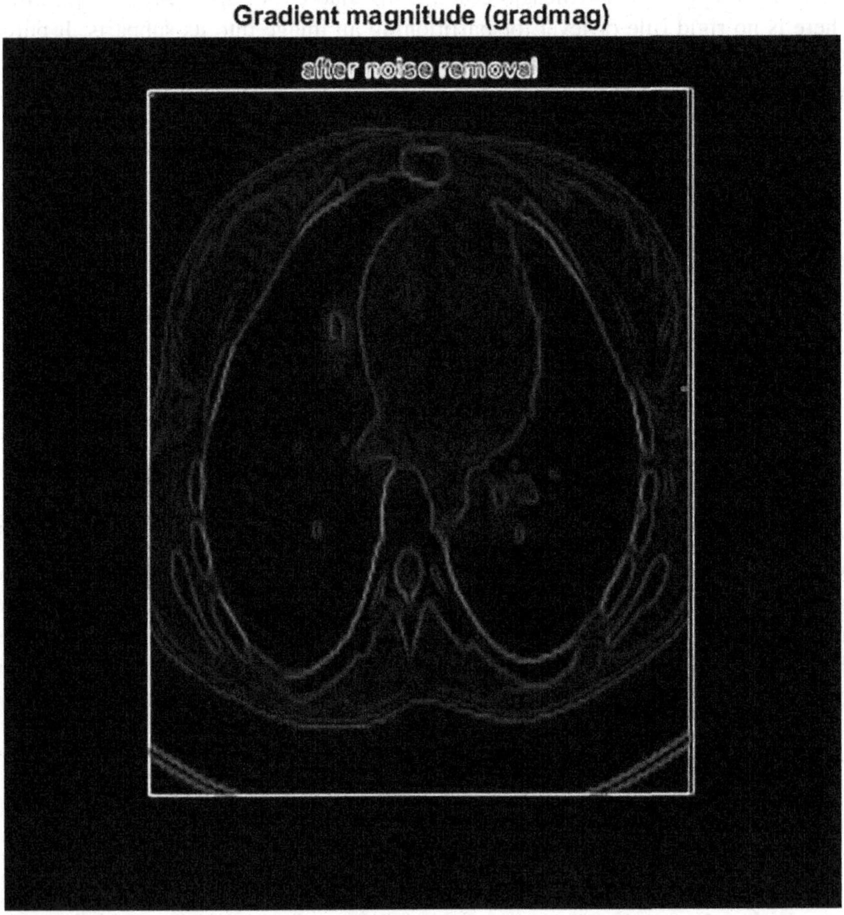

**FIGURE 8.12** Watershed Segmentation Result of a Normal Lung Image.

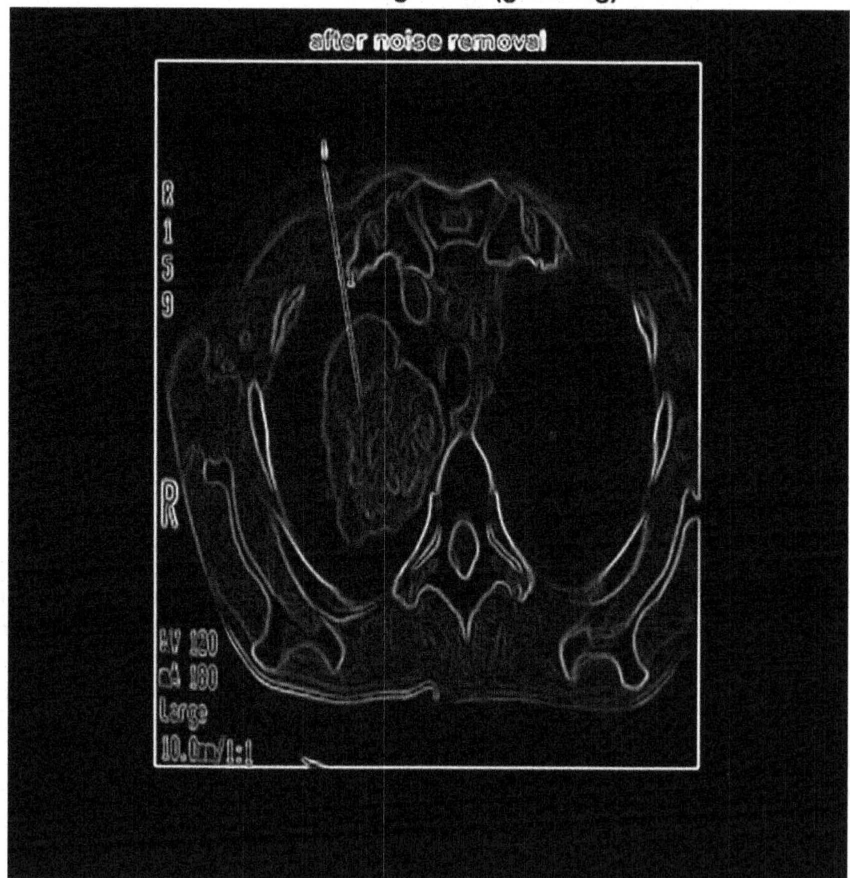


FIGURE 8.13 Watershed Segmentation Result of Cancerous Lung Image.

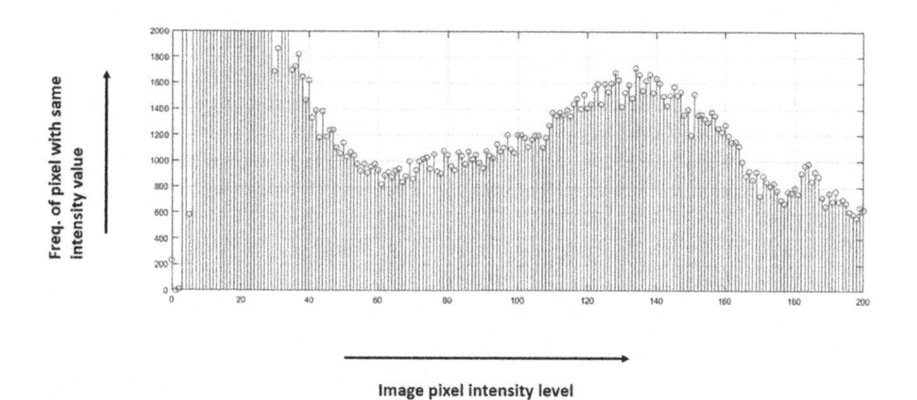


FIGURE 8.14 Output Curve of a Normal Lung Cell after AHE.

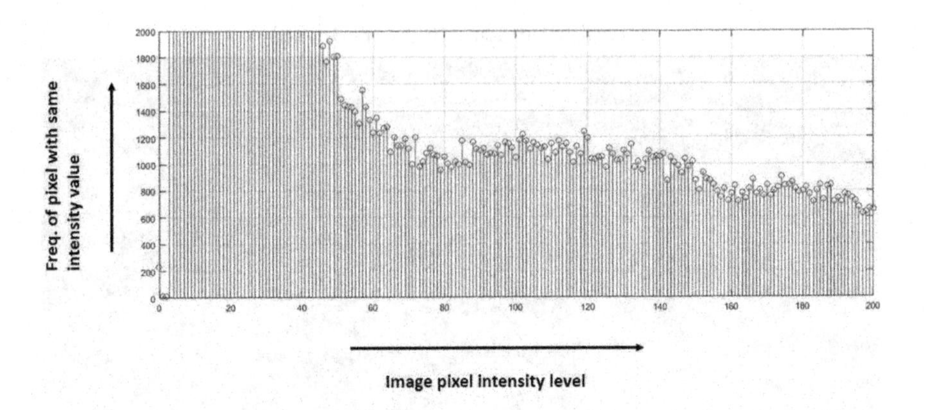


**FIGURE 8.15** Output Curve of an Affected Lung Cell after AHE.

## 8.6 CONCLUSION

The proposed set of algorithms tries to conclude an accurate and fast method for detecting cancer-affected cells. In each step, starting with image enhancement through histogram equalization, the input images are used as the output of previous steps (e.g., in watershed segmentation technique, the input image is used as the output of adaptive histogram equalization) to get accurate results. After reaching the end stage of our proposed algorithm, output curves are compared with normal and cancerous lung images. The proposed algorithm is beneficial for the detection of diseases like lung cancer, diabetic retinopathy, and the like. It is also beneficial for reducing human efforts in a real-life scenario with less time consumed than in classical techniques. In the future scope of this system, PET images and X-ray images could also be used. The future model could also be focused on different machine learning, deep learning algorithms for better accuracy.

## REFERENCES

Alam, J., Alam, S., & Hossan, A. (2018). Multi-stage lung cancer detection and prediction using multi-class svm classifier. In *2018 International Conference on Computer, Communication, Chemical, Material and Electronic Engineering (IC4ME2)*, pp. 1–4. https://10.1109/IC4ME2.2018.8465593.

Al-Amri, S. S., Kalyankar, N. V., & Khamitkar, S. D. (2010). Image segmentation by using edge detection. *International Journal on Computer Science and Engineering*, 2(3), 804–807.

Al-Tarawneh, M. S. (2012). Lung cancer detection using image processing techniques. *Leonardo Electronic Journal of Practices and Technologies*, 11(21), 147–158.

American Cancer Society. Facts & Figures. (2019). *American Cancer Society*. Atlanta, GA. https://www.cancer.org/, last accessed 2019/01/07.

Amoda, N., et al. (2013). Image segmentation and detection using watershed transform and region-based image retrieval. *International Journal of Emerging Trends & Technology in Computer Science (IJETTCS)*, 2(2).

Chander, M. P., Rao, M. V., & Rajinikanth, T. V. (2017). Detection of lung cancer using digital image processing techniques: A comparative the study. *International Journal of Medical Imaging*, 5(5), 58–62. https://doi.org/10.11648/j.ijmi.20170505.12

Chaudhary, A., & Singh, S. S. (2012). Lung cancer detection on CT images by using image processing. *2012 International Conference on Computing Sciences*. https://doi.org/10.1109/iccs.2012.43

Choi, H., & Mazzone, P. (2014, September). Radon and lung cancer: Assessing and mitigating the risk. *Cleveland Clinic Journal of Medicine*, 81(9), 567–575.

Chouhan, S. S., Kaul, A., & Singh, U. P. (2018). Image segmentation using computational intelligence techniques: Review. *Archives of Computational Methods in Engineering*, 26, 533–596. https://doi.org/10.1007/s11831-018-9257-4

De Geus-Oei, L. F., van der Heijden, H. F., Visser, E. P., et al. (2007). Chemotherapy response evaluation with 18F-FDG PET in patients with non-small cell lung cancer. *The Journal of Nuclear Medicine*, 48(10), 1592–1598.

Govindan, R., Page, N., Morgenstern, D., et al. (2006). Changing epidemiology of small-cell lung cancer in the United States over the last 30 years: Analysis of the surveillance, epidemiologic, and end results database. *Journal of Clinical Oncology*, 24, 4539–4544.

Gujral, H. (2015). A review of techniques for lung cancer detection. *International Journal of Current Engineering and Technology*, 5(3), 1597–1602.

Hore, S., et al. (2016). An integrated interactive technique for image segmentation using stack-based seeded region growing and thresholding. *International Journal of Electrical and Computer Engineering*, 6(6), 2773.

Hossain, F., & Alsharif, M. R. (2007). Image enhancement based on logarithmic transform coefficient and adaptive histogram equalization. *2007 International Conference on Convergence Information Technology*, pp. 1439–1444. 10.1109/ICCIT.2007.258. https://ieeexplore.ieee.org/abstract/document/4420457

Jiao, L., et al. (2010). Natural and remote sensing image segmentation using memetic computing. *IEEE Computational Intelligence Magazine*, 5, 78–91. https://doi.org/10.1109/MCI.2010.936307

Jothi, J. A. A., & Rajam, V. M. A. (2017). A survey on automated cancer diagnosis from histopathology images. *Artificial Intelligence Review*, 48, 31–81. https://doi.org/10.1007/s10462-016-9494-6

Kalemkerian, G. P., & Schneider, B. J. (2017). Advances in small cell lung cancer. *Hematology/Oncology Clinics of North America*, 31(1), 143–156.

Khandelwal, R. (2013). Various image enhancement techniques: A critical review. *International Journal of Advanced Research in Computer and Communication Engineering*, 2(3), 267–274.

Khuder, S. A. (2001). Effect of cigarette smoking on major histological types of lung cancer: A meta-analysis. *Lung Cancer*, 31(2–3), 139–148. https://doi.org/10.1016/s0169- 5002(00)00181-1

Krishna, C. R., Dutta, M., & Kumar, R. (Eds.). (2019). A comparative analysis of various image segmentation techniques. *Proceedings of 2nd International Conference on Communication, Computing, and Networking*. Lecture Notes in Networks and Systems. https://doi.org/10.1007/978-981-13-1217-5

Kumar, S. A., Ramesh, J., Vanathi, P. T., & Gunavathi, K. (2011). Robust and automated lung nodule diagnosis from CT images based on fuzzy systems. In *2011 International Conference on Process Automation, Control and Computing*, pp. 1–6. https://10.1109/PACC.2011.5979050. https://ieeexplore.ieee.org/abstract/document/5979050

Logesh Kumar, S., Swathy, M., Sathish, S., Sivaraman, J., & Rajasekar, M. (2016). Identification of lung cancer cell using watershed segmentation on CT images. *Indian Journal of Science and Technology*, 9(1). https://doi.org/10.17485/ijst/2016/v9i1/85765

Lung Cancer Statistics. www.wcrf.org/dietandcancer/cancer-trends/lung-cancer-statistics, last accessed 2019/01/07.

Mac Manus, M. P., Hicks, R. J., Matthews, J. P., et al. (2003). Positron emission tomography is superior to computed tomography scanning for response-assessment after radical radiotherapy or chemoradiotherapy in patients with non-small-cell lung cancer. *Journal of Clinical Oncology*, 21(7), 1285–1292.

Mesejo, P., Ibanez, O., Cordon, O., & Cagnoni, S. (2016). A survey on image segmentation using metaheuristic-based deformable models: State of the art and critical analysis. *Applied Soft Computing*, 44, 1–29. https://doi.org/10.1016/j.asoc.2016.03.004

Minaee, S., Boykov, Y., Porikli, F., Plaza, A., Kehtarnavaz, N., & Terzopoulos, D. (2020). Image segmentation using deep learning: A survey. *arXiv preprint arXiv:2001.05566.*

Open Database of Lung Cancer Detection Cases with CT Images. https://github.com/ddhaval04/Lung-Cancer-Detection, last accessed 2019/01/07.

Peto, R., Lopez, A. D., Boreham, J., et al. (2006). *Mortality from Smoking in Developed Countries 1950–2000: Indirect Estimates from National Vital Statistics.* Oxford: Oxford University Press. ISBN 978-0-19-262535-9. Archived from the original on 5 September 20.

Senthilkumaran, N., & Rajesh, R. (2009). Edge detection techniques for image segmentation – A survey of soft computing approaches. *International Journal on Recent Trends in Engineering*, 1(2), 250–254.

Shakeel, P. M., Burhanuddin, M. A., & Desa, M. I. (2019). Lung cancer detection from CT image using improved profuse clustering and deep learning instantaneously trained neural networks. *Measurement*, 145, 702–712.

Sim, K. S., Tso, C. P., & Tan, Y. Y.(2007). Recursive sub-image histogram equalization applied to grayscale images. *Pattern Recognition Letters*, 28(10), 1209–1221.

Slotman, B., Faivre-Finn, C., Kramer, G., Rankin, E., Snee, M., Hatton, M., Senan, S., et al. (2007). Prophylactic cranial irradiation in extensive small-cell lung cancer. *New England Journal of Medicine*, 357(7), 664–672. https://doi.org/10.1056/nejmoa071780

Thun, M. J., Hannan, L. M., Adams-Campbell, L. L., Boffetta, P., Buring, J. E., Feskanich, D., Samet, J. M., et al. (2008). Lung cancer occurrence in never-smokers: An analysis of 13 cohorts and 22 cancer registry studies. *PLoS Medicine*, 5(9), e185.

Vas, M., & Dessai, A. (2017). Lung cancer detection system using lung CT image processing. *2017 International Conference on Computing, Communication, Control and Automation (ICCUBEA)*, Pune, India, pp. 1–5. https://doi.org/10.1109/ICCUBEA.2017.8463851

Wang, Z., & Tao, J. (2006). A fast implementation of adaptive histogram equalization. *2006 8th International Conference on Signal Processing.* https://doi.org/10.1109/icosp.2006.345602. https://ieeexplore.ieee.org/abstract/document/4129083

Yabuuchi, H., Hatakenaka, M., Takayama, K., Matsuo, Y., Sunami, S., Kamitani, T., Honda, H., et al. (2011). Non – small cell lung cancer: Detection of early response to chemotherapy by using contrast-enhanced dynamic and diffusion-weighted MR imaging. *Radiology*, 261(2), 598–604. https://doi.org/10.1148/radiol.11101503

Yadav, N., Yadav, A., Bansal, J. C., Deep, K., & Kim, J. H. (Eds.). (2019). Harmony Search and Nature Inspired Optimization Algorithms. *Advances in Intelligent Systems and Computing*, Springer, Singapore. https://doi.org/10.1007/978-981-13-0761-4. https://link.springer.com/book/10.1007/978-981-13-0761-4

Yang, I. A., Holloway, J. W., & Fong, K. M. (2013). Genetic susceptibility to lung cancer and co-morbidities. *Journal of Thoracic Disease*, 5(Suppl 5), S454–S462.

# Section III

## Computational Genomics

# 9 Improved Prediction of Gene Expression of Epigenomics Data of Lung Cancer Using Machine Learning and Deep Learning Models

*Kalpdrum Passi, Zhengxin Shi, and Chakresh Kumar Jain*

**CONTENTS**



DOI: 10.1201/9781003142751-12

## 9.1 INTRODUCTION

Cancer is a deadly disease caused due to genetic and epigenetic changes that happen at gene levels that are manifested in the form of their phenotypic characteristics (Shen & Abate-Shen, 2010). These alterations are reflected at the genomic level and noticeable in cancerous or tumor cells as compared to normal cells. Although there has been a lot of advancement and research over the last 50 years in the field of cancer, still, very little is known about the mechanism of action and molecular genetics of human cancer. This triggers the urgent attention of the scientific community for more holistic research at genomic, proteomics, and transcriptomics levels. Currently, the research on cancer genetics is going through a revolution (Risch & Plass, 2008), and combination therapy, along with chemotherapy, is the possible treatment for lung cancer (Krushkal et al., 2020). Lung cancer is one of the important causes of several deaths. Around 1.8 million deaths (18%) have been reported due to lung cancer worldwide (Risch & Plass, 2008), (Li et al., 2015). Basically, it has been divided into two types, that is, small cell lung cancer (SCLC) and non–small cell lung cancer (NSCLC). NSCLC includes squamous cell carcinoma, adenocarcinoma, and large cell carcinoma and is found to be higher in scale in lung cancer.

Epigenetics refers to the study of genetic factors associated with gene expression that regulates gene function. Alterations in these factors may reflect on dysregulation of genes; lead to altered cell division, cellular process, cell development, cell proliferation, and cell death; and have been reported as the main cause behind tumor biogenesis. The chromatin structure is highly packaged with 147bp of a DNA molecule wrapped by four units of histone proteins (H2A, H2B, H3, H4) and arranged inside the nucleus. The expression of genes depends on the accessibility of the DNA. In chromatin structure, the euchromatin part is less dense compared to heterochromatin and is ready to express. The pattern of gene regulation is unique in individuals where the cell function is programed according to inheritance of epigenetic changes held at the DNA level. On the other side, nongenetic alterations are not heritable and remain present within the organism (Russo et al., 1996; Sadakierska-Chudy & Filip, 2015). Primarily, the epigenetic alteration process is mediated by DNA methylation, acetylation, and histone and nonhistone modification events.

DNA methylation is an important event under epigenetic regulation of genes in which the methyl group is used to bind with DNA in presence of DNA methyltransferase enzyme. This process leads to silencing tumor suppressor and DNA repair gene, thus making space for tumor formation and metastasis through continuous cell growth, cell development, and uncontrolled regulation. Therefore, the quantified methylation pattern that is person-specific is one of the important modes of cancer diagnosis (Greer & Shi, 2012). The triggering of the methylation process is often due to connected process and mutations in oncogenes.

The gene expression mechanism, which is mainly governed by histone modification whereby the histone protein binding is altered or modified with DNA molecule, is very complex. This protein is basic in nature and responsible for stabilizing the DNA and finally controlling gene regulation. The genome silencing or expression is the main process behind the protein function. The activation/deactivation of transcription followed by translation, chromosome packaging, unpackaging, chromatic modelling, and modifications are the major processes effecting gene regulation. The mutation or damage of DNA may lead to changes, thus altering the overall machinery functioning, although an inbuilt DNA repair mechanism is present within the cell. Hence, the degree of alteration in DNA/histone is the inherent cause behind altered phenotypes (Stephens et al., 2015). Histone modifiers and chromatic bound proteins are the main players in altering transcription machinery. There are five main types of histone modifications: methylation, phosphorylation, acetylation, ubiquitylation, and simulation.

Primarily in the acetylation process that is reversible in nature; two enzymes, that is, histone acetyltransferases and histone deacetylases, play an important role in acetylation (attachment of acetyl group to lysine residues of histone H3 and H4) and deacetylation (detachment of acetyl group) and play an important role in chromatic dynamics, gene silencing and expression, and the like (Shen & Abate-Shen, 2010). Most of the modifications used to occur at H3 unit of histone, which conclusively helped in the determination of heterochromatin versus euchromatin and governs the states (active/inactive) of various elements such as promoters, enhancers, and gene bodies of the genome.

In methylation, one, two, or three methyl groups used to bind with histone protein and alter the structure, which, in turn, effects gene expression. The methylation leads to switching off the gene expression while demethylation leads to switching on the gene, thereby facilitating the access of the DNA molecule for transcription factors. Similarly, ubiquitylation and phosphorylation processes also affect histone modification through epigenetic alteration, leading to gene expression.

### 9.1.1 Software and Tools

Various R packages were used for data processing (Li et al., 2015). Weka 3 data mining software was used for feature selection and classification (www.cs.waikato.ac.nz/ml/weka/index.html), and Python 3 was used for constructing a convolutional neural network model.

## 9.2 DATA SETS AND PROCESSING

In total, four types of data were used to extract features and perform classification, including CpG methylation data and histone marker data for metastasis and normal phase, human genome data, and RNA-Seq expression data. These datasets have been further processed with the help of several available software packages, R-scripts and packages, and libraries, such as Corrplot, for the generation of the correlation matrix.

### 9.2.1 DNA Methylation Data

The methylation features were extracted from the Cancer Genome Atlas (TCGA) Methylation data based on Illumina's Infinium HumanMethylation450 Beadchip (Illumina 450K). Since the genomic location of CpG, the exons and coding regions, coding DNA sequences, and transcripts are identified from an annotation file, it has been reannotated with the protein-coding genes using the Illumina Genome hg19 Refeseq as followed in Li et al. (2015).

### 9.2.2 Histone Data

The raw histone dataset has been downloaded from UCSC genome browser (http://genome.ucsc.edu) of relevant cell lines and further processed through normalization and alignment procedure with maintaining the consistency and non-redundancy of the dataset using tools such as Samtool toolkit, bowtie2 software (Li et al., 2009; Love et al., 2014). The alignment is performed against hg19; after that, the aligned reads are intersected with the relevant segments of transcript by using the Bedtools toolkit (Li et al., 2009).

### 9.2.3 Human Genome Data

Regarding the human genome dataset, the FASTA files (http://genome.ucsc.edu) of hg19 genome have been downloaded, and three species, such as vertebrates, primates, and placental animals, were considered. The conservation scores of conserved elements that govern the importance from the evolutionary functional point of view have been computed with the help of phastCons software (Siepel et al., 2005; Li et al., 2015).

### 9.2.4 RNA-Seq Data

RNA-Seq data represent the transcriptomics data of all types of RNA, such as mRNA, TRNA, rRNA, and so on, within the cell that demonstrate the gene expression pattern for lung cancer. These datasets have been collected from the TCGA network (http://cancergenome.nih.gov) and further could be analyzed using DESeq2 package based on R platform (Adetiba & Olugbara, 2015). Gene expression for upregulation and downregulation is catagorized in a binary form of 1 and 0 following Li et al. (2015).

## 9.3 FEATURE EXTRACTION

All the features were extracted from three gene types, CpG methylation features were extracted from HumanMethylation 450 Beadchip (Illumina 450k) data, Histone marker modification features are extracted from histone marker data, Nucleotide features and conservation features were extracted from human genome data.

### 9.3.1 CpG Methylation Feature

The CpG methylation feature is very important for understanding the epigenetics of gene expression and can be analyzed using limma library on an R platform, where the Toptable function is used to compute the log fold change (logFC) between the tumor and the normal tissues and to compute the average methylation (avgMval) of each CpG site (Li et al., 2015). This value of logFC is further used to catagorize the gene as in either a hypermethylation or a hypomethylation state. Whereas the positive value of logFC refers to hypermethylation, a negative logFC indicates hypomethylation (Li et al., 2015; Adetiba & Olugbara, 2015; Quinlan & Hall, 2010). Hypomethylation contributes to cancer development and progression. Apart from the hypermethylation state also increasing genomic instability and activating proto-oncogenes, site-specific hypermethylation thus contributes to tumorigenesis. Several features of CpG methylation can be computed as given in Li et al. (2015).

### 9.3.2 Histone Marker Change Feature

This feature is very important as it reveals the pattern of the binding of histone with DNA; therefore, any change in the binding pattern may decipher the change in phenotypic characteristics. This feature could be computed using the multicov function from the BEDTools package on to each transcript after aligning the raw histone marker data with transcript and their intercept region. Before that, the histone reads are normlized per 1000bp length of each segment per 1 million aligned read library. The remaining calculation is the same based on segment-by-segment calculations as followed in the CpG methylation feature. Generally, the histone data are represented in the form of a certain nomenclature like H3K36me3, where H3 is histone, k refers to lysine residue, and position number 36 is where trimethylation takes place. The same could be interpreted for H3K27me3 (Pages et al., 2009).

### 9.3.3 Nucleotide Feature

The nucleotide features (nucleotide compositions) of each segment of the transcript can be computed using the Biostrings library in R by processing the Hg19 genome as a reference (Li et al., 2015; Saeys et al., 2007; Pages et al., 2009).

### 9.3.4 Conservative Feature

This can be calculated/extracted through the UCSC genome browser for different comparative species, that is, vertebrate, primate, and placental animals, on the basis of the arithmetic mean of the conservation score per nucleotide in the segment. The model proposed by Li et al. (2015) was obtained with 120 features composed of four types of genomic and epigenomic data. The 67 features were selected for the best model development and were considered to contribute to the correct gene expression prediction. In these selected features, histone modification features are composed of 32 features, followed by the 15 methylation features.

Those two types of data contribute most toward the correct gene expression prediction. Moreover, the same type of data tends to cluster together, and the relationship with the same type is closer than the relationship between different data types. In our best model, the original 67 features are included with additional 53 features for a total of 120 features. In the additional 53 features, 33 features are histone modification features, 11 features are the methylation features, 4 features are the nucleotide features, and 2 features are conservation features. In total, there are 65 histone modification features, 26 methylation features, 16 nucleotide features, 7 conservation features, 6 features are the element length.

## 9.4 FEATURE SELECTION

The object of feature selection is to find the optimal feature subset. Irrelevant features can be removed by feature selection methods, thereby reducing the number of features that generally results in the improvement in classification performance and makes the prediction more efficient (Chandrashekar & Sahin, 2014). In our research, we used four different types of feature selection methods, including principal component analysis (PCA), correlation-based feature selection (CFS), gain ratio, and ReliefF.

### 9.4.1 PCA

PCA finds a subset of variables from a larger set using a linear combination that retains the information from the original and therefore is referred as a dimension-reduction technique (Peng et al., 2005). The variables that are correlated are linearly combined into a set of variables that are reduced but still keep the information from the original variables. High dimensionality often results in redundancy and noise in the data, which causes errors in actual applications such as image recognition, and therefore, it is important that the number of features is reduced in order to reduce the error and the accuracy of recognition is improved. In many algorithms, dimensionality reduction algorithms have become part of data preprocessing. For some high-dimension data, it is hard to obtain good results without reducing the dimensions. There are four main steps in the PCA process, which are feature normalization, covariance matrix computation, dimensionality reduction, and reconstruction.

### 9.4.2 ReliefF

ReliefF is an efficient algorithm for data with strong dependencies between attributes. It is an improved version of the Relief algorithm proposed by Kira and Rendell (1992a, 1992b) to solve binary classification problems. The Relief algorithm assigns weights to the features based on the correlation between features. A threshold on the weight determines which features are to be retained for classification purposes. The correlation between features influences the ability to discriminate data into classes. A data tuple is randomly picked along with two nearest neighbors, one that is similar and one that is dissimilar to the tuple. Every feature is assigned a weight based on the nearness of the tuple to the similar and dissimilar neighbors. If the distance of the tuple to the similar neighbor is less than the distance to the dissimilar neighbor, the weight of the feature is increased; otherwise, it is decreased. Similar neighbors closest to each other belong to the same class. The average weight of a feature is computed over a number of iterations. A feature that has a larger weight has a greater influence on the classification of the data tuple. The Relief algorithm has a high running efficiency as the increase in the number of features has a linear relationship with execution time. The Relief algorithm is relatively simple but has a high operating efficiency, so it is widely used. However, it can only handle two types of data.

Kononenko (1994) proposed an ReliefF algorithm that can handle multiclass problems. Regression techniques are used for continuous valued attributes, and ReliefF is one such algorithm. For multiclass problems, ReliefF picks a tuple randomly from the training set, finds k-nearest neighbors from the sample set similar to the tuple, and updates the weight of each feature. The ReliefF algorithm runs very efficiently and has no restrictions on data types. The features that are highly correlated get higher weights, and the redundant features are removed effectively (Urbanowicz et al., 2018).

### 9.4.3 Gain Ratio

In machine learning, gain ratio refers to the information gain rate, which represents the ratio of the node's information to the node split information metric. The information gain is modified to reduce its bias. The attribute with the maximum gain ratio is selected as the splitting attribute. Attributes that have large values influence the information gain measure. The gain ratio corrects the information gain by normalization (Priyadarsini et al., 2011). Entropy was originally a physical concept and was later used in mathematics to describe randomness, disorder, or uncertainty. Therefore, the higher the information entropy of a system, the more disordered it is; the lower the information entropy, the less disordered it is. The expected information or the entropy of a tuple X is defined as

$$H(X) = -\sum_{i=1}^{n} p_i \log(p_i), \tag{9.1}$$

where $n$ represents the type of all samples and $p$ represents the proportion of the $i$th sample, when $p = 0$, $H(X) = 0$.

Different decision tree algorithms are mainly determined by the way of dividing the nodes. Three algorithms commonly used are ID3, C4.5, and classification and regression tree (CART). ID3 is mainly applied in information gain, C4.5 is applied in gain ratio, and CART is used in the Gini Index. In our study, we only used gain ratio for feature selection, so we only focus on C4.5 algorithm, which is an improvement of ID3. Suppose in a dataset, there are many categories of a feature; as a result, each sample is divided into an independent node on the decision tree, which maximizes the information gain. It will affect the decision tree because we want the decision tree to focus more on the attributes that the samples have in common so that the new samples can be classified by the trained decision tree. To balance the information gain and the number of attributes, the information gain ratio was proposed, the formula is as follows:

$g_R(D, A) = \dfrac{g(D, A)}{H_A(D)}$; $H_A(D)$ is the empirical entropy given by

$$H_A(D) = -\sum_{i=1}^{n} \frac{|D_i|}{|D|} log_2 \frac{|D_i|}{|D|}. \tag{9.2}$$

The information gain is multiplied by a penalty to obtain the gain ratio. When there are more features, the penalty parameter is smaller; when there are fewer features, the penalty parameter is larger. The Penalty parameter is the inverse of entropy $H_A(D)$.

$$\frac{1}{H_A(D)} = \frac{1}{-\sum_{i=1}^{n} \frac{|D_i|}{|D|} log_2 \frac{|D_i|}{|D|}} \tag{9.3}$$

ID3 uses the information gain as the evaluation standard, and it tends to take more valued features. The information gain reflects the degree of uncertainty reduction after a given condition. More feature values mean higher certainty, smaller entropy, and greater information gain. This is a flaw of entropy. C4.5 improved this situation by using a gain ratio. The features with more values are punished to avoid overfitting.

### 9.4.4 CFS

CFS is a classic filtering feature selection algorithm. The merit of a feature subset is evaluated by a search algorithm and a function. CFS finds the correlation between the attributes to predict the class label with the most relevant features (Hall & Smith, 1997). CFS selects a feature subset by assigning merits to the features. The merits are assigned according to the following formula:

$$Merit_S = \frac{k\bar{r_{cf}}}{\sqrt{k + k(k-1)\bar{r_{ff}}}}. \tag{9.4}$$

Here, $Merit_S$ is an evaluation of a subset of k features, $\bar{r_{cf}}$ is the average correlation between attributes and categories, and $\bar{r_{ff}}$ is the correlation between attributes.

The training set is used to calculate the correlation matrices of features which is then reduced to find a subset of features using the "best first" approach. Forward selection and backward elimination are other search methods that are often used. In forward selection, the best features are selected and added at each step, whereas in backward elimination, the reverse process is used by eliminating the irrelevant features one by one to obtain the most relevant feature set. Both these methods are similar to the "best first" approach. An initial set *S* is taken that would be empty in case of forward select, and it would be complete in the case of backward elimination. Features are selected on the basis of $merit_s$ and the features added to set S with largest value, then the second feature with the second-largest value, and so on. Features having the higher merit score are kept, and those with lower scores are eliminated.

## 9.5 CLASSIFIERS

Data classification is the classification of data so that they can be used most efficiently and to ensure their efficiency is the highest. It uses a basic method to store computer data that may be classified according to its critical value or the frequency it needs to be accessed. These critical or frequently used data are stored in a medium that is fast, while other data may be stored in a slower (and less expensive) storage medium. This type of classification aims to optimize the use of multipurpose data storage—technical optimization. In our research, we used seven different classifiers, namely, random forests (RFs), logistic regressions, naïve Bayes, Gaussian SVMs, linear SVMs, neural networks, and convolutional neural networks (CNNs).

### 9.5.1 Random Forest

RF is an ensemble of CART (Breiman et al., 1984), which form binary trees by selecting the most appropriate attribute to split the tree at every step where the leaves give the prediction outcome. The idea of a decision tree was first proposed by Breiman et al. (1984), whereby a decision tree is formed by selecting the attribute with the least cost to split the data into two branches. Breiman (2001) proposed RFs by combining decision trees. The forest consists of multiple independent decision trees (Breiman, 2001; Sarica et al., 2017; Fawagreh et al., 2014). To assign a class label to a new sample, all decisions trees independently calculate the class label. Majority vote is used to select the class label based on all the votes from the decision trees.

Random samples are extracted from the training set, and decision trees are formed from the samples to create an RF. Each decision tree outputs the class label, and the final class label is selected based on the majority vote. Since RFs consist of many decision trees, their performance is better when compared to the performance of a single decision tree. Each sample governs the formation of the decision tree within the forest. The correlation in the sample influences the classification error in each decision tree although the samples have the same distribution. The features are selected based on the classification error generated by individual decision trees.

The decision tree is the keystone of an RF. The non-leaf nodes are decision attributes that branch out to subtrees by applying some criterion to split the data, and the leaf nodes represent the class labels. For each decision tree, the root node is selected to split the data into two branches. At each step, the best attribute is selected for splitting the features till the leaf nodes are reached. New samples representing the test data are run through the independent decision trees and the class label is determined by the majority vote on the class labels found by individual decision trees.

### 9.5.2 Logistic Regression

Logistic regression is a technique in statistics that models a binary classification using a logistic function, has several extensions to multiple classes, and is an efficient method for classifying microarray data for a particular cancer (Sperandei, 2014). Logistic regression is commonly used to estimate the likelihood of an event. Some examples can be "What is the likelihood that a student would purchase a laptop?" and "What is the likelihood that a patient has lung cancer?" Note that the "probability" here is different from the mathematical "probability". A weighted summation of features is used in the computation. Logistic regression and linear regression fall in the category of generalized linear models. Logistic regression obeys the Bernoulli distribution, while Gaussian distribution is used in linear regression. Therefore, there are many similarities between the two methods (Liao & Chin, 2007). However, there are also differences between logistic regression and linear regression. Logistic regression yields a discrete result, but linear regression yields a continuous result. For example, the house price predicting model returns continuous results. The value varies depending on parameters such as the size or position of the house. Discrete results are always binary: Either a patient has cancer or not. The activation function of logistic regression is given by the following formula:

$$h_\theta(x) = \frac{1}{1+e^{-\theta^T x}}. \tag{9.5}$$

This function is also called sigmoid function, also known as the logistic function whose value is between [0,1]. It is particularly suitable for binary problems. The logistic function can be rewritten as

$$h_\theta(x) = g(\theta^T x), g(z) = \frac{1}{1+e^{-z}}, \tag{9.6}$$

where $x$ is the input and $\theta$ is the parameter.

The parameter values have a certain set of values that determine the range of values that are mapped between 0 and 1. Conditions can be defined on the logistic function to characterize the mode as

$$P(y=1|x;\theta) = g(\theta^T x) = \frac{1}{1+e^{-\theta^T x}}, \tag{9.7}$$

where $P$ is the conditional probability of $y = 1$ given $x$ and $\theta$.

### 9.5.3 Naive Bayes

Naive Bayes is based on a statistical technique that performs probabilistic prediction and has been derived from Bayes' theorem. Naive Bayes is one of the widely used classification algorithms. Naive Bayes classifier is considered a supervised learning algorithm and is used for prediction purposes such as "customers will increase or decrease", "investing will be profitable or not", and "credit rating is good or bad". Naive Bayes is a simple algorithm and performs well, and it is comparable to other classifiers for certain data types. However, this algorithm is based on a simple assumption that the attributes are conditionally independent and for continuous variables, Gaussian distribution is used (Langarizadeh & Moghbeli, 2016; Ray, 2017). Assume that the probability of a random event A is $P(A)$ and the probability of a random event B is $P(B)$. Then, under the condition that the probability of event B occurs given probability of event A is

$$P(B|A) = \frac{P(A|B)P(B)}{P(A)}. \tag{9.8}$$

If event A and event B are independent of each other, then $P(A|B)\,P(B) = P(AB)$. We get

$$P(B|A) = \frac{P(AB)}{P(A)}, \tag{9.9}$$

where $P(A|B)$, $P(B)$, and $P(A)$ are called prior probabilities and $P(B|A)$ is called the posterior probability.

There are several reasons we choose this classifier in our research; it is a computationally efficient and low-cost algorithm. It is robust and efficient for big datasets. The practical application of naïve Bayes is accomplished by assuming the attributes to be conditionally independent and the performance of naive Bayes is comparable to other algorithms as it can train the model with small data sample.

### 9.5.4 Support Vector Machine

SVM maps the training data to a higher dimension with a nonlinear function. SVM is applied to linear, as well as nonlinear, data (Huang et al., 2018). A linear hyperplane in the higher dimension separates the data into two classes. SVM and related kernel methods are extremely good at solving epigenetics problems (Huang et al., 2018). The support vectors define the margins of a hyperplane. In SVM, the sample of the smallest distance from the hyperplane is called the support vector. The distance between the margins of the support vectors is maximized to find the hyperplane. The equation of hyperplane can be written as

$$w^T x + b = 0, \tag{9.10}$$

where $w$ is the lateral quantity and $x$ is the longitudinal quantity.

When the data are linearly inseparable, nonlinear SVMs are created. The first step is to use a nonlinear function to map the data into a higher-dimensional space. Then, a linear hyperplane has to be found in the higher-dimensional space. The problem is formulated as a quadratic optimization problem that is solved by maximizing the quadratic function of several variables with linear constraints. The maximal marginal hyperplane (MMH) is found in the higher-dimensional space is obtained through a nonlinear function on the lower dimension. The non-linear function used in the transformation of training tuples is expressed as a kernel function. Some common kernel functions include linear, polynomial, radial basis function (RBF), and standard RBF (SRBF). We used both linear kernel and RBF (Gaussian) kernel functions in our experiments.

### 9.5.5 Neural Networks

Neural networks are an old technique but have acquired importance in recent years, and they also have a foundation in deep learning. Neural networks are learning algorithms that work on a set of connected input and output nodes, where each connection has an associated weight. The input nodes would represent a gene for the epigenomics data and the connections within the network would represent the interactions between the genes (Knocklein, 2019). In this research, we used multilayer perception (MLP), which means it has at least one or more hidden layers. The network learns by modifying the weights while training the model until the error of prediction is minimized. An MLP is a multilayer feed-forward neural network that performs nonlinear regression. When a neural network is trained, backpropagation is most commonly used to modify the weights in the backward direction. The main idea of backpropagation is to find the error at the output and moving backward, keep modifying the weights till the error at the output is minimized. The training stops when the weights converge, and the mean squared error is below a threshold value.

### 9.5.6 CNN

CNN is an extension of a neural network that has a number of hidden layers and convolutional layers. CNN is a type of deep network with a number of layers and the process of training involves finding the optimal weights in the fully connected layers to minimize the loss function. Filters are applied to different layers of the CNN that determine the parameters. The ability of the CNNs to learn and distinguish between image-based patterns makes them highly specific for image recognition and image classifications (Kumar et al., 2017). A CNN consists of mainly four types of layers: a convolutional layer, a pooling layer, a Rectified Linear Unit correction layer, and a fully connected layer. By stacking these layers together, a complete CNN can be constructed.

## 9.6 RESULTS

The epigenomics data was compiled using four sources, Illumina 450K CpG methylation data, histone H3 marker CHIP-Seq data, human genome data, and RNA-Seq gene expression data. In total, 1,424 features were calculated. Feature selection was performed with four different feature selection methods namely, ReliefF, gain ratio, CFS, and PCA. After the feature selection, seven classifiers were trained using 10-fold cross-validation. The classifiers used in this study include neural networks, RFs, logistic regression, naive Bayes, Gaussian SVM, linear SVM, and CNN. In order to find the best performance, different ratios of features were selected, starting at 95% features and reducing to 50% features. Furthermore, a fixed number of features were selected to compare with the results of (Li et al., 2015) as they get the best performance for 67 features. The models were trained by splitting the dataset into different training-to-testing ratios, with training data 90%, 80%, 70% and 60%, respectively. In order to deal with the problem of overfitting, 10-fold cross validation was used (Refaeilzadeh et al., 2009). The feature selection methods with the classifiers were tested for accuracy and area under the curve (AUC) of the receiver operating characteristics (ROC).

The purpose of this research was to use and test the performance of deep learning using a CNN with the feature selection methods. The results show that the CNN gave the best results for all but a few feature selection methods. An improved AUC of 0.998 (+13.4%) as compared to the AUC of 0.864 in Li et al. (2015) was significant. It further consolidates the argument that there is an interaction between different types of data due to the nonlinearity of the CNN and other classifiers.

Among the different percentages (95%–50%) of the features selected by ReliefF method, amounting to 1353–712 features (from a total 1,424 features), CNN gives the highest AUC of 0.88 for 1281 (90%) features for a training-to-testing ratio 80:20. The best accuracy of 84.6% was also obtained by CNN for 1,210 (85%) features for the same training-to-testing ratio. By selecting a much smaller number of features (50–120), the highest AUC of 0.998 was achieved with 120 features using the CNN classifier for a training-to-testing ratio 90:10. This is the highest AUC value achieved among all the four feature selection methods. The best accuracy of 95.4% was achieved with 100 features using a CNN classifier for a training-to-testing ratio of 60:40.

Among the different percentages (95%–50%) of the features selected by gain ratio, the CNN gives the highest AUC of 0.869 for 1,068 (75%) features for a training-to-testing ratio of 80:20. The best accuracy of 84.6% was obtained by the RF for 783 (55%) features for the same training-to-testing ratio. By selecting a much smaller number of features (50–120), the CNN gave the highest AUC of 0.991 with 120 features with 60% training data. The best accuracy of 95.6% was achieved with 120 features using CNN classifier with 60% training data.

Among the different percentages (95%–50%) of the features selected by Correlation Feature Selection (CFS), CNN gives the highest AUC of 0.88 for 1068 (75%) features for a training-to-testing ratio 80:20. The best accuracy of 85.8% was obtained by CNN for 712 (50%) features for the training-to-testing ratio 90:10. By selecting a much smaller number of features (50–120), the highest AUC of 0.962 was achieved with 100 features and a CNN classifier with 80% training data. The best accuracy of 96.4% was achieved with 70 features and a CNN classifier with 90% training data. This is the highest accuracy achieved among all the feature selection methods.

Among the different percentages (95–50%) of the features selected by PCA, the CNN gives the highest AUC of 0.861 for 1,352 (95%) features for a training-to-testing ratio of 80:20. The best accuracy of 83.8% was obtained by logistic regression for 1,352 (95%) features for the training-to-testing ratio of 90:10. By selecting a much smaller number of features (50–120), the highest AUC of 0.851 was achieved with 120 features and a CNN classifier with 90% training data. The best accuracy of 80.2% was achieved with 120 features and a CNN classifier with 90% training data.

Figure 9.1 shows the AUC results of all seven classifiers with the four feature selection methods for 120 features. It is clear from the figure that CNN performs the best among all the classifiers while ReliefF (0.998) and gain ratio (0.991) feature selection methods give the best combined results with CNN. The next best performance is by RF and Gaussian SVM in combination with gain ratio, CFS, and ReliefF feature selection methods while PCA gives the lowest performance. Higher AUC values were obtained with other classifiers as well, such as RF gave an AUC of 0.912 with the CFS method for 70 features and 90% training data. Similarly, Gaussian SVM gave an AUC of 0.88 with a gain ratio feature selection method for 120 features with 90% training data.

Figure 9.2 shows the accuracy results of the seven classifiers with four feature selection methods for 120 selected features. The figure shows that CNN gives the best accuracy with gain ratio (95.6%), ReliefF (95.4%), and CFS (95.1%). The next best accuracy is by a neural networks classifier in combination with the CFS and gain ratio feature selection methods.

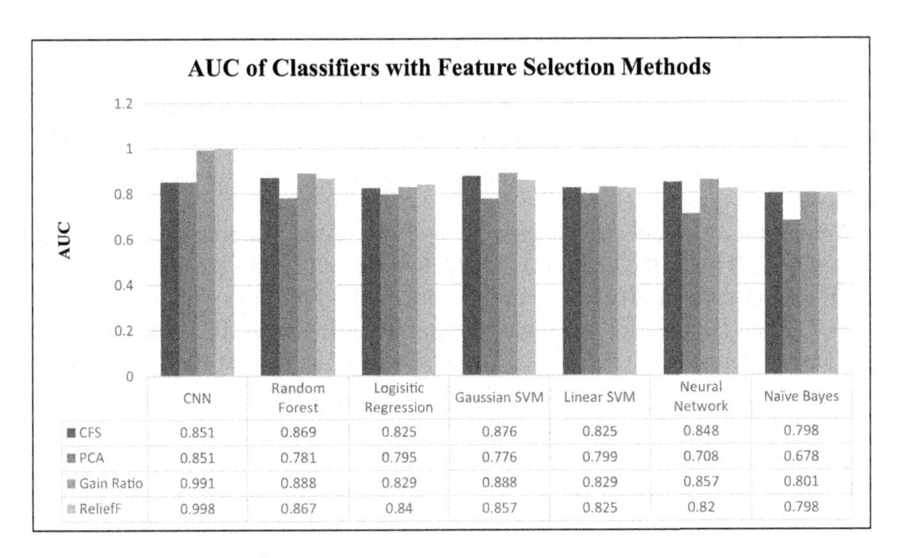


| | CNN | Random Forest | Logisitic Regression | Gaussian SVM | Linear SVM | Neural Network | Naïve Bayes |
|---|---|---|---|---|---|---|---|
| CFS | 0.851 | 0.869 | 0.825 | 0.876 | 0.825 | 0.848 | 0.798 |
| PCA | 0.851 | 0.781 | 0.795 | 0.776 | 0.799 | 0.708 | 0.678 |
| Gain Ratio | 0.991 | 0.888 | 0.829 | 0.888 | 0.829 | 0.857 | 0.801 |
| ReliefF | 0.998 | 0.867 | 0.84 | 0.857 | 0.825 | 0.82 | 0.798 |

**FIGURE 9.1** The AUC Results of Classifiers for Each Feature Selection Method.

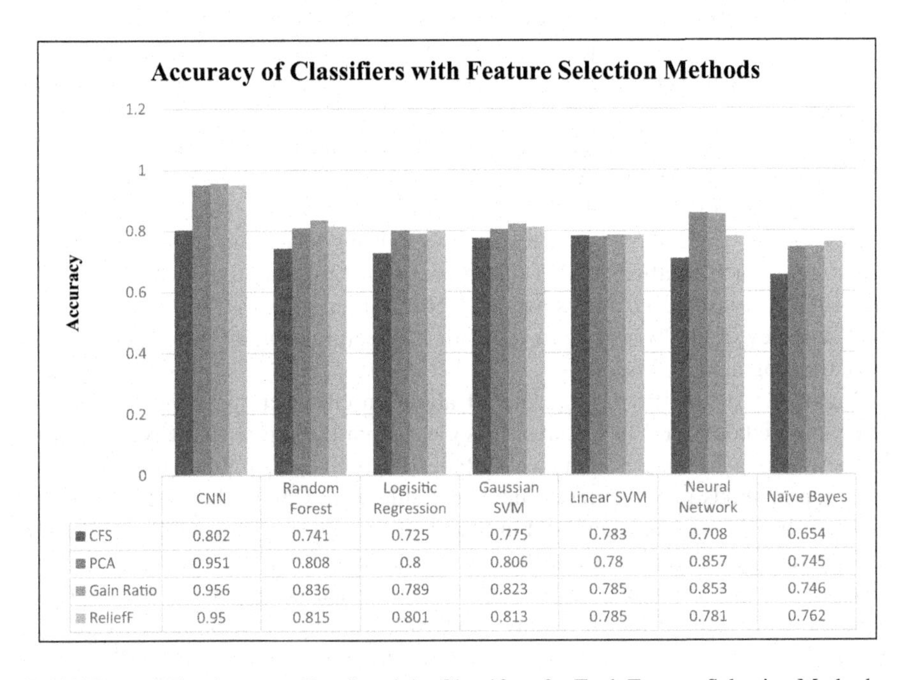


| | CNN | Random Forest | Logisitic Regression | Gaussian SVM | Linear SVM | Neural Network | Naïve Bayes |
|---|---|---|---|---|---|---|---|
| CFS | 0.802 | 0.741 | 0.725 | 0.775 | 0.783 | 0.708 | 0.654 |
| PCA | 0.951 | 0.808 | 0.8 | 0.806 | 0.78 | 0.857 | 0.745 |
| Gain Ratio | 0.956 | 0.836 | 0.789 | 0.823 | 0.785 | 0.853 | 0.746 |
| ReliefF | 0.95 | 0.815 | 0.801 | 0.813 | 0.785 | 0.781 | 0.762 |

**FIGURE 9.2** The Accuracy Results of the Classifiers for Each Feature Selection Method.

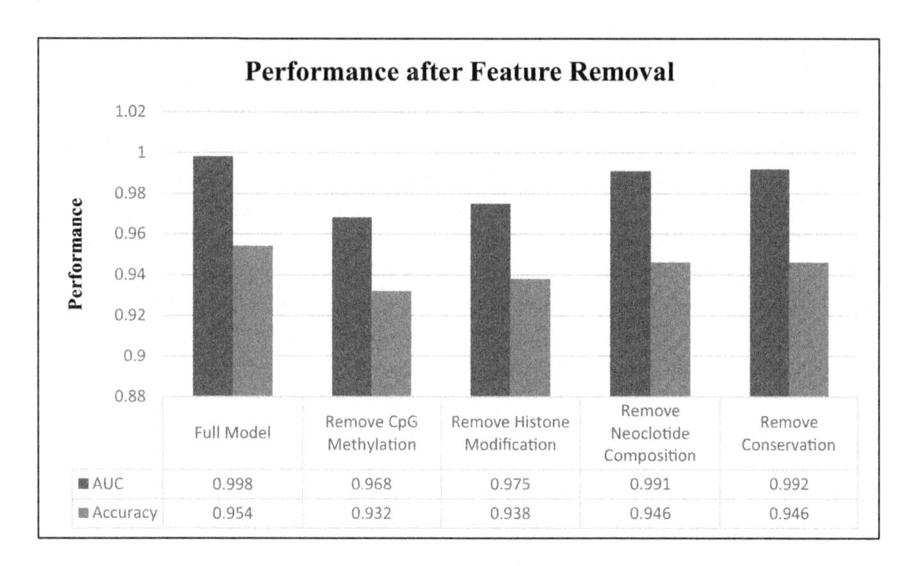


**FIGURE 9.3** Performance Comparison after Feature Set Removal.

In our best model, the original 67 features are included with additional 53 features for a total of 120 features. In the additional 53 features, there are 33 histone modification features, 11 methylation features, 4 nucleotide features, and 2 conservation features. In total, there are 65 histone modification features, 26 methylation features, 16 nucleotide features, 7 conservation features, 6 features are the element length. Then, we dropped off each data type to see the performance of the prediction. Figure 9.3 shows that after CpG methylation data type features were removed, the AUC dropped from 0.998 to 0.968, and the accuracy dropped from 0.954 to 0.932, which reduced the accuracy most within the four types of data. Histone modification comes the second; after removal of the histone modification features, the AUC dropped from 0.998 to 0.975, and the accuracy dropped from 0.954 to 0.938. Nucleotide composition features come next; the AUC dropped to 0.991, and the accuracy dropped to 0.946. The last is the conservation features; the AUC dropped to 0.992, and the accuracy drooped to 0.946. As a result, the greatest effect was seen by removing methylation features among individual feature sets. It was also observed that CpG methylation features were the most highly correlated features in the gene expression. In comparison, we confirmed that CpG methylation data contribute the most toward correct gene expression prediction. Since some features in the Illumina 450K model were hand coded, there might be a bias in the number of methylation features.

## 9.7 CONCLUSION AND FUTURE WORK

The purpose of this study is to test and improve the AUC that (Li et al., 2015) have achieved, and to compare which classification method gives the best results. We have compared four feature selection methods (CFS, ReliefF, gain ratio, PCA) with different training data sizes and seven different classifiers (RFs, neural networks, logistical regression, SVMs, CNNs). We also compared the performance by 10-fold cross-validation with each classifier. The results show that 120 features represent the best set of features in the data. A CNN with gain ratio feature selection method improves the AUC and accuracy significantly. The neural network comes the second, and naive Bayes with PCA gives the lowest AUC. In this research, only three common histone markers were used, namely, H3k4Me3, H3K27Me3, and H3K36Me3. The heterogeneity of the sample resources could influence the model's accuracy. In the future, we can include more histone marker data to obtain a stronger model. For classification, we only tried one deep learning method; there are many other deep learning methods we have not tested yet, such as RNN and DBN. In the future, we can also test how the other deep learning models work with epigenetics data.

## REFERENCES

Adetiba, E., & Olugbara, O. O. (2015). Lung cancer prediction using neural network ensemble with histogram of oriented gradient genomic features. *The Scientific World Journal*, 2015.

Breiman, L. (2001). Random forests. *Machine Learning*, 45(1), 5–32. https://doi.org/10.1023/a:1010933404324

Breiman, L., Friedman, J. H., Olshen, R. A., & Stone, C. J. (1984). *Classification and Regression Trees* (1st ed.). New York/Boca Raton, FL: Chapman and Hall/CRC.

Chandrashekar, G., & Sahin, F. (2014). A survey on feature selection methods. *Computers & Electrical Engineering*, 40(1), 16–28.

Fawagreh, K., Gaber, M. M., & Elyan, E. (2014). Random forests: From early developments to recent advancements. *Systems Science & Control Engineering*, 2(1), 602–609. https://doi.org/10.1080/21642583.2014.956265

Greer, E. L., & Shi, Y. (2012). Histone methylation: A dynamic mark in health, disease and inheritance. *Nature Reviews Genetics*, 13(5), 343.

Hall, M. A., & Smith, L. A. (1997). Feature subset selection: A correlation-based filter approach. In *1997 International Conference on Neural Information Processing and Intelligent Information Systems* (pp. 855–858). Berlin: Springer.

Huang, S., Cai, N., Pacheco, P. P., Narrandes, S., Wang, Y., & Xu, W. (2018). Applications of support vector machine (SVM) learning in cancer genomics. *Cancer Genomics Proteomics*, 15(1), 41–51. https://doi.org/10.21873/cgp.20063

Kira, K., & Rendell, L. A. (1992a). A practical approach to feature selection. In Sleeman, D., & Edwards, P. (eds.), *Machine Learning Proceedings* (pp. 249–256). Burlington, MA: Morgan Kaufmann. ISBN 9781558602472. https://doi.org/10.1016/B978-1-55860-247-2.50037-1

Kira, K., & Rendell, L. A. (1992b). The feature selection problem: Traditional methods and a new algorithm. In *Proceedings of the Tenth National Conference on Artificial Intelligence (AAAI'92)* (pp. 129–134). San Jose, CA: AAAI Press.

Knocklein, O. (2019, June 15). *Classification Using Neural Networks.* Retrieved November 25, 2019, from https://towardsdatascience.com/classification-using-neural-networks-b8e98f3a904f

Kononenko, I. (1994). *Estimating Attributes: Analysis and Extensions of RELIEF.* Machine Learning: ECML-94. Lecture Notes in Computer Science. 784. Berlin, Heidelberg: Springer, pp. 171–182. ISBN 978–3540578680. https://doi.org/10.1007/3-540-57868-4_57

Krushkal, J., Silvers, T., Reinhold, W. C., et al. (2020). Epigenome-wide DNA methylation analysis of small cell lung cancer cell lines suggests potential chemotherapy targets. *Clinical Epigenetics*, 12, 93.

Kumar, N., Verma, R., Arora, A., Kumar, A., Gupta, S., Sethi, A., & Gann, P. H. (2017, March). Convolutional neural networks for prostate cancer recurrence prediction. In *Medical Imaging 2017: Digital Pathology* (Vol. 10140, p. 101400H). Orlando, FL: International Society for Optics and Photonics.

Langarizadeh, M., & Moghbeli, F. (2016). Applying naive bayesian networks to disease prediction: A systematic review. *Acta Informatica Medica: AIM: Journal of the Society for Medical Informatics of Bosnia & Herzegovina: casopis Drustva za medicinsku informatiku, BiH*, 24(5), 364–369. https://doi.org/10.5455/aim.2016.24.364-369

Li, H., Handsaker, B., Wysoker, A., Fennell, T., Ruan, J., Homer, N., Marth, G., Abecasis, G., & Durbin, R. (2009). Genome project data processing S: The sequence alignment/map format and SAMtools. *Bioinformatics*, 25(16), 2078–2079. https://doi.org/10.1093/bioinformatics/btp352

Li, J., Ching, T., Huang, S., et al. (2015). Using epigenomics data to predict gene expression in lung cancer. *BMC Bioinformatics*, 16(S10). https://doi.org/10.1186/1471-2105-16-S5-S10

Liao, J. G., & Chin, K.-V. (2007). Logistic regression for disease classification using microarray data: Model selection in a large *p* and small *n* case. *Bioinformatics*, 23(15), 1945–1951. https://doi.org/10.1093/bioinformatics/btm287

Love, M. I., Huber, W., & Anders, S. (2014). Moderated estimation of fold change and dispersion for RNA-Seq data with DESeq2. *Genome Biology*, 15(12), 550.

Pages, H., Aboyoun, P., Gentleman, R., & DebRoy, S. (2009). *String objects representing biological sequences, and matching algorithms*. R package version 2.38.4.

Peng, H., Long, F., & Ding, C. (2005). Feature selection based on mutual information: Criteria of max-dependency, max-relevance, and min-redundancy. *IEEE Transactions on Pattern Analysis and Machine Intelligence*, 27, 1226–1238. https://doi.org/10.1109/TPAMI.2005.159

Priyadarsini, R. P., Valarmathi, M. L., & Sivakumari, S. (2011). Gain ratio-based feature selection method for privacy preservation. *ICTACT Journal on Soft Computing*, 1(4), 201–205.

Quinlan, A. R., & Hall, I. M. (2010). BEDTools: A flexible suite of utilities for comparing genomic features. *Bioinformatics*, 26(6), 841–842. https://doi.org/10.1093/bioinformatics/btq033

Ray, S., Business Analytics. (2017, September). *6 Easy Steps to Learn Naive Bayes Algorithm (with code in Python)*. Retrieved from www.analyticsvidhya.com/blog/2017/09/naive-bayes-explained/

Refaeilzadeh, P., Tang, L., & Liu, H. (2009). Cross-validation. In Liu, L., & Özsu, M. T. (eds.), *Encyclopedia of Database Systems*. Boston, MA: Springer.

Risch, A., & Plass, C. (2008). Lung cancer epigenetics and genetics. *International Journal of Cancer*, 123(1), 1–7.

Russo, T. A., Jodush, S. T., Brown, J. J., & Johnson, J. R. (1996). Identification of two previously unrecognized genes (guaA and argC) important for uropathogenesis. *Molecular Microbiology*, 22(2), 217–229.

Sadakierska-Chudy, A., & Filip, M. (2015). A comprehensive view of the epigenetic landscape. Part II: Histone post-translational modification, nucleosome level, and chromatin regulation by ncRNAs. *Neurotoxicity Research*, 27(2), 172–197.

Saeys, Y., Inza, I., & Larrañaga, P. (2007). A review of feature selection techniques in bioinformatics. *Bioinformatics*, 23(19), 2507–2517.

Sarica, A., Cerasa, A., & Quattrone, A. (2017). Random forest algorithm for the classification of neuroimaging data in Alzheimer's disease: A systematic review. *Frontiers in Aging Neuroscience*, 9, 329. https://doi.org/10.3389/fnagi.2017.00329

Shen, M. M., & Abate-Shen, C. (2010). Molecular genetics of prostate cancer: New prospects for old challenges. *Genes & Development*, 24(18), 1967–2000.

Siepel, A., Bejerano, G., Pedersen, J. S., Hinrichs, A. S., Hou, M., Rosenbloom, K., Clawson, H., Spieth, J., Hillier, L. W., & Richards, S. (2005). Evolutionarily conserved elements in vertebrate, insect, worm, and yeast genomes. *Genome Research*, 15(8), 1034–1050. https://doi.org/10.1101/gr.3715005

Sperandei, S. (2014). Understanding logistic regression analysis. *Biochemia Medica (Zagreb)*, 24(1), 12–18. https://doi.org/10.11613/BM.2014.003

Stephens, Z. D., Lee, S. Y., Faghri, F., Campbell, R. H., Zhai, C., Efron, M. J., & Robinson, G. E. (2015). Big data: Astronomical or genomical? *PLoS Biology*, 13(7), e1002195.

Urbanowicz, R. J., Meeker, M., La Cava, W., Olson, R. S., & Moore, J. H. (2018). Relief-based feature selection: Introduction and review. *Journal of Biomedical Informatics*, 85, 189–203. https://doi.org/10.1016/j.jbi.2018.07.014

Weka Software. www.cs.waikato.ac.nz/ml/weka/index.html

# 10 Genetic Study of Schizophrenia and Role of Computational Genomics in Mental Healthcare

*Namrata Jawanjal and Indranath Chatterjee*

**CONTENTS**



DOI: 10.1201/9781003142751-13

## 10.1 INTRODUCTION

Schizophrenia is a brain disorder that affects a small percentage of the U.S. population. Signs and symptoms can include delusions, hallucinations, disorganized speech, difficulty with questions, and a lack of motivation when schizophrenia is active (Chatterjee et al., 2018). Nevertheless, unfortunately, there is no cure for schizophrenia. Artificial intelligence (AI) systems are well adapted for the resolution and definition of clinical diagnostic tasks useful in medical genetics and genomics, genome annotation and variant classification, and phenotype-to-genotype correspondence.

### 10.1.1 Types and Symptoms

#### 10.1.1.1 Paranoid Schizophrenia

Symptoms are included like hallucinations and/or delusions, but in this type of schizophrenia, one's speaking ability and emotions may not be affected (Chatterjee & Mittal, 2020).

#### 10.1.1.2 Hebephrenic Schizophrenia

Symptoms include disorganized speaking patterns, behaviors, and thoughts, alongside short-lasting delusions and hallucinations.

#### 10.1.1.3 Catatonic Schizophrenia

In this type of schizophrenia, a person behaves like being very active but still, not talking and imitating others' voices and gestures.

#### 10.1.1.4 Non-Differentiated Schizophrenia

Undifferentiated schizophrenia displays specific anxiety symptoms and hebephrenic or catatonic schizophrenia, but it does not fall into either of these forms alone.

#### 10.1.1.5 Residual Schizophrenia

It has some negative symptoms (slow movement, poor memory, a lack of concentration, and poor hygiene).

#### 10.1.1.6 Simple Schizophrenia

It is rarely diagnosed in the United Kingdom.

#### 10.1.1.7 Cenesthopathic Schizophrenia

People have unusual bodily sensations in this type of schizophrenia.

#### 10.1.1.8 Unspecified Schizophrenia

Symptoms of unspecified schizophrenia meet the general conditions for a diagnosis but do not fit into any preceding categories.

### 10.1.2 Evolution of Schizophrenia and Its Occurrence

Recently, Banerjee et al. (2018) explained that the presence of methylated regions varies between us. Primates and Denisova hominins are abundant with single nucleotide polymorphisms (SNPs) believed to be associated with the disorder. The physical health problems of schizophrenia patients are common, give excess chances of mortality rate, and decrease the quality of life (QoL). Meanwhile, we can say that the physical health of schizophrenia patients is poor (von Hausswolff-Juhlin et al., 2009). According to researchers, it affects both men and women, with schizophrenia in men presenting between 15 and 24 years while in women between 55 to 64 years (Chatterjee et al., 2020b). There are two terms early-onset schizophrenia (EOS) and childhood-onset schizophrenia (COS). In EOS, it occurs before the age of 18 years and in COS before the age of 13 years.

#### 10.1.2.1 Cardiovascular Health

Weight gain and obesity: In schizophrenia patients, this condition is ordinary. According to the survey, schizophrenia patients have a 42% body mass index (BMI) while normal persons have a 27% BMI. In the last decade, obesity was considered a vital physical concern in these patients. In the development of diabetes, metabolic syndrome, and cardiovascular disease, this factor is essential. Patients with a low diet, with increased fat and reduced fiber, are eating too many calories and lacking exercise. It can be a significant problem in potential weight gain for schizophrenia patients. According to the National Cholesterol Education Program (NCEP)–ATPIII description, it becomes a metabolic syndrome. Abdominal obesity includes type 2 diabetes, an abnormal amount of lipids, and hypertension, leading to uterine, prostate gland, kidney, and gallbladder cancer (Kurzthaler & Fleischhacker, 2001).

Metabolic syndrome: It increases the risk of mortality. According to the NCEP-ATPIII, it involves abdominal obesity, high triglycerides, low levels of high-density lipoprotein (HDL), high blood pressure, and high fasting blood glucose.

#### 10.1.2.2 Lifestyle Factors

##### *10.1.2.2.1 Exercise: Lack of Exercise*

Smoking: Of people with schizophrenia, 75–92% are smokers. They can smoke heavily; it increases the percentage of nicotine in the bloodstream (Connolly & Kelly, 2005). It includes deeper inhalation or genetic factors linking with schizophrenia

(Olincy et al., 1997). Heavy smoking reduces the plasma concentrations of any antipsychotic medications.

Diet: Factors are as discussed in cardiovascular health.
Alcohol: The hereditary pattern influences the dopamine system.
Effects related to antipsychotics: Side effects are related to cardiovascular and physical health.

### 10.1.3 Parameters Influencing the Growth of the Disorder

Some risk factors influence the growth of a schizophrenia-like combination of genetic, physical, psychological, and environmental factors:

- Genetics: Schizophrenia is hereditary, but sometimes no single gene is responsible. This means that a different combination of genes in the patient does not mean they will develop schizophrenia. Identical twins share the same genes, and schizophrenia is partly inherited. An identical twin has a 1 in 2 chance of being diagnosed with schizophrenia while in non-identical twin has a 1 in 8 chance and the general population 1 in 100 chance.
- Neurotransmitters: Neurotransmitters are a type of chemicals that help carry messages between brain cells. Schizophrenia is caused by the change in two neurotransmitters like dopamine and serotonin.
- Pregnancy and birth complications such as low weight at birth, premature labor, and a lack of oxygen (asphyxia) during birth can be factors.
- Stressful life events like loneliness, out of work, financial problems, homeless, and/or losing someone close to you contribute to schizophrenia.
- Drug: While no drug directly causes schizophrenia, research shows that drug addiction increases the disorder's risk of occurrence. Thus, drug abuse cannot be wiped out as a potential factor in the causation of the disorder.

## 10.2 WHAT IS A GENE?

Genes are nothing but physical and functional units of heredity, elements of heredity transmitted from parents to offspring in reproduction.

DNA is an inheritance molecule, and a gene is a nucleotide sequence present in the DNA or RNA. For making molecules called proteins, some genes are listed. Several genes do not code proteins. The size of genes ranges from a few hundred to more than 2 million DNA bases. The human genome project has worked to determine the human genome sequence and to classify the genes. Humans have about 20,000 to 25,000 genes. Each gene has two copies in every individual, from which one is inherited from each of the parents. In all individuals, most genes are the same (<1% of the total).

### 10.2.1 Relationship between Genes and Schizophrenia

Schizophrenia is phenotypic heterogeneity divided into positive, hostile, and cognitive symptoms (Arlington & American Psychiatric Association, 2013). Positive symptoms include hallucinations and delusions. In a patient with schizophrenia, adverse selection is more prominent than in children in the general population. Defects in mindfulness, focus, selective attention, and critical analysis can be seen in cognitive functions. This disease heritability is calculated as greater than 80% (McGue & Gottesman, 1991). Representing the nonheritable fraction of the risk of developing schizophrenia are a diverse array of environmental factors. These include migrant status, urban environment (Vassos et al., 2012), maternal malnutrition, and birth month (Davies et al., 2003) all play an important role. One example of environmental factors with genetic variation is at CTNNA3 gene replication of cytomegalovirus infection has been reported (Avramopoulos et al., 2015). Genome-wide association studies (GWASs) support highly polygenic architecture and, due to adverse selection, common schizophrenia alleles show low odd ratios.

### 10.2.2 Importance of Studying the Genetic Variability and Its Link to Schizophrenia

1. **High-Penetrance Genetic Variation:**
   a. **Chromosomal abnormalities and copy number variation:** Schizophrenia and deletion (chromosome band 22q11.2) syndrome (Velo-cardio-facial syndrome) have received increased attention, and a strong link between them has been shown but has been unable to provide the desired result (Chow et al., 1994). It is crucial to study the genetics of the disease schizophrenia. In autism, gains and losses of genetic material are observed. In some unique cases, the rearrangement of the chromosome on the DISC1 locus has been linked to schizophrenia. There is a genetic link between the disease and translocation of two genes in the Scottish pedigree, chromosome 1 DISC1 and chromosome 11 DISC2.
   b. **Rare and de novo variation:** Gene SETD1A demonstrated by Takata et al. (2016) in this gene found two loss-of-function and one synonymous de novo mutation. Transcriptional co-expression and protein interaction networks are involved in synaptic transmission, neuronal migration, cellular transport, signaling, and transcriptional regulation. Rare, segregating, and potentially functional variants were explained in sequencing. These will be the significant risk factors for schizophrenia (Gulsuner et al., 2013).
2. **Common Low-Penetrance Variants and GWASs:**
   GWASs are a more convenient way of mapping the genes of schizophrenia than are genome-wide linkage studies (Risch & Merikangas, 1996). Genome-wide association study reported three genomic loci: human leukocyte antigen (HLA) on chromosome 6, TCF4, and NRGN genes on 18 and 11 chromosomes, respectively (Stefansson et al., 2009). The International Schizophrenia Consortium has also described some genetic overlap with bipolar disorder (BD) and HLA association (Consortium, 2009). Scientists found enrichment in brain tissue enhancers and those tissues that play an essential role in immunity (CD19 and CD20 B cells). In 2007, the Psychiatric Genomics Consortium (PGC) was established.

There are some Cross-Disorder Group of the PGC, and they overlapped across five disorders like schizophrenia, autism, attention-deficit hyperactivity disorder, blood pressure (BP), and major depressive disorder (MDD), showing the strongest overlaps among schizophrenia, BP, and MDD (Lee et al., 2013).

### 10.2.3 The Approaches for Computational Genomics

Computational genomics, also known as computational genetics, is a branch of science dealing with computational and statistical analysis to understand gene data's internal biology. The gene data may include both DNA and RNA. In addition to the computational and statistical approaches, it is also pursued to understand genes' roles from the post-genomic data. This domain is a subdomain of bioinformatics under the area of computational biology. However, it relies entirely on studying genes and their structures to comprehend the biology principles from the molecular level of DNA of a species. Recently, plenty of massive biological datasets have become available, which boosts computational studies and subsequently popularizing the study of computational genomics. Rapid growing methods and research in microarray and sequencing are noticeable advanced computational approaches in genomics. With biomedical help, the results make it possible to improve diagnosis and care through genetic and genomic studies. The application of computational genomics is to diagnose and manage human disorders (Wei et al., 2015). A mathematical algorithm (FARMS) was explored to specify the outcome from optimal prediction and high-dimensional dataset of more than 800 individuals and proved this method is more systematic than other methods—a method (low-rank and sparse decomposition, or LRSDec) to identify gene modules and genetic interactions between them. We

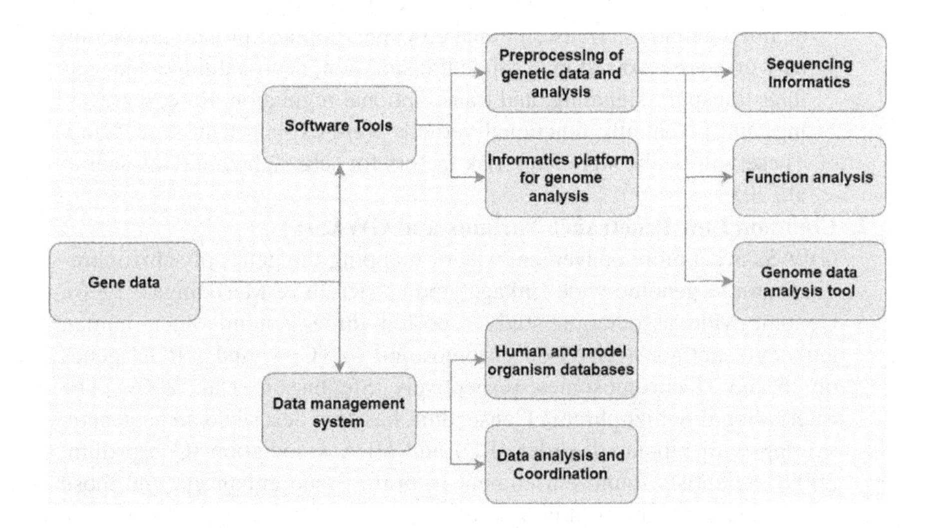


**FIGURE 10.1** Sample Stepwise Approaches for Computational Genomics.

can use genetic interaction widely in data analysis and image processing (Wang et al., 2015). Polycystic ovary syndrome in women with the association between follicle-stimulating hormone receptor (FSHR) polymorphisms and polycystic ovaries' anatomy (Du et al., 2015). The use of quantitative genetics by genotypic data and determine the sample size impact, trait, and amount of SNPs to check the result's accuracy (de Vlaming & Groenen, 2015).

### 10.2.4 State-of-the-Art AI and Machine Learning Techniques Used in Computational Biology

A program powered by artificial intelligence can help promote research into big data analytics among scientists without specialized expertise, specifically in computation biology. With the invention of vast amounts of datasets being available from different research labs worldwide, big data analytics plays a pivotal role (Gautam & Chatterjee, 2020). Machine learning is a branch of computer science knowledge in that we can learn computational methods from data (Kumar & Chatterjee, 2016; Chatterjee, 2018). There are two kinds of machine learning methods: supervised and unsupervised. The supervised learning approach describes the collection of objects with a classmark called the training set. Unsupervised learning methods define that it does not depend on prior knowledge availability (Bhattacharyya & Kalita, 2013). In bioinformatics, machine learning methods are monotonous and equivalent, network security, healthcare, banking and finance, and transportation, machine learning is a potent and applicable technique (Chatterjee et al., 2019).

- "DeepGestalt": The advantages of this technique are the analysis and processing of images and deep learning algorithms, suggesting a genetic syndrome (Gurovich et al., 2018).
- Machine learning in genetics and genomics is to handle complex genomic data with the algorithm's help.
- FDNA is a Boston-based company. A smartphone-based research framework has been developed, Face2Gene, for facial images to categorize typical facial attributes of people with congenital and neurodevelopment disorders. This tool's use assists medical genetics in clinical and laboratory practices and cancer diagnoses and in the sequencing process can help improve accuracy, probe design, and predict DNA binding rates from sequence data and also helps in early detection of infections or diseases (Gurovich et al., 2018).
- Big data is used to reduce the cost of input–outputs and to maximize monotonous processing.
- High-performance technology, such as fast data capture tools and very high-resolution satellite data recording tools, is used in biomedical research.
- There are five types of big data in bioinformatics: (a) data on gene expression; (b) data on DNA, RNA, and protein sequence; (c) data on protein–protein interaction (PPI); (d) data on pathways; and (e) ontology of the gene.
- RNA sequencing: To use more accurate and quantitative gene expression measurements.

There are some machine learning properties to analyze big data (Kashyap et al., 2015):

a. Scalable to high volume: It can handle a large chunk of data.
b. High velocity robust: A method with low time complexity that can ingest and process stream data.
c. Transparent to diversity: In principle, big data may be semistructured or unstructured and handle data from various sources.
d. Incremental: It can manage whole databases and information at the minimum cost.
e. Distributed: This process is dependent on partial data and the fusion of partial outcomes.

### 10.2.5 Recent Advancement in Schizophrenia for Computational Genomics

There are varieties of symptoms observed in schizophrenia using Bayesian models, also called a psychopathology model. As we know, Bayesian models are widely used as a tool for understanding the cognitive capacity of the human brain. It is also used to compare and determine the suitable model for the selection of psychopathological analysis. Other than these, there are models involving abnormalities of neuromodulation and/or imbalance in receptors. This model acts as a connectionist and works in the neural network process. We can analyze understanding behavioral versus neurobiological abnormalities (Valton et al., 2017). Research tools like computational and mathematical models showed that it is beneficial in neural computational exploration, and the interconnection between neural systems and function were investigated three different analytical complementary levels (Montague et al., 2012). The level of computation ("What" does the brain measure and "why"?), the level of an algorithm ("Which" representation and algorithms will explain these computations?), and the physical level ("How" are these algorithms neutrally implemented?) (Dayan & Abbott, 2005). Two separate forms of models: top-down and bottom-up. Top-down models are related to computational levels, while bottom-up models are related to the physical substrate (Colombo & Seriès, 2012). A recent computational psychiatric study enlarged the convergence between computational neuroscience and psychiatry (Huys et al., 2016, 2011; Montague et al., 2012).

### 10.2.6 Importance of the Study of Computation Genomics in Schizophrenia

Schizophrenia is a long-lasting mental illness requiring care and significant advances. Worldwide, schizophrenia patients are between 15 and 20 per 100,000 per year. A patient suffers from chronicity for their whole life. That is why it is crucial to study and discuss associated genes, common elements, differences, and cooperation with the environment. It will help in the prevention, care, and management of individual patients. We study the genes involved in normal brain function, how they play a role, and the extent to which the brain grows. For schizophrenia, we explain which genes affect normal brain function and turn into mental disorders like schizophrenia, their functions, characteristics, and how they affect a person's lifestyle and genetic structure. We already discussed the study of computational genomics in schizophrenia.

## 10.3 GENETIC STUDY OF SCHIZOPHRENIA BRAIN

### 10.3.1 Genes Involved in Average Human Brain Growth

Although the total number of genes involved in normal brain growth and function is still uncertain, here, we explain a few genes and their functions that are probably capable of altering brain behavior in normal human beings. Around the age of 20 or 21, vigorous changes continuously occur in every individual's brain structure. In average brain growth, it is essential to understand which genes are expressed.

#### 10.3.1.1 Role of Genes in Normal Human Brain Growth

**Apolipoprotein E (apoE):** Astrocytes and microglia are responsible for producing the brain's apoE gene. When the excitotoxic injury occurs, neurons express the apoE gene. Cholesterol and phospholipid movement between cells is promoted by apoE (Huang & Mahley, 2014).

**Growth hormone 1 (GH1):** The development of GH1 (growth hormone 1) in a brain region called the hypothalamus. It is developed in the somatotropic growth-stimulating cells of the pituitary gland. This gene promotes the development and maturation of cells (Mullis, 2005).

**Brain-derived neurotrophic factor (BDNF):** Brain-derived neurotrophic factor (BDNF) offers the instructions for producing a protein called BDNF that is generally found in the brain and spinal cord. By playing a role in the growth, maturation (differentiation), and maintenance of these cells, this protein promotes nerve cells' survival (neurons). This protein is active in the brain at the junctions between nerve cells (synapses) where cell-to-cell contact occurs. In reaction to the experience, the synapses will alter and adapt over time, a synaptic plasticity trait. The BDNF protein, which is essential for learning and memory, helps regulate synaptic plasticity. In regions of the brain that regulate food, drinking, and body weight, the BDNF protein is found; the protein likely contributes to controlling those functions (Sears et al., 2011).

**Notch homolog 2 N-terminal-like (NOTCH2NL):** Three members of the NOTCH2NL family proteins could have been included in developing the big cortex of humans (Suzuki et al., 2018).

**Prune exopolyphosphatase (PRUNE):** PRUNE is necessary for normal human cortical growth. It is a member of the phosphodiesterase protein DHH (Asp-His-His) superfamily of molecules essential for cell motility and cancer progression (Zollo et al., 2017).

**Homeobox gene:** In a temporally and spatially limited way, homeobox genes' expression is regulated, and homeoproteins serve as a critical factor in defining and sustaining cell types. In developing and adult brains, both in the hypothalamus and pituitary gland, various homeobox genes are expressed (Vollmer & Clerc, 1998).

**TABLE 10.1**
**Genes Involved in Normal Brain Function**

| Sr. no | Genes | |
|---|---|---|
| 1 | apoE | Apolipoprotein E |
| 2 | GH1 | Growth hormone 1 |
| 3 | BDNF | Brain-derived neurotrophic factor |
| 4 | NOTCH2NL | Notch homolog 2 N-terminal-like |
| 5 | PRUNE | Prune exopolyphosphatase |
| 6 | Homeobox gene | Homeobox gene |

#### 10.3.1.2 Role of Genetic and Environmental Factors for Structural Brain Growth in Normal Human Brain

Brain growth arises from the correlation between genetics and environmental factors. A pair of monozygotic twins shared their genetic material 100%, while dizygotic (DZ) twins 50%. As compared to DZ twins, their trait variation influences genetic factors, while DZ twins showed their effect on shared environment (Neale & Cardon, 2013). In some areas, such as the frontal and superior temporal lobes, the disparity between cortical thickness indicates a significant genetic contribution, whereas environmental factors are noticeable in other regions. (In, 2010; Rimol et al., 2001; Thompson et al., 2001). The study observed some stronger environmental influences in topological features such as gyrification (White et al., 2002). Genetic influences over childhood and adolescence were enhanced in both gray and white matter quantities. However, white matter volume is decreased in environmental variance, while in the gray matter, it increased, showed increased heritability in white matter, and decreased in gray matter (Chatterjee et al., 2020a). Changes in heritability of cortical thickness differ by brain areas as age increases. We may determine related or environmental variables with multivariate tests, which leads to a brain structure and cognitive phenotype. Mechanisms of increasing genetic contribution and development are not well understood. The study of Stead et al. (2006) and Sun et al. in 2006 proved that animal brain gene expression is well established. Dopamine transporter–linked genes, BDNF, its receptor, tyrosine kinase B (trkB), and some growing gene expression changes in the human brain are being recorded (Webster et al., 2006; Weickert et al., 2007). Some changes in heritability are because of the changing interactivity of genes and the environment. There are three different processes: passive, active, and evocative of gene–environmental interrelation. In the passive type, the offspring's genes and environment correspond to each other due to the parents' genes and shape their climate. In the second gene–environmental climate correlation, the child creates an environment by choosing their activities. Evocative genotype correlation occurs when the child's genetically induced traits induce unique responses in others that again affect the child's environment. As the child's age increases, heritability and the ability to choose their environment increase, as proposed by Scarr and McCartney in 1983 (McGue 2010; Rutter 2007).

### 10.3.2 Role of Genes in Schizophrenia

Ruth Whelan et al. (2018) confirmed 108 genetic loci interrelated with schizophrenia with a GWAS. Among 108 genes, we have listed only 42 genes in Table 10.2.

**TABLE 10.2**
**Genes Involved in Schizophrenia Disorder**

| Sr.no. | Genes | |
|---|---|---|
| 1 | apoE | Apolipoprotein E |
| 2 | BDNF | Brain-derived neurotrophic factor, or abrineurin |
| 3 | CHRNA7 | Neuronal acetylcholine receptor subunit alpha-7 or nAChRα7 |
| 4 | COMT | Catechol-O-methyltransferase |
| 5 | DISC1 | Disrupted in schizophrenia 1 |
| 6 | DRD2 | Dopamine receptor D2 or D2R |
| 7 | HTR2A | 5-Hydroxytryptamine Receptor 2A |
| 8 | NRG1 | Neurolregulin 1 |
| 9 | DAO | D-Amino Acid Oxidase |
| 10 | DAOA | D-Amino Acid Oxidase Activator |
| 11 | DTNBP1 | Dystrobrevin-binding protein 1 or dysbindin |
| 12 | RGS4 | Regulator of G protein signaling 4 |
| 13 | AKT1 | AKT serine/threonine kinase 1 |
| 14 | DRD3 | Dopamine D3 receptor or D3R |
| 15 | DRD4 | Dopamine D4 receptor or D4R |
| 16 | GRM3 | Glutamate metabotropic receptor 3 |
| 17 | KCNN3 | Potassium calcium–activated channel subfamily N member 3 |
| 18 | MTHFR | Methylenetetrahydrofolate reductase |
| 19 | NOTCH4 | Notch Receptor 4 |
| 20 | PPP3CC | Protein phosphatase 3 catalytic subunit gamma |
| 21 | PRODH | Proline dehydrogenase |
| 22 | SLC6A3 | Solute carrier family 6 member 3 |
| 23 | SLC6A4 | Solute carrier family 6 member 4 |
| 24 | TNF | Tumor necrosis factor |
| 25 | ZDHHC8 | Zinc finger DHHC-type palmitoyltransferase 8 |
| 26 | TAAR6 | Trace amine receptor 4 |
| 27 | CAPON | Carboxyl-terminal PDZ ligand of neuronal nitric oxide synthase |
| 28 | EPSIN4 | Epsin 4 |
| 29 | GJA5 | Gap junction protein alpha 5 |
| 30 | TCF | Transcription factor |
| 31 | NRGRN | Neugrin, neurite outgrowth associated |
| 32 | TPH1 | Tryptophan hydroxylase 1 |
| 33 | AHI1 | Abelson helper integration site 1 |
| 34 | RELN | Reelin |
| 35 | TRKA | Tropomyosin receptor kinase A |
| 36 | NRXN1 | Neurexin 1 |
| 37 | BCL9 | BCL9 transcription coactivator |

*(Continued)*

**TABLE 10.2** (Continued)

| Sr.no. | Genes | |
|---|---|---|
| 38 | CYF1R1 | Cytoplasmic FMR1 interacting protein 1 |
| 39 | GJA8 | Gap junction protein alpha 8 |
| 40 | SNAP29 | Synaptosome associated protein 29 |
| 41 | CTNNA3 | Catenin alpha 3 |
| 42 | TRANK1 | Tetratricopeptide repeat and ankyrin 1 |

#### 10.3.2.1 The Primary Functions and Characteristics of Each Gene Responsible for Schizophrenia

**Proline dehydrogenase (PRODH):** The gene codes of L-Proline, an amino acid that is immediately involved in glutamatergic transmission, metabolize the enzyme. At the 3′ end of the gene, haplotypic overexpression has been observed on the 22q11 locus (Liu et al., 2002; Renick et al., 1999).

**Cathecol-O-methyltransferase (COMT):** Situated on the 22q11 region between two anterior genes. It regulates an enzyme involving breaking down dopamine and an alternative that balances the enzyme activity (in the 158 codons, Val-high and Met-low activity; Karayiorgou & Gogos, 2006).

**ZDHHC8:** The locus is the same 22q11 and is observed in the same LD screen for the PRODH gene (Liu et al., 2002).

**Dystrobrevin-binding protein 1 or dysbindin (DTNBP1):** This is another gene observed with a fine mapping of 6p24–22. Dysbindin is a representation of the dystrophin protein complex and biogenesis of a lysosome-related organelle complex (BLOC; Benson et al., 2001; Li et al., 2003).

**Neurolregulin 1 (NRG1):** Chromosome is located on 8p12–2. There was no stability in the repetition of haplotypes samples and showed essential heterogeneity of linkage disequilibrium (Li et al., 2004; Zhao et al., 2004).

**Disrupted in schizophrenia 1 (DISC1):** Gene discovered in linkage analysis and a 1q422 positional nominee. It is elaborated in cellular functions like cell migration, microtubule, receptor membrane trafficking, neurite outgrowth, mitochondrial activity, and phosphodiesterase signaling (Sawa & Snyder, 2005).

**Trace amine receptor 4 (TAAR6):** It is known as a positional candidate for schizophrenia on the 6q23.2 and 13q32–34 locus. Glutamatergic signaling is caused by the enzymatic activity of the D-amino acid oxidase (DAAO) activity of G72 (Chumakov et al., 2002).

**Carboxyl-terminal PDZ ligand of neuronal nitric oxide synthase (CAPON):** This showed an irregularity of carboxyl-terminal PDZ ligand (Mulle et al., 2005).

**Epsin 4:** The location is on chromosome 5q33, and four haplotypes revealed the presence of LD with schizophrenia (Pimm et al., 2005). Proteins involved in the gene play a function in the transport and stabilization of vesicles of neurotransmitters in synapses.

**Tetratricopeptide repeat and ankyrin 1 (TRANK1):** 3, 3p22.2 is the locus of this gene on the chromosome's short arm. It assists in controlling the immune response mediated by interferon (IFN) and in combating the development of schizophrenia. Due to the activation STAT-JAK pathway in the liver, IFN-α could express upregulation in the TRANK1 gene (Whelan et al., 2018; Irudayam et al., 2015).

Some genes are initiated at serotoninergic (HTR2A, SLC6A4, and TPH1) and dopaminergic (DRD2, DRD3, and DRD4) mechanisms while meta-analysis studies. Another 12 regions of genes have been identified by researchers (2p, 5q, 3p, 11q, 2q, 1q, 22q, 8p, 6p, 20p, 13q, and 14q; Badner et al., 2002). A polymorphism in zinc finger protein 804A gene (ZNF804) was identified in a genome-wide study (O'Donovan et al., 2008). Significant research on the interrelation between Transcription factor 4 (TCF), genes of neurogranin (NRGRN), and the areaof the major histocompatibility complex (MHC; Stefansson et al., 2009). Luo et al. (2014) studied genetic interrelation and PPI, and they observed the genes NRXN1, CHRNA7, BCL9, CYFIP1, GJA8, NDEI, SNAP29, and GJA5 in schizophrenia by disordering CNVs (Copy Number Variants) study.

### 10.3.3 The Effect of Genes Influencing the Growth of Schizophrenia

Proline dehydrogenase gene is located on 22q11 chromosome and consortium in between schizophrenia and hemizygous deletion of the 22q11 locus. In another study of 360 Iranian, there were three polymorphisms (757C/T, 1766A/G, AND 1852G/A) of the PRODH gene that were allied with an increased chance of schizophrenia (Ghasemvand et al., 2015). Catechol-O-methyltransferase shows the high- and low-activity allele and is responsible for increasing the risk of schizophrenia (Shifman et al., 2002; Tsai et al., 2006; Egan et al., 2001; Ho et al., 2005). ZDHHC8 increases the risk of disease 1.5-fold times (Mukai et al., 2004). Talbot et al. (2004) and Weickert et al. (2004) found a decreased level of the DTNBP1 gene increases the risk factor in the dorsolateral prefrontal cortex and hippocampus schizophrenia patients. The DISC1 gene is involved in schizophrenia for allelic heterogeneity (Falola et al., 2017; Hennah et al., 2003; Hodgkinson et al., 2004). The trace amine receptor 4 (TAAR6) is a positional candidate for schizophrenia on the 6q23.2 and 13q32–34 locus. Glutamatergic signaling is caused by the enzymatic activity of the D-amino acid oxidase (DAAO) activity of G72 (Chumakov et al., 2002). In the Epsin 4 gene, researchers (Pimm et al., 2005) found some evidence of LD with schizophrenia. Genes like AHI1, MTHFR, RELN, and TRKA directly affect neurodevelopment. In the human genome, some linked genes for schizophrenia were observed (Badner & Gershon, 2002).

## 10.4 TREATMENT AND PSYCHOSOCIAL INTERVENTION OF SCHIZOPHRENIA

As far we have discovered in the paper that schizophrenia can be caused due to transmission of genetic influence in the form of prenatal or perinatal insult. It is evident that only psychosocial intervention is not an effective way of treating schizophrenia. Clinical intervention is necessary (Chatterjee & Mittal, 2020).

The treatment process in this disorder can be broadly divided into few types:

1. Pharmacological therapy
2. Electroconvulsive therapy
3. Non-pharmacological therapy (Chatterjee & Mittal, 2020)
   a. Individual therapy
      i. Counseling
      ii. Personal therapy
   b. Group level
      i. Social therapy
   c. Cognitive therapy
      i. Cognitive behavioral therapy
      ii. Cognitive enhancement therapy

Due to social taboo and exclusion criteria for patients with mental disorders, most schizophrenia patients remain untreated. This disorder's mortality rate is not so much due to the illness but instead due to the high suicidal tendency and social exclusion impact. However, good clinical intervention and social therapy can pose impactful treatment mechanisms for the disorder.

## 10.5 DISCUSSION

We can evaluate millions of SNPs and the genetic concept of schizophrenia with the help of a GWAS. From the earliest observations of twin pairs (meta-analyses of a GWAS), we can go through testing the genetic fabric of schizophrenia. CNVs are nothing but structural genomic variants related to duplications, deletions, insertions, and translocations. There are three studies on next-generation sequencing. Gulsuner et al. (2013) studied genes that are after sequencing 105 patients with schizophrenia, participants in fetal prefrontal cortex neurogenesis were observed. Fromer et al. (2014) studied genetic overlapping of neurodevelopmental disorders like schizophrenia, autism, and mental retardation. Purcell et al. (2014) were observed the existence of CNVs in ARC (activity-regulated cytoskeleton), NMDAR, fragile X mental retardation (FMRP)—complexes of protein–protein—and calcium channels. Grey matter volume in the dorsolateral prefrontal cortex follows an inverted-U developmental path while white matter volumes increase. Electroencephalographic, functional magnetic resonance imaging, postmortem, and neuropsychological changes are the anatomic changes that increase connectivity in brain development. Several neuroscience concepts are related to "connectivity". During adolescence, all types of connectivity increase. There are two conditions of psychological tests: "cold cognition"—hypothetical low-emotion scenarios—and "hot cognition"—strong anticipation of peer pressure and real implications. Hallmark is defined as the brain's ability to adapt and change according to environmental demands (Giedd & Rapoport, 2010). Against peptide antigens, schizophrenia patients had IgG antibodies, and schizophrenia-associated genes encode that. Schizophrenia patients showed an increased level of IgG in 6 of 18 antigens. Increased antibodies contribute to an immune system failure, causing an autoimmune response (Goldsmith & Rogers, 2008). According to guidelines for physical monitoring in schizophrenia, the patient should undergo regular check-ups like blood pressure, cardiovascular status, weight, blood glucose/diabetes, lipid profiles, and drug-related adverse-event monitoring. Schizophrenia patients need desegregated care.

The current issues and challenges are searching and getting all information on 108 genes related to schizophrenia disorder. The complexity of the phenotype, heterogeneity, and lack of biological marker are the reasons for difficulty finding genes. There are some association studies that have flaws. First, put light on a large number of candidate genes. Second, cases and control are equal to the genetic background of the population. Third, the associated decision between two probabilities is difficult: (a) a risk for disease collision by the marker alleles and (b) a marker allele, which is near to a genetic variant is a fundamental determinant. Some inherent problems are also found in schizophrenia disorder, such as (a) a magnetic effect of the gene, (b) the population linkage signals from the power of magnitude, and (c) small size samples having genes with mild effects. It has doubts while using the linkage method, the use of some new informative marker systems will help scientists localize particular genes with their effects. Multiple candidate loci and genes have been targeted by genetic linkage and association research, which has failed to specify specific gene variants. A combination of genes has some specific role in causing schizophrenia. There are 108 genes responsible to cause schizophrenia. But still some genetic variations have limited evidences.

## 10.6 CONCLUSION AND SCOPE OF FUTURE WORK

The PGC sheds light on all genetic heterogeneity and 80% heritability reported by Bulik-Sullivan et al. (2015). As we have focused on a genetic study in this chapter, human genome sequencing is a challenging field in the genetics of schizophrenia. The study of cerebral mechanisms might be showing some genetic faults in psychosis or mental disorders like schizophrenia. For future studies, we have to collect specific information related to the human schizophrenia brain. Future studies of phenotypic assessment could examine its relationship with genetic liability and heterogeneity. The whole method generates a full sequence map with markers. The markers are incontrovertibly mapped; discovering and mapping the remaining novel genes have contributed to the Human Genome Project's advanced development of any biological marker related to disorders such as neurocognitive dysfunction, brain dysmorphology, and neurochemical anomalies. The scope of future studies on psychiatric disorders such as schizophrenia will be in molecular biology, cognitive science, brain imaging, and new insights on the origin of disorders and brain function.

## REFERENCES

Arlington, V. A., & American Psychiatric Association. (2013). Diagnostic and statistical manual of mental disorders. *American Psychiatric Association*, *5*, 612–613.

Avramopoulos, D., Pearce, B. D., McGrath, J., Wolyniec, P., Wang, R., Eckart, N., . . . & Coneely, K. (2015). Infection and inflammation in schizophrenia and bipolar disorder: A genome wide study for interactions with genetic variation. *PLoS ONE*, *10*(3), e0116696.

Badner, J. A., & Gershon, E. S. (2002). Meta-analysis of whole-genome linkage scans of bipolar disorder and schizophrenia. *Molecular Psychiatry, 7*(4), 405–411.

Banerjee, N., Polushina, T., Bettella, F., Giddaluru, S., Steen, V. M., Andreassen, O. A., & Le Hellard, S. (2018). Recently evolved human-specific methylated regions are enriched in schizophrenia signals. *BMC Evolutionary Biology, 18*(1), 1–11.

Benson, M. A., Newey, S. E., Martin-Rendon, E., Hawkes, R., & Blake, D. J. (2001). Dysbindin, a novel coiled-coil-containing protein that interacts with the dystrobrevins in muscle and brain. *Journal of Biological Chemistry, 276*(26), 24232–24241.

Bhattacharyya, D. K., & Kalita, J. K. (2013). *Network anomaly detection: A machine learning perspective.* CRC Press.

Bulik-Sullivan, B., Finucane, H. K., Anttila, V., Gusev, A., Day, F. R., Loh, P.-R., Duncan, L., Perry, J. R. B., Patterson, N., Robinson, E. B., Daly, M. J., Price, A. L., & Neale, B. M. (2015). An atlas of genetic correlations across human diseases and traits ReproGen Consortium 8, Psychiatric Genomics Consortium 8, Genetic Consortium for Anorexia Nervosa of the Wellcome Trust Case Control Consortium 3 HHS Public Access. *Nature Genetics, 47*(11), 1236–1241. https://doi.org/10.1038/ng.3406.An

Chatterjee, I. (2018). Mean deviation based identification of activated voxels from time-series fMRI data of schizophrenia patients. *F1000Research, 7*(1615). https://doi.org/10.12688/f1000research.16405.2

Chatterjee, I., Agarwal, M., Rana, B., Lakhyani, N., & Kumar, N. (2018). Bi-objective approach for computer-aided diagnosis of schizophrenia patients using fMRI data. *Multimedia Tools and Applications, 77*(20), 26991–27015. https://doi.org/10.1007/s11042-018-5901-0

Chatterjee, I., Kumar, V., Rana, B., Agarwal, M., & Kumar, N. (2020a). Identification of changes in grey matter volume using an evolutionary approach: An MRI study of schizophrenia. *Multimedia Systems, 26*, 383–396. https://doi.org/10.1007/s00530-020-00649-6

Chatterjee, I., Kumar, V., Rana, B., Agarwal, M., & Kumar, N. (2020b). Impact of ageing on the brain regions of the schizophrenia patients: An fMRI study using evolutionary approach. *Multimedia Tools and Applications, 79*(33), 24757–24779. https://doi.org/10.1007/s11042-020-09183-z

Chatterjee, I., Kumar, V., Sharma, S., Dhingra, D., Rana, B., Agarwal, M., & Kumar, N. (2019). Identification of brain regions associated with working memory deficit in schizophrenia. *F1000Research, 8*(124). https://doi.org/10.12688/f1000research.17731.1

Chatterjee, I., & Mittal, K. (2020). A concise study of schizophrenia and resting-state fMRI data analysis. *Qeios, 414*(599711.2). https://doi.org/10.32388/599711.2

Chow, E. W., Bassett, A. S., & Weksberg, R. (1994). Velo-cardio-facial syndrome and psychotic disorders: Implications for psychiatric genetics. *American Journal of Medical Genetics, 54*(2), 107–112.

Chumakov, I., Blumenfeld, M., Guerassimenko, O., Cavarec, L., Palicio, M., Abderrahim, H., . . . & Puech, A. (2002). Erratum: Genetic and physiological data implicating the new human gene G72 and the gene for D-amino acid oxidase in schizophrenia (Proceedings of the National Academy of Sciences of the United States of America (October 15, 2002) 99: 21 (13675–13680)). *Proceedings of the National Academy of Sciences of the United States of America, 99*(26), 17221.

Colombo, M., & Seriès, P. (2012). Bayes in the brain – on Bayesian modelling in neuroscience. *The British Journal for the Philosophy of Science, 63*(3), 697–723.

Connolly, M., & Kelly, C. (2005). Lifestyle and physical health in schizophrenia. *Advances in Psychiatric Treatment, 11*(2), 125–132.

Consortium, I. S. (2009). Common polygenic variation contributes to risk of schizophrenia that overlaps with bipolar disorder. *Nature, 460*(7256), 748.

Davies, G., Welham, J., Chant, D., Torrey, E. F., & McGrath, J. (2003). A systematic review and meta-analysis of Northern Hemisphere season of birth studies in schizophrenia. *Schizophrenia Bulletin, 29*(3), 587–593.

Dayan, P., & Abbott, L. (2005). *Theoretical neuroscience: Computational and mathematical modelling of neural systems. Computational neuroscience series.* Cambridge, MA: MIT Press, 2001.

de Vlaming, R., & Groenen, P. J. (2015). The current and future use of ridge regression for prediction in quantitative genetics. *BioMed Research International, 2015.* https://doi.org/10.1155/2015/143712

Du, T., Duan, Y., Li, K., Zhao, X., Ni, R., Li, Y., & Yang, D. (2015). Statistical genomic approach identifies association between FSHR polymorphisms and polycystic ovary morphology in women with polycystic ovary syndrome. *BioMed Research International, 2015.* https://doi.org/10.1155/2015/483726

Egan, M. F., Goldberg, T. E., Kolachana, B. S., Callicott, J. H., Mazzanti, C. M., Straub, R. E., . . . & Weinberger, D. R. (2001). Effect of COMT Val108/158 Met genotype on frontal lobe function and risk for schizophrenia. *Proceedings of the National Academy of Sciences, 98*(12), 6917–6922.

Falola, O., Osamor, V. C., Adebiyi, M., & Adebiyi, E. (2017). Analyzing a single nucleotide polymorphism in schizophrenia: A meta-analysis approach. *Neuropsychiatric Disease and Treatment, 13*, 2243.

Fromer, M., Pocklington, A. J., Kavanagh, D. H., Williams, H. J., Dwyer, S., Gormley, P., . . . & Carrera, N. (2014). De novo mutations in schizophrenia implicate synaptic networks. *Nature, 506*(7487), 179–184.

Gautam, A., & Chatterjee, I. (2020). Big data and cloud computing: A critical review. *International Journal of Operations Research and Information Systems (IJORIS), 11*(3), 19–38. https://doi.org/10.4018/IJORIS.2020070102

Ghasemvand, F., Omidinia, E., Salehi, Z., & Rahmanzadeh, S. (2015). Relationship between polymorphisms in the proline dehydrogenase gene and schizophrenia risk. *Genetics and Molecular Research, 14*(4), 11681.

Giedd, J. N., & Rapoport, J. L. (2010). Structural MRI of pediatric brain development: What have we learned and where are we going? *Neuron, 67*(5), 728–734.

Goldsmith, C. A. W., & Rogers, D. P. (2008). The case for autoimmunity in the etiology of schizophrenia. *Pharmacotherapy: The Journal of Human Pharmacology and Drug Therapy, 28*(6), 730–741.

Gulsuner, S., Walsh, T., Watts, A. C., Lee, M. K., Thornton, A. M., Casadei, S., . . . & Calkins, M. E. (2013). Spatial and temporal mapping of de novo mutations in schizophrenia to a fetal prefrontal cortical network. *Cell, 154*(3), 518–529.

Gurovich, Y., Hanani, Y., Bar, O., Fleischer, N., Gelbman, D., Basel-Salmon, L., . . . & Gripp, K. W. (2018). DeepGestalt-Identifying rare genetic syndromes using deep learning. *arXiv preprint arXiv:1801.07637.*

Hennah, W., Varilo, T., Kestilä, M., Paunio, T., Arajärvi, R., Haukka, J., . . . & Meyer, J. (2003). Haplotype transmission analysis provides evidence of association for DISC1 to schizophrenia and suggests sex-dependent effects. *Human Molecular Genetics, 12*(23), 3151–3159.

Ho, B. C., Wassink, T. H., O'leary, D. S., Sheffield, V. C., & Andreasen, N. C. (2005). Catechol-O-methyl transferase Val 158 Met gene polymorphism in schizophrenia: Working memory, frontal lobe MRI morphology and frontal cerebral blood flow. *Molecular Psychiatry, 10*(3), 287–298.

Hodgkinson, C. A., Goldman, D., Jaeger, J., Persaud, S., Kane, J. M., Lipsky, R. H., & Malhotra, A. K. (2004). Disrupted in schizophrenia 1 (DISC1): Association with schizophrenia, schizoaffective disorder, and bipolar disorder. *The American Journal of Human Genetics, 75*(5), 862–872.

Huang, Y., & Mahley, R. W. (2014). Apolipoprotein E: Structure and function in lipid metabolism, neurobiology, and Alzheimer's diseases. *Neurobiology of Disease*, *72*, 3–12.

Huys, Q. J., Maia, T. V., & Frank, M. J. (2016). Computational psychiatry as a bridge from neuroscience to clinical applications. *Nature Neuroscience*, *19*(3), 404.

Huys, Q. J., Moutoussis, M., & Williams, J. (2011). Are computational models of any use to psychiatry? *Neural Networks*, *24*(6), 544–551.

Irudayam, J. I., Contreras, D., Spurka, L., Subramanian, A., Allen, J., Ren, S., . . . & French, S. W. (2015). Characterization of type I interferon pathway during hepatic differentiation of human pluripotent stem cells and hepatitis C virus infection. *Stem Cell Research*, *15*(2), 354–364.

Karayiorgou, M., & Gogos, J. A. (2006). Schizophrenia genetics: Uncovering positional candidate genes. *European Journal of Human Genetics*, *14*(5), 512–519.

Kashyap, H., Ahmed, H. A., Hoque, N., Roy, S., & Bhattacharyya, D. K. (2015). Big data analytics in bioinformatics: A machine learning perspective. *arXiv preprint arXiv:1506.05101*.

Kumar, A., & Chatterjee, I. (2016). Data mining: An experimental approach with WEKA on UCI Dataset. *International Journal of Computer Applications*, *138*(13). https://doi.org/10.5120/ijca2016909050

Kurzthaler, I., & Fleischhacker, W. W. (2001). The clinical implications of weight gain in schizophrenia. *The Journal of Clinical Psychiatry*, *62*(Suppl 7), 32–37.

Lee, S. H., Ripke, S., Neale, B. M., Faraone, S. V., Purcell, S. M., Perlis, R. H., . . . & Absher, D. (2013). Genetic relationship between five psychiatric disorders estimated from genome-wide SNPs. *Nature Genetics*, *45*(9), 984.

Li, T., Stefansson, H., Gudfinnsson, E., Cai, G., Liu, X., Murray, R. M., . . . & Ingason, A. (2004). Identification of a novel neuregulin 1 at-risk haplotype in Han schizophrenia Chinese patients, but no association with the Icelandic/Scottish risk haplotype. *Molecular Psychiatry*, *9*(7), 698–704.

Li, W., Zhang, Q., Oiso, N., Novak, E. K., Gautam, R., O'Brien, E. P., . . . & Jenkins, N. A. (2003). Hermansky-Pudlak syndrome type 7 (HPS-7) results from mutant dysbindin, a member of the biogenesis of lysosome-related organelles complex 1 (BLOC-1). *Nature Genetics*, *35*(1), 84–89.

Liu, H., Abecasis, G. R., Heath, S. C., Knowles, A., Demars, S., Chen, Y. J., . . . & Karayiorgou, M. (2002). Genetic variation in the 22q11 locus and susceptibility to schizophrenia. *Proceedings of the National Academy of Sciences*, *99*(26), 16859–16864.

Liu, H., Heath, S. C., Sobin, C., Roos, J. L., Galke, B. L., Blundell, M. L., . . . & Gogos, J. A. (2002). Genetic variation at the 22q11 PRODH2/DGCR6 locus presents an unusual pattern and increases susceptibility to schizophrenia. *Proceedings of the National Academy of Sciences*, *99*(6), 3717–3722.

Luo, X., Huang, L., Han, L., Luo, Z., Hu, F., Tieu, R., & Gan, L. (2014). Systematic prioritization and integrative analysis of copy number variations in schizophrenia reveal key schizophrenia susceptibility genes. *Schizophrenia Bulletin*, *40*(6), 1285–1299.

McGue, M. (2010). The end of behavioral genetics? *Behavior Genetics*, *40*(3), 284–296.

McGue, M., & Gottesman, I. I. (1991). The genetic epidemiology of schizophrenia and the design of linkage studies. *European Archives of Psychiatry and Clinical Neuroscience*, *240*(3), 174–181.

Montague, P. R., Dolan, R. J., Friston, K. J., & Dayan, P. (2012). Computational psychiatry. *Trends in Cognitive Sciences*, *16*(1), 72–80.

Mukai, J., Liu, H., Burt, R. A., Swor, D. E., Lai, W. S., Karayiorgou, M., & Gogos, J. A. (2004). Evidence that the gene encoding ZDHHC8 contributes to the risk of schizophrenia. *Nature Genetics*, *36*(7), 725–731.

Mulle, J. G., Chowdari, K. V., Nimgaonkar, V., & Chakravarti, A. (2005). No evidence for association to the G72/G30 locus in an independent sample of schizophrenia families. *Molecular Psychiatry*, *10*(5), 431–433.
Mullis, P. E. (2005). Genetic control of growth. *European Journal of Endocrinology*, *152*(1), 11–31.
Neale, M. C. C. L., & Cardon, L. R. (2013). *Methodology for genetic studies of twins and families* (Vol. 67). Amsterdam, Netherlands: Springer. https://10.1007/978-94-015-8018-2.
O'donovan, M. C., Craddock, N., Norton, N., Williams, H., Peirce, T., Moskvina, V., . . . & Dwyer, S. (2008). Identification of loci associated with schizophrenia by genome-wide association and follow-up. *Nature Genetics*, *40*(9), 1053–1055.
Olincy, A., Young, D. A., & Freedman, R. (1997). Increased levels of the nicotine metabolite cotinine in schizophrenic smokers compared to other smokers. *Biological Psychiatry*, *42*(1), 1–5.
Pimm, J., McQuillin, A., Thirumalai, S., Lawrence, J., Quested, D., Bass, N., . . . & Badacsonyi, A. (2005). The Epsin 4 gene on chromosome 5q, which encodes the clathrin-associated protein enthoprotin, is involved in the genetic susceptibility to schizophrenia. *The American Journal of Human Genetics*, *76*(5), 902–907.
Purcell, S. M., Moran, J. L., Fromer, M., Ruderfer, D., Solovieff, N., Roussos, P., . . . & Duncan, L. (2014). A polygenic burden of rare disruptive mutations in schizophrenia. *Nature*, *506*(7487), 185–190.
Renick, S. E., Kleven, D. T., Chan, J., Stenius, K., Milner, T. A., Pickel, V. M., & Fremeau, R. T. (1999). The mammalian brain high-affinity L-proline transporter is enriched preferentially in synaptic vesicles in a subpopulation of excitatory nerve terminals in rat forebrain. *Journal of Neuroscience*, *19*(1), 21–33.
Rimol, L. M., Panizzon, M. S., Fennema-Notestine, C., Eyler, L. T., Fischl, B., Franz, C. E., . . . & Perry, M. E. (2010). Cortical thickness is influenced by regionally specific genetic factors. *Biological Psychiatry*, *67*(5), 493–499.
Risch, N., & Merikangas, K. (1996). The future of genetic studies of complex human diseases. *Science*, *273*(5281), 1516–1517.
Rutter, M. (2007). Gene-environment interdependence. *Developmental Science*, *10*(1), 12–18.
Sawa, A., & Snyder, S. H. (2005). Two genes link two distinct psychoses. *Science*, *310*(5751), 1128–1129.
Scarr, S., & McCartney, K. (1983). How people make their own environments: A theory of genotype→ environment effects. *Child Development*, *54*(2), 424–435.
Sears, C., Markie, D., Olds, R., & Fitches, A. (2011). Evidence of associations between bipolar disorder and the brain-derived neurotrophic factor (BDNF) gene. *Bipolar Disorders*, *13*(7–8), 630–637.
Shifman, S., Bronstein, M., Sternfeld, M., Pisanté-Shalom, A., Lev-Lehman, E., Weizman, A., . . . & Schiffer, R. (2002). A highly significant association between a COMT haplotype and schizophrenia. *The American Journal of Human Genetics*, *71*(6), 1296–1302.
Stead, J. D., Neal, C., Meng, F., Wang, Y., Evans, S., Vazquez, D. M., . . . & Akil, H. (2006). Transcriptional profiling of the developing rat brain reveals that the most dramatic regional differentiation in gene expression occurs postpartum. *Journal of Neuroscience*, *26*(1), 345–353.
Stefansson, H., Ophoff, R. A., Steinberg, S., Andreassen, O. A., Cichon, S., Rujescu, D., . . . & Sigurdsson, E. (2009). Common variants conferring risk of schizophrenia. *Nature*, *460*(7256), 744–747.
Suzuki, I. K., Gacquer, D., Van Heurck, R., Kumar, D., Wojno, M., Bilheu, A., . . . & Detours, V. (2018). Human-specific NOTCH2NL genes expand cortical neurogenesis through Delta/Notch regulation. *Cell*, *173*(6), 1370–1384.

Takata, A., Ionita-Laza, I., Gogos, J. A., Xu, B., & Karayiorgou, M. (2016). De novo synonymous mutations in regulatory elements contribute to the genetic etiology of autism and schizophrenia. *Neuron*, *89*(5), 940–947.

Talbot, K., Eidem, W. L., Tinsley, C. L., Benson, M. A., Thompson, E. W., Smith, R. J., . . . & Blake, D. J. (2004). Dysbindin-1 is reduced in intrinsic, glutamatergic terminals of the hippocampal formation in schizophrenia. *The Journal of Clinical Investigation*, *113*(9), 1353–1363.

Thompson, P. M., Cannon, T. C., Narr, K. L., Van Erp, T., Poutanen, V.-P., Huttunen, M., . . . Toga, A. W. (2001). Genetic influences on brain structure. *Nature Neuroscience*, *4*(12), 1253–1258.

Tsai, S. J., Hong, C. J., Hou, S. J., & Yen, F. C. (2006). Lack of association of catechol-O-methyltransferase gene Val108/158Met polymorphism with schizophrenia: A family-based association study in a Chinese population. *Molecular Psychiatry*, *11*(1), 2–3.

Valton, V., Romaniuk, L., Steele, J. D., Lawrie, S., & Seriès, P. (2017). Comprehensive review: Computational modelling of schizophrenia. *Neuroscience & Biobehavioral Reviews*, *83*, 631–646.

Vassos, E., Pedersen, C. B., Murray, R. M., Collier, D. A., & Lewis, C. M. (2012). Meta-analysis of the association of urbanicity with schizophrenia. *Schizophrenia Bulletin*, *38*(6), 1118–1123.

Vollmer, J. Y., & Clerc, R. G. (1998). Homeobox genes in the developing mouse brain. *Journal of Neurochemistry*, *71*(1), 1–19.

von Hausswolff-Juhlin, Y., Bjartveit, M., Lindström, E., & Jones, P. (2009). Schizophrenia and physical health problems. *Acta Psychiatrica Scandinavica*, *119*, 15–21.

Wang, Y., Yang, D., & Deng, M. (2015). Low-rank and sparse matrix decomposition for genetic interaction data. *BioMed Research International*, *2015*.

Webster, M. J., Herman, M. M., Kleinman, J. E., & Weickert, C. S. (2006). BDNF and trkB mRNA expression in the hippocampus and temporal cortex during the human lifespan. *Gene Expression Patterns*, *6*(8), 941–951.

Wei, Z., Chang, X., & Wang, J. (2015). Advanced computational approaches for medical genetics and genomics. *BioMed Research International*, *2015*.

Weickert, C. S., Straub, R. E., McClintock, B. W., Matsumoto, M., Hashimoto, R., Hyde, T. M., . . . & Kleinman, J. E. (2004). Human dysbindin (dtnbpl) gene expression innormal brain and in schizophrenic prefrontal cortex and midbrain. *Archives of General Psychiatry*, *61*(6), 544–555.

Weickert, C. S., Webster, M. J., Gondipalli, P., Rothmond, D., Fatula, R. J., Herman, M. M., . . . & Akil, M. (2007). Postnatal alterations in dopaminergic markers in the human prefrontal cortex. *Neuroscience*, *144*(3), 1109–1119.

Whelan, R., St Clair, D., Mustard, C. J., Hallford, P., & Wei, J. (2018). Study of novel autoantibodies in schizophrenia. *Schizophrenia Bulletin*, *44*(6), 1341–1349.

White, T., Andreasen, N. C., & Nopoulos, P. (2002). Brain volumes and surface morphology in monozygotic twins. *Cerebral Cortex*, *12*(5), 486–493.

Zhao, X., Shi, Y., Tang, J., Tang, R., Yu, L., Gu, N., . . . & Zhao, S. (2004). A case control and family based association study of the neuregulin1 gene and schizophrenia. *Journal of Medical Genetics*, *41*(1), 31–34.

Zollo, M., Ahmed, M., Ferrucci, V., Salpietro, V., Asadzadeh, F., Carotenuto, M., . . . & Mojarrad, M. (2017). PRUNE is crucial for normal brain development and mutated in microcephaly with neurodevelopmental impairment. *Brain*, *140*(4), 940–952.

# 11 Prediction of Disease–lncRNA Associations via Machine Learning and Big Data Approaches

*Mariella Bonomo, Armando La Placa, and Simona Ester Rombo*

**CONTENTS**



## 11.1 CODING AND NON-CODING RNA

According to the central dogma of molecular biology, ribonucleic acid (RNA) is the transition element from deoxyribonucleic acid (DNA) to proteins. RNA, DNA, proteins, carbohydrates, and lipids constitute the main macromolecules essential for cell life.

RNA is a polymeric molecule involved in various biological roles of coding, decoding, regulation, and gene expression. Messenger RNA (mRNA) is a type of RNA that encodes and carries information during transcription from DNA to protein synthesis sites for translation. The short life of an mRNA molecule begins with transcription and ends with degradation. During its life, an mRNA molecule can also be examined, modified, and transported prior to translation. Cellular organisms use mRNA to transmit genetic information (using the nitrogenous bases guanine, uracil,

DOI: 10.1201/9781003142751-14

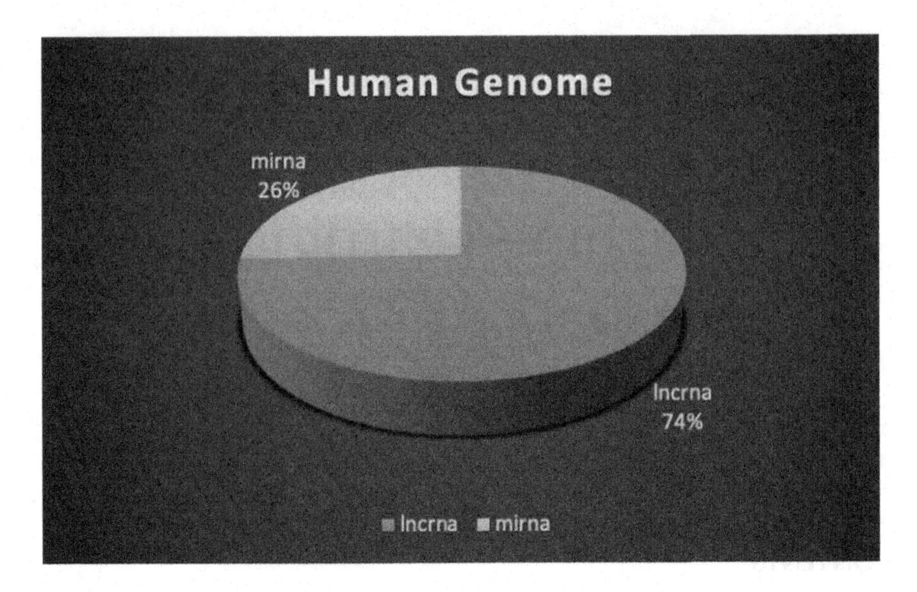


**FIGURE 11.1** Distribution of miRNA and lncRNA in the Human Genome.

adenine, and cytosine, indicated by the letters G, U, A, and C, respectively) that direct the synthesis of specific proteins. Many viruses encode their genetic information using an RNA genome.

Most of the genome is not encoded; that is, the information it contains is not used for protein synthesis. For a long while, such "non-coding" regions of the genome have been considered "junk DNA". However, it is now well recognized that also DNA sequences that do not give rise to proteins, also known as *non-coding RNA*, may be important for specific cell functions. Non-coding RNAs differ in the length of nucleotides. There are two types of ncRNAs:

- Long non-coding RNA (lncRNA)—molecules made up of more than 200 nucleotides that perform functions in the regulation of gene transcription, epigenetic regulation, tissue aging, and the onset of certain diseases such as cancer
- MicroRNA (miRNA)—small molecules characterized by approximately 20 to 22 nucleotides that are particularly active in the regulation of gene expression at transcriptional and post-transcriptional levels

In Figure 11.1, the distribution of miRNA and lncRNA in the human genome is shown.

## 11.2 lncRNAs AND THEIR IMPLICATIONS IN DISEASES

lncRNAs are molecules emerging as key regulators of various critical biological processes, and their alterations and dysregulations have been associated with many important complex diseases (Chen G. et al., 2013); we illustrate lncRNAs associated with disease in Figure 11.2.

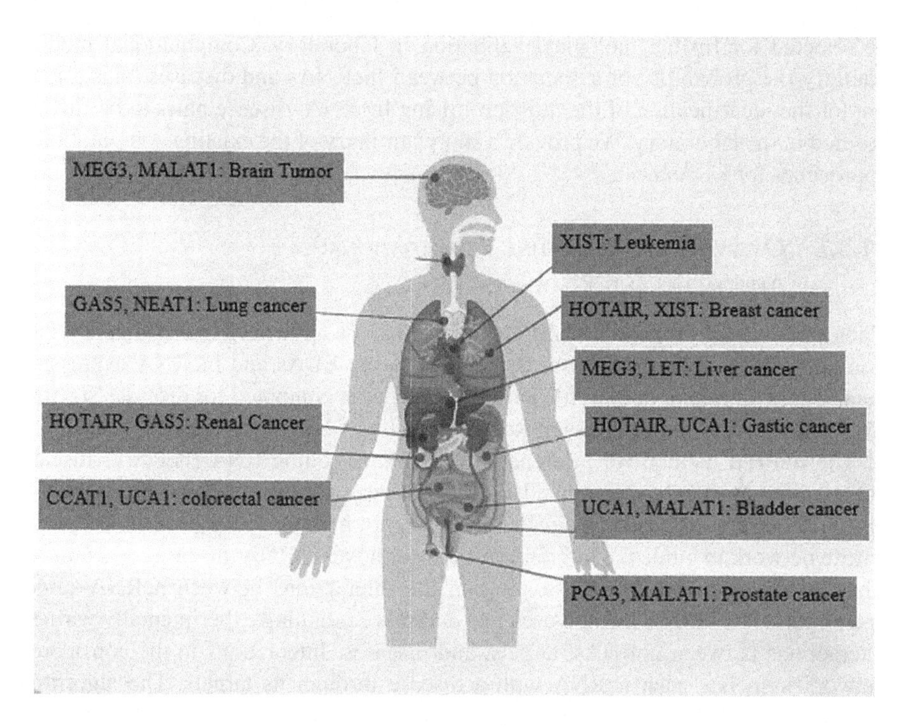


**FIGURE 11.2** Representation lncRNA Associated with Diseases.

Many studies have demonstrated that the true catalog of RNAs encoded within the genome (the "transcriptome") is more extensive and complex than previously thought. As an example, lncRNAs are emerging as new players in the cancer paradigm, demonstrating potential roles in both oncogenic and tumor-suppressive pathways (Gibb et al., 2011). Recently, lncRNAs have attracted much attention from researchers due to the increasing evidence that they play important roles in different biological processes (Zhao et al., 2014). In addition, it has been shown that many human diseases may be related to both mutations and dysregulations of lncRNAs (Li et al., 2014; Taft et al., 2010), and there are more than 200 diseases associated with various lncRNAs and more than 300 lncRNAs, which play critical roles in various human complex diseases (Chen G. et al., 2013).

## 11.3 PREDICTION OF lncRNA–DISEASE ASSOCIATIONS (LDAs)

The discovery of novel LDAs may provide valuable input to the understanding of disease mechanisms at the RNA level, as well as to the detection of biomarkers for disease diagnosis, treatment, prognosis, and prevention. Unfortunately, the experimental confirmation of disease-related lncRNAs is much expensive and requires a long time. Therefore, the number of experimentally verified associations between lncRNAs and diseases is very limited. Computational approaches for the prediction of potential LDAs are important in order to identify only those candidate pairs to be selected for further biological validation in laboratory. Computational models quantify the probability of association between lncRNAs and diseases, thus allowing for the identification of the most promising lncRNA–disease pairs to be further verified in the laboratory. We provide a short summary of the existing computational approaches for LDAs next.

### 11.3.1 Overview of the Existing Computational Approaches and Resources

Chen et al. (2013) propose **LRLSLDA**, which can be applied to interactions prediction between lncRNAs and diseases by integrating LDAs and lncRNA expression data. Gaussian interaction profile kernel similarity is computed for diseases starting from known LDAs and Laplacian Regularized Least Squares is applied.

The **ncPred** method for inference of novel non-coding RNA (ncRNA)–disease association based on a recommendation technique has been proposed by Alaimo et al. (2014). ncPred is based on a resource propagation methodology, which uses a tripartite network to guide the inference process of novel ncRNA–disease associations. The tripartite network permits exploiting the interactions between ncRNA–target and target–disease. The method uses two datasets containing experimentally verified interactions between ncRNAs, targets, and diseases. Interactions in the considered network associate each ncRNA with a disease through its targets. The algorithm is based on a multilevel resource transfer technique, which computes the weights between each ncRNA–disease pair and, at each step, considers the resource transferred from the previous step. The area under the curve (AUC) was equal to 0.76.

Chen (2015) proposed a model based on the hypergeometric distribution in order to predict associations between lncRNAs and diseases, with the particularity that associations between miRNAs and diseases are used as bridge. In the same article, the authors propose a model for the computation of functional similarity between lncRNAs based on disease semantic similarity, miRNA–disease associations (MDAs), and miRNA–lncRNA interactions. The AUC was equal to 0.76.

An integrative framework, **IntNetLncSim**, has been presented by Cheng et al. (2016) to infer lncRNA functional similarity by modelling the information flow in an integrated network that comprises both lncRNA-related transcriptional and post-transcriptional information. The AUC performed by IntNetLncSim was equal to 0.73.

Lan et al. (2016) proposed a web server for implementing a method for predicting LDAs. The input of the web server is a lncRNA sequence or a text file with multiple sequences in the FASTA format. A weighted support vector machine is used to build classifiers and discriminate positive samples from each subsample. The performed AUC was equal to 0.83.

Yu et al. (2017) developed a model that uses bi-random walks to predict novel LDAs. In particular, they constructed heterogeneous networks encoding the functional similarity of lncRNAs and diseases, which are used to build a directed bi-relational. Then, bi-random walks were applied to such a network to predict potential associations. The resulting AUC was equal to 0.79 for this method.

The method presented by Shi et al. (2017) used graph regression in a unified framework working on lncRNAs that have no previously known disease association and diseases that have no known association with any lncRNAs. The resulting AUC value was equal to 0.88.

The **GMCLDA** (Geometric Matrix Completion LDA) method was proposed by Lu et al. (2020) to infer underlying associations based on geometric matrix completion. Using association patterns among functionally similar lncRNAs and phenotypically similar diseases, GMCLCA is performed in five steps: computing disease semantic similarity according to the disease ontology hierarchy, computing lncRNA Gaussian interaction profile kernel similarity on the basis of known LDAs, calculating lncRNA sequence similarity by using Needleman–Wunsch algorithm, constructing an association profile for a new lncRNA using k-nearest neighbors, and recovering the association matrix with geometric matrix completion by incorporating the proximity of lncRNA association patterns and disease association patterns. GMCLDA achieved an AUC value of 0.85.

Xie et al. (2019) propose a method that constructs a lncRNA–disease correlation matrix based on the known LDAs. Then, it calculates the similarity between lncRNAs and that between diseases, according to specific metrics, and integrates such data. Finally, a predicted LDA matrix is obtained by the Laplacian Regularized Least Squares method. The obtained AUC value was equal to 0.90.

In Wei et al. (2019), a method is proposed based on convolution neural networks (CNNs) able to integrate different data sources by a supervised algorithm. Directed acyclic graphs (DAGs) are used to describe diseases, and disease semantic information is obtained from disease ontology. Since similar diseases have many common miRNAs to perform identical molecular pathogenic mechanisms, the hypergeometric test is adopted to measure whether diseases are significantly associated with common miRNAs. The obtained AUC score was equal to 0.80.

A novel local random walk–based prediction model is proposed by Li et al. (2019). They built a heterogeneous network, and local random walk was applied for the prediction of novel LDAs, also using Gaussian interaction profile kernel similarity constructed between the diseases. They achieved a reliable AUC of 0.83.

In Ou-Yang et al. (2019), a two-side sparse self-representation algorithm is proposed using the estimated representations of lncRNAs and diseases to predict potential LDAs, based on the known LDAs. It obtained an AUC score of 0.87.

Sumathipala et al. (2019) applied a network diffusion algorithm over a multilevel network built using lncRNA–protein, protein–protein, and protein–disease associations. They achieved an AUC value of 0.968 for cardiovascular diseases, 0.919 for cancer, and 0.902 for neurological diseases.

Zhu et al. (2019) built an LDA network and represented it by the corresponding adjacency matrix. They decomposed the known lncRNA–disease correlation matrix into two characteristic matrices and then defined an optimization function using disease semantic similarity, lncRNA functional similarity and known LDAs, and, finally, solved two optimal feature matrices by the least squares method. Their AUC was 0.95.

Fan et al. (2019) aimed at searching for candidate LDAs based on data integration and random walk, through the construction of biological networks from both lncRNA and disease datasets. Then, they implement the random walk with restart algorithm on these similarity networks for extracting the topological similarities that are fused with positive pointwise mutual information to build a large-scale lncRNA–disease heterogeneous network. The AUC was equal to 0.86.

A network-based model is proposed by Li et al. (2019) in which two novel lncRNA–disease weighted networks are built. An lncRNA–lncRNA weighted matrix and a disease–disease weighted matrix are obtained as well, based on a resource allocation strategy of unequal allocation and unbiased consistency. Finally, a label propagation algorithm is applied to predict associated lncRNAs for the investigated diseases. The average value of AUC by negative binomial linear discriminant analysis was 0.83.

A computational model is proposed by Wang et al. (2019) that collects associations between lncRNA from different data sources, and then a collaborative filtering approach is applied to compute functional association between lncRNAs. Their average AUC was 0.78.

In Zhang J. et al. (2019), a deep learning approach is proposed based on the topological similarity of heterogeneous tripartite networks, including lncRNA–disease, lncRNA–miRNA, and MDAs. Vertices in the networks are vectorized to reveal the topology features of the nodes, with the aim of computing their similarities. A rule-based inference method is applied to calculate a relevance score for each lncRNA-disease pair. The inferred association scores are used to calculate the AUC value, which was 0.93.

The model proposed by Cui et al. (2020) predicts novel LDAs from the lncRNA side and the disease side, respectively. A support vector machine is used as a classifier to train data. The obtained AUC score was 0.94.

The model proposed by Wang et al. (2020) is based on unknown human LDAs combined with clinical data. In particular, lncRNA expression and clinical data for prostate cancer are considered and used to select the characteristic variables of the LDAs. The tumor clinical stage is used to predict lncRNAs with a significant impact on the prognosis of cancer patients. The authors used random forests to order the importance of variables with a large impact on the performance and efficiency of the algorithm. The resulting AUC of this approach applied to prostate cancer was 0.67.

In Deng et al. (2020), a three-layer neural network model is used together with lncRNA similarity networks, disease similarity networks, and miRNA similarity networks. The associations between them are used to build a further network, and a gradient-boosting tree classifier is built with the feature vectors to predict LDAs. The obtained AUC score was equal to 0.97.

In Wang et al. (2020), a computational approach is proposed that considers disease-associated lncRNAs' identification as a recommendation system problem. The method consists of three main steps: (1) Calculate the disease similarity matrix, based on known LDAs and disease semantic information. (2) Compute the similarity matrix of lncRNAs, based on known LDAs. (3) Compute the disease similarity, based on Gaussian interaction profile kernel. In particular, the semantic similarities between different diseases can be computed using DAG based on the Mesh database, where nodes represent diseases and edges represent the association between diseases. After that, the affinity graphs for lncRNAs and diseases are constructed using a p-nearest neighbor graph. Second, an LDA adjacency matrix is reconstructed using the weighted k-nearest known neighbor (WKNKN) interaction profiles. Potential LDAs are predicted using the graph regularized nonnegative matrix factorization.

The LDAs prediction based on graph regularized nonnegative matrix factorization (LDGRNMF) achieved an AUC of 0.89.

Zhang et al. (2020) propose an approach that change a Boolean network of known LDAs into the weighted networks via combining all the global information (e.g., disease semantic similarities, lncRNA functional similarities, and known LDAs). They obtain this way the space projection scores via vector projections of the weighted networks to form the final prediction scores without biases. The approach has shown an AUC value of 0.91 for inferring diseases, 0.88 for inferring new lncRNAs (whose associations related to diseases are unknown), 0.75 for inferring isolated diseases (whose associations related to lncRNAs are unknown).

In Bonomo et al. (2020), an approach based on neighborhood analysis in tripartite graphs is presented. We provide more details on this approach in the following sections, where big data–related issues are also discussed.

Table 11.1 summarizes the main resources available for storing lncRNAs and their relationships with diseases, also providing further information about the availability of such resources and specifying which of the reviewed approaches use them.

**TABLE 11.1**
**Availability with Resources**

| Database Name | Availability | Database Content | ApproachesUsing It |
|---|---|---|---|
| Consortium TR. RNAcentral (Wang et al., 2019) | https://rnacentral.org/ | Non-coding RNA sequences | Wang et al. (2019) |
| Disease ontology (Kibbe et al., 2015) | https://disease-ontology.org/ | Formal ontology of human disease | Wang et al. (2019); Cheng et al. (2016) |
| DisGeNet (Schriml et al., 2012) | www.disgenet.org/downloads | disease-gene associations | Shi et al. (2017); Fan et al. (2019) |
| EVLncRNAs (Zhou et al., 2018) | http://biophy.dzu.edu.cn/EVLncRNAs/ | curated database for lncRNA validated by esperiment | Sumathipala et al. (2019) |
| GeneRIF (Lu et al., 2020) | https://ftp.ncbi.nlm.nih.gov/gene/ | functional association of genes | Yu et al. (2017); Fan et al. (2019); Lu et al. (2020); |
| HMDD (Li et al., 2014) | Availability in the Supplementary Material of (Li et al., 2014) | miRNA disease associations | Fan et al. (2019); Liu et al. (2019); Zhang J. et al. (2019) |
| iLncRNAdis-FB (Wei et al., 2019) | http://bliulab.net/iLncRNAdis-FB/ | lncRNA disease associations | Wei et al. (2019) |
| Lnc2Cancer (Ning et al., 2016) | www.bio-bigdata.net/lnc2cancer | lncRNA cancer associations | Lan et al. (2016; Shi et al. (2017); Fan et al. (2019); Liu et al. (2019); Zhang J. et al. (2019); Wei et al. (2019) |

*(Continued)*

**TABLE 11.1 (Continued)**

| Database Name | Availability | Database Content | ApproachesUsing It |
|---|---|---|---|
| LncRNADisease (Chen et al., 2013) | Available in the Supplementary Material of (Chen et al., 2013) | lncRNA disease associations | Alaimo et al. (2014); Lan et al. (2016); Shi et al. (2017); Yu et al. (2017); Fan et al. (2019); Wang et al. (2019); Xie et al. (2019); Zhang J. et al. (2019b); Bonomo et al. (2020); |
| | | | Cheng et al. (2016); Cui et al. (2020); Liu et al. (2019); Lu et al. (2020); Zhu et al. (2019); Wang et al. (2020); |
| LncRNADisease (Chen, 2015) | www.cuilab.cn/lncrnadisease | lncRNA disease associations | Liu et al. (2019); Xuan et al. (2019) |
| LncRNADisease (Chen et al. 2016) | www.cuilab.cn/lncrnadisease | lncRNA disease associations | Chen et al. (2013); Li et al. (2019); Liu et al. (2019); Deng et al. (2020); |
| | | | Wei et al. (2019) |
| lncRNAtor (Park et al., 2014) | Availability in the Supplementary Material of (Park et al., 2014) | Gene expression from RNA seq and protein lncRNA interaction | Wang et al. (2020) |
| lncRInter (Liu et al., 2017) | Availability in the Supplementary Material of (Liu et al., 2017) | lncRNA interactions | Sumathipala et al. (2019) |
| LncRNA EMBLEBI (E-MTAB-5214) NPInter v3.0, RAID v2.0 (Fan et al., 2019) | Availability in the Supplementary Material of (Xiao et al., 2009) www.rna-society.org/raid/ | lncRNA-protein interactions | Fan et al. (2019) |
| MeSH (Chen et al., 2013) | www.ncbi.nlm.nih.gov/mesh | Gene sequence | Yu et al. (2017); Cui et al. (2020); Zhu et al. (2019) |
| miR2Disease (Jiang et al., 2009) | www.mir2disease.org | miRNA disease interactions | Zhang J. et al. (2019) |
| miRCancer (Xie et al., 2019) | Availability in the Supplementary Material of (Xie et al., 2019) | miRNA cancer interactions | Zhang J. et al. (2019) |
| miRBase (Release 21) (Kozomara, 2014) | www.mirbase.org/ | miRNA sequences and annotation | Cheng et al. (2016) |
| miRNA–lncRNA (Li et al., 2014) | Availability in the Supplementary Material of (Li J.-H., 2014) – starbase v2.0 | miRNA lncRNA interactions | Fan et al. (2019); Wang et al. (2019); Cheng et al. (2016), |

| Database Name | Availability | Database Content | ApproachesUsing It |
|---|---|---|---|
| miRNA–miRNA HPRD (Cheng et al., 2016) | https://hprd.org/ | miRNA miRNA interactions | Cheng et al. (2016) |
| miRmine (Liao et al., 2011) | Availability in the Supplementary Material of (Liao et al., 2011) | human miRNAs | Deng et al. (2020) |
| MNDR (Wang et al., 2019) | www.rna-society.org/mndr/ | lncRNA disease associations | Liu et al. (2019) |
| NPInter (Hao et al., 2016) | Availability in the Supplementary Material of (Hao et al., 2016) | lncRNA interactions | Sumathipala et al. (2019) |
| Nucleotide (Cui et al., 2020) | www.ncbi.nlm.nih.gov/nuccore | lncRNA gene sequence | Cui et al. (2020) |
| starBase (Lu et al., 2018) | http://starbase.sysu.edu.cn/ | miRNA–lncRNA interactions | Chen (2015); Bonomo et al. (2020); Chou et al. (2016) |
| (TarBase v6.0, miRTarBase v4.5, miRecords v. 4) (Xiao et al., 2009) | www.microrna.gr/tarbase http://mirtarbase.cuhk.edu.cn/php/index.php<br>Availability miRecords in the Supplementary Material of (Xiao et al., 2009) | human miRNAs | Cheng et al. (2016)<br>Vergoulis et al. (2012) |
| The STRING database in 2017 (Wang et al., 2019) | https://string-db.org/ | Protein–protein interactions | Wang et al. (2019) |

The approach in Zhang J. et al. (2019) utilizes a flow propagation algorithm to integrate multiple networks based on a variety of biological information including lncRNA similarity, protein–protein interactions, disease similarity, and the associations between them to infer LDAs. The obtained AUC value was equal to 0.84.

A novel model (Ping et al., 2019) based on a bipartite network based on known LDAs proposes for inferring potential LDAs, the analysis of properties of the bipartite network and found that it closely followed a power-law distribution. To evaluate the performance, a leave-one-out cross-validation (LOOCV) framework was implemented, and the simulation results of a computational model significantly outperformed previous state-of-the-art models, with AUCs of 0.8825, 0.9004, and 0.9292 for known LDAs obtained from the LncRNADisease database, Lnc2Cancer database, and MNDR database, respectively.

In Zhao et al. (2020), a multilayer network was designed by integrating the similarity networks of lncRNAs, diseases, and genes and the known association networks of lncRNA–disease, lncRNAs–gene, and disease–gene. The model developed for predicting the lncRNA–disease potential associations is based on random walk with restart and had an AUC value of 0.91 with LOOVC.

Guo et al. (2019) integrate and analyze the associations among miRNA, lncRNA, protein, drug, and disease to predict potential new associations. A random forest classifier is trained to classify and predict new interactions or associations between biomolecules. They obtained an AUC equal to 0.97.

In Wang et al. (2019), the authors propose an improved diffusion model for predicting LDAs applied in a bipartite graph in order to solve the correlation between lncRNAs and diseases. This model relies on the assumption that phenotypically similar diseases are often related to functionally similar lncRNA and vice versa. Two ensemble similarities are defined, one for diseases and one for lncRNA, then are applied in the model to compute the correlation score. The model achieved an AUC of 0.95.

Tan et al. (2020) built a consensus graph using several similarity matrices for lncRNAs and diseases. At each iteration of the learning, the predicted association probability for lncRNA and disease was optimized. The AUC obtained was larger than 0.9.

In Yu et al. (2018), a quadruple global network was constructed using miRNA–lncRNA, lncRNA–disease, miRNA–disease, and gene–lncRNA associations. Then, the authors defined a probabilistic model based on a naive Bayesian classifier and used it to predict LDAs. The AUC performed by the classifier was 0.88 on a network that includes information regarding disease semantic similarity.

Chen et al. (2016) included information related LDAs, disease semantic similarity, and various lncRNA similarity measures, such as the expression similarity. Then they apply random walk, for a resulting AUC equal to 0.78.

## 11.4 AN APPROACH BASED ON TRIPARTITE GRAPHS

We now describe in more detail an approach that relies on the analysis of lncRNA-related information stored in public databases, as well as their interactions with other types of molecules (Bonomo et al., 2020). In particular, large amounts of lncRNA–miRNA interactions (LMIs) have been collected in public databases, and plenty of experimentally confirmed MDAs are available as well. Therefore, the prediction of LDAs may be based on known LMIs, and MDAs. In the considered approach, the problem of LDA prediction is modeled as a neighborhood analysis performed on tripartite graphs (see Figure 11.3), in which the three sets of vertices represent lncRNAs, miRNAs, and diseases, and the vertices are linked according to LMIs and MDAs. Due to the fact that similar lncRNAs interact with similar diseases (Lu et al., 2018), we aim at identifying novel LDAs by analyzing the behavior of neighbor lncRNAs in terms of their intermediate relationships with miRNAs. A score is assigned to each LDA $(l, d)$ by considering both their respective interactions with common miRNAs and the interactions with miRNAs shared by the considered disease $d$ and other lncRNAs in the neighborhood of $l$. Significant predictions for candidate LDAs to be proposed for further laboratory validation are computed by a statistical test performed through a Monte Carlo test.

**Problem Statement:** Let $L = \{l_1, l_2, \ldots, l_n\}$ be a set of lncRNAs and $D = \{d_1, d_2, \ldots, d_k\}$ be a set of diseases. The goal is to return a set $P = \{(l_x, d_y)\}$ of predicted LDAs.

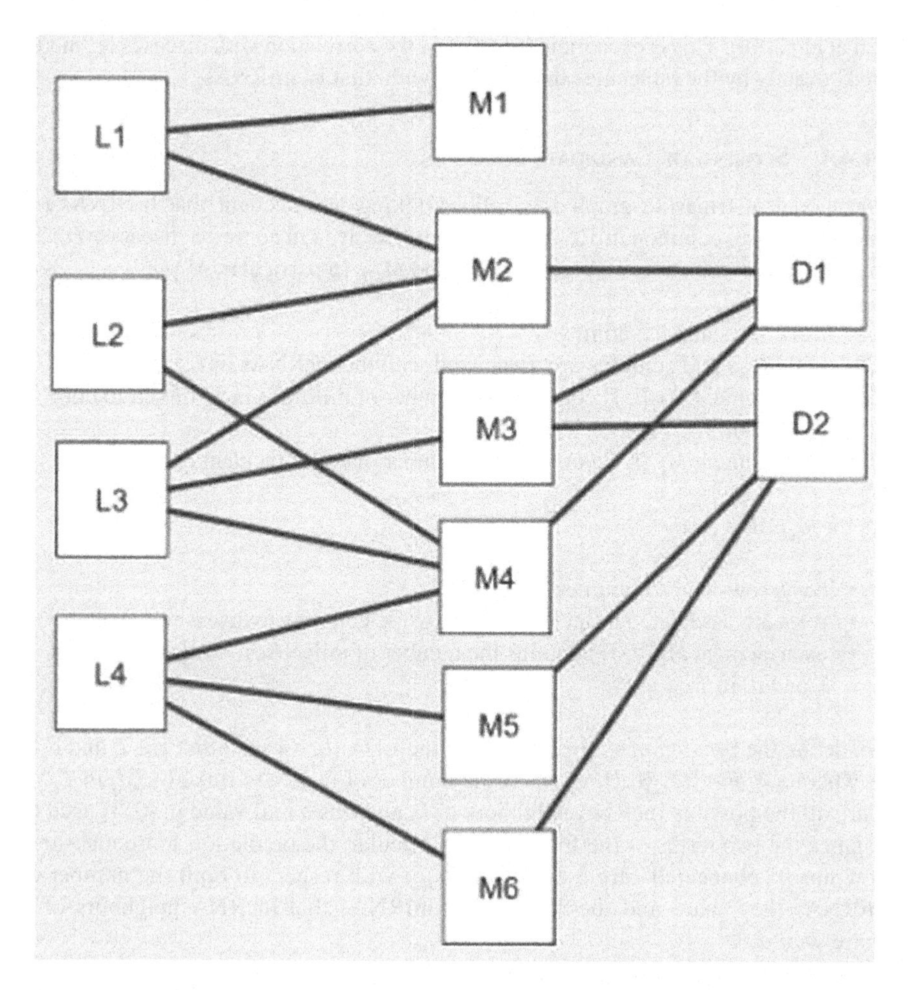


**FIGURE 11.3** Tripartite Graph with $L$ (lncRNAs), $M$ (miRNAs), $D$ (diseases).

Let $T_{LMD}$ be a tripartite graph defined on the three sets of disjoint vertices $L$, $M$, and $D$, which can also be represented as $T_{LMD} = ((l, m), (m, d))$, where $(l, m)$ are edges between vertices in $L$ and $M$ and $(m, d)$ are edges between vertices in $M$ and $D$. In the proposed approach, $L$ is a set of lncRNAs, $M$ is a set of miRNAs and $D$ is a set of diseases. In such a context, edges of the type $(l, m)$ represent molecular interactions between lncRNAs and miRNAs, experimentally validated in the laboratory; edges of the type $(m, d)$ correspond to known associations between miRNAs and diseases, according to the existing literature. In both cases, interactions and associations suitably annotated and stored in public databases are considered.

A commonly recognized assumption is that lncRNAs with similar behavior in terms of their molecular interactions with other molecules may also reflect this similarity in their involvement in the occurrence and progress of disorders and diseases

(Lu et al., 2018). This is even more effective if the correlation with diseases is "mediated" exactly by the molecules they interact with, that is, miRNAs.

### 11.4.1 Scoring of Candidate LDAs

The model of tripartite graph $T_{LMD}$ allows taking into account that lncRNAs ($L$) interacting with common miRNAs ($M$) may be involved in common diseases ($D$). To this aim, let us consider two matrixes $M_{LL}$ and $M_{LD}$. In particular, $M_{LL}$

- has $h$ rows and $h$ columns,
- both rows and columns are associated with the lncRNAs in $L$,
- each element $M_{LL}[i, j]$ contains the number of miRNAs in $M$ linked to both $l_i$ and $l_j$ in $T_{LMD}$, and
- each element $M_{LL}[i, i]$ contains the number of edges incident onto $l_i$.

As for $M_{LD}$, it

- has $h$ rows and $k$ columns;
- rows are associated to lncRNAs in $L$, while columns to disease in $D$; and
- each element $M_{LD}[i, j]$ contains the number of miRNAs in $M$ linked to both $l_i$ and $d_j$ in $T_{LMD}$.

We define the *prediction score* $S(l_i, d_j)$ for the LDA $(l_i, d_j)$ such that $l_i \in L$ and $d_j \in D$, where $n = min\ (M_{LL}[i, i], n_j)$, $n_j$ is the number of miRNAs linked to $d_j$ in $T_{LMD}$, $x$ are all the possible lncRNA neighbors of $l_i$, and $\alpha$ is a real value in [0, 1] used to balance the two terms of the formula. In particular, the prediction score measures how much "connected" are $l_i$ and $d_j$ on $T_{LMD}$, with respect to both the number of miRNAs they share and the numbers of miRNAs that lncRNA neighbors of $l_i$ share with $d_j$.

***Example.*** Consider the $T_{LMD}$ represented in Figure 11.3. The two matrixes $M_{LL}$ and $M_{LD}$ are shown in Tables 11.2 and 11.3. The prediction score for the pair $(l_1, d_1)$ is $S(l_1, d_1) = 0.5\left(\frac{1}{2}\right) + (1-0.5)\left(\frac{(1\cdot 2)+(1\cdot 2)}{(2\cdot 3)+(3\cdot 3)}\right) = 0.25 + 0.13 = 0.38.$

### 11.4.2 Prediction of Significant LDAs

Given a set of LDAs scored according to the prediction score computed as described earlier, it is necessary to select the only associations that are statistically significant to produce the output predictions. To establish the statistical significance of the considered LDAs, a hypothesis test via a Monte Carlo simulation is proposed (Giancarlo et al., 2015; Gordon, 1996). The null hypothesis is that lncRNAs and diseases have been associated with chance. It is important to focus on the importance that the intermediate miRNAs have in the prediction score computation and, more generally, in the measure of how much similar the behavior of different lncRNAs with respect

**TABLE 11.2**
**$M_{LL}$ for Prediction Score Computation and Disease**

| | *L* 1 | *L* 2 | *L* 3 | *L* 4 |
|---|---|---|---|---|
| *L* 1 | 2 | 1 | 1 | 0 |
| *L* 2 | 1 | 2 | 2 | 1 |
| *L* 3 | 1 | 2 | 3 | 1 |
| *L* 4 | 0 | 1 | 1 | 3 |

**TABLE 11.3**
**$M_{LD}$ for Prediction Score Computation**

| | *D* 1 | *D* 2 |
|---|---|---|
| *L* 1 | 1 | 0 |
| *L* 2 | 2 | 0 |
| *L* 3 | 3 | 1 |
| *L* 4 | 1 | 2 |

to the occurrence of diseases is. In particular, in the adopted model, interactions with miRNAs are the key factors in order to determine the association between a lncRNA and a disease. Let then $(l, m)$ be the association lncRNA–miRNA in a set $A$ and shuffle them for 100 times by producing 100 new sets of pairs $A_i$. The meaning is to interchange the associations between lncRNAs and miRNAs while still maintaining the same number of interactions. The test to reject the null hypothesis consists of comparing the prediction score $S(l, d)$ of an association $(l, d)$ with the maximum value of the prediction score $\hat{S}(l,d)$ obtained using the associations in the 100 $A_i$. The null hypothesis is rejected if $S(l,d) > \hat{S}(l,d)$.

## 11.5 BIG DATA TECHNOLOGIES FOR LDAs

The approach described in the previous section has been implemented by big data technologies, and a framework for the computational prediction of LDAs has been designed and developed. In the following, we first provide some details on the exploited technologies, then we describe the implemented system, and finally we discuss the results obtained by applying it on real datasets.

### 11.5.1 Adopted Technologies

We have based the implementation of our system on Apache Spark. This framework provides the tools to parallelize the computation and to distribute large amount of data into several resilient and fault-tolerant clusters, named Resilient Distributed Dataset (RDD; Zaharia et al., 2016). Apache Spark is a master–slave architecture

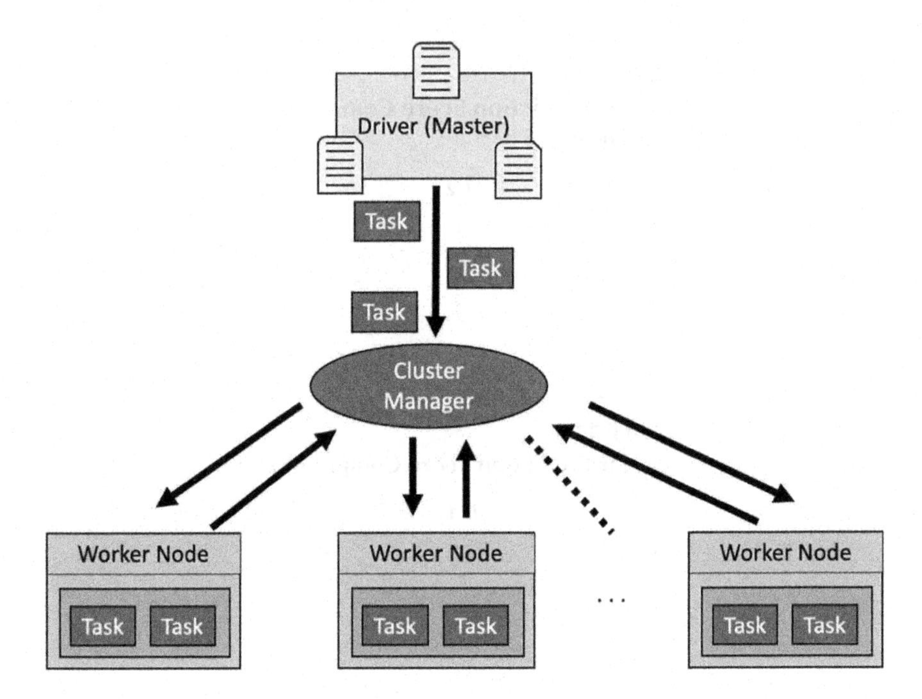


**FIGURE 11.4** Master–Slave Architecture. The driver receives the applications from the users and then sends the tasks to the cluster manager to be assigned to the workers.

framework, in which a *driver* (master) works as an interface between the user's program and the workers (Figure 11.4). A cluster manager is used by the driver to allocate the workers' resources. In our case, we used a standalone cluster manager for local execution, but it is also possible to parallelize the computation over nodes in different networks using other cluster managers such as YARN or Mesos. This is possible due to the fact that Apache Spark is agnostic to the type of cluster manager.

In order to execute the user application, the program needs to be sent to the workers. In this regard, the application is first split into *jobs*. Each job is then divided into *stages*, containing a set of operations that depend on one another. In this phase, the computational boundaries of each operation are taken into account for the subdivision of jobs. Finally, each stage is split into *tasks*, which are the smallest unit of computation of the RDDs executed by the worker's nodes. Sometimes, executors need to access one variable several times in the driver context within several tasks. Apache Spark permits to define shared variables, called *broadcast variables*, which are not sent every time with the task but are cached in the worker's cache memory.

Two types of operations can be applied over RDDs: *actions* and *transformations*. The transformations produce new RDDs, while the actions return the values to the driver. The actions are the most expensive, since Apache Spark follows a lazy approach and all the transformations are applied when an action is called for. Any further optimizations would imply the reduction of actions, in favor of transformations.

Another data abstraction provided by Apache Spark is the *dataset*. Datasets are an evolution of RDDs that organize the data in columns, integrating the SQL relational approaches to the RDDs. They permit to increment the performance of execution since provide a better compression of the objects (storing data organized as columns is more efficient than storing objects). Another advantage of using datasets is the possibility to store the partial results in a column-based file called *parquet*. This functionality is exploited several times by our system, allowing quick reloads of data instead of reiterating computations.

RDD is used in each of the five steps described in the next section. For instance, once all the lncRNAs and diseases are extracted from LncRNADisease and Human microRNA Disease Database (HMDD) datasets and put in RDDs, it is possible to produce all the possible combinations of lncRNA–disease by calling the *cartesian* transformation. As for the results of each of the five steps, they are converted into datasets and stored in *parquet files*, allowing an easy and quick access for future analysis.

### 11.5.2 System Overview

We developed an Apache Spark–based system that aims to efficiently compute graph-based algorithms, such as neighborhood analysis, over a large tripartite graph (Figure 11.5).

The following steps are executed by the system (Figure 11.6):

1. Parsing and graph building
2. Score computation
3. Generation of 100 random lncRNA–miRNA's associations
4. Prediction based on Monte Carlo simulation
5. Results analysis

Each step's result is converted into a dataset and then saved as *parquet*, in order to provide easy and fast access for future uses. Several optimizations were made in order to speed up the time execution.

The first step of the system (Figure 11.7) consists of parsing the input data from the considered databases, extracting all the different miRNAs, lncRNAs, and diseases.

Following the graph generation, there is the score computation step (Figure 11.8) over all the associations generated in the previous step. A service is available that receives in input the LDAs and the associations between lncRNA and miRNA and then returns in output a tuple of raw data containing the following:

- lncRNA
- Disease
- Computed score
- number of miRNAs in common with both lncRNA and disease
- miRNA associated with the lncRNA
- miRNA associated with the disease
- First term of the equation
- Second term of the equation
- A Boolean with a value of True, if the association belongs to the golden standard and False otherwise

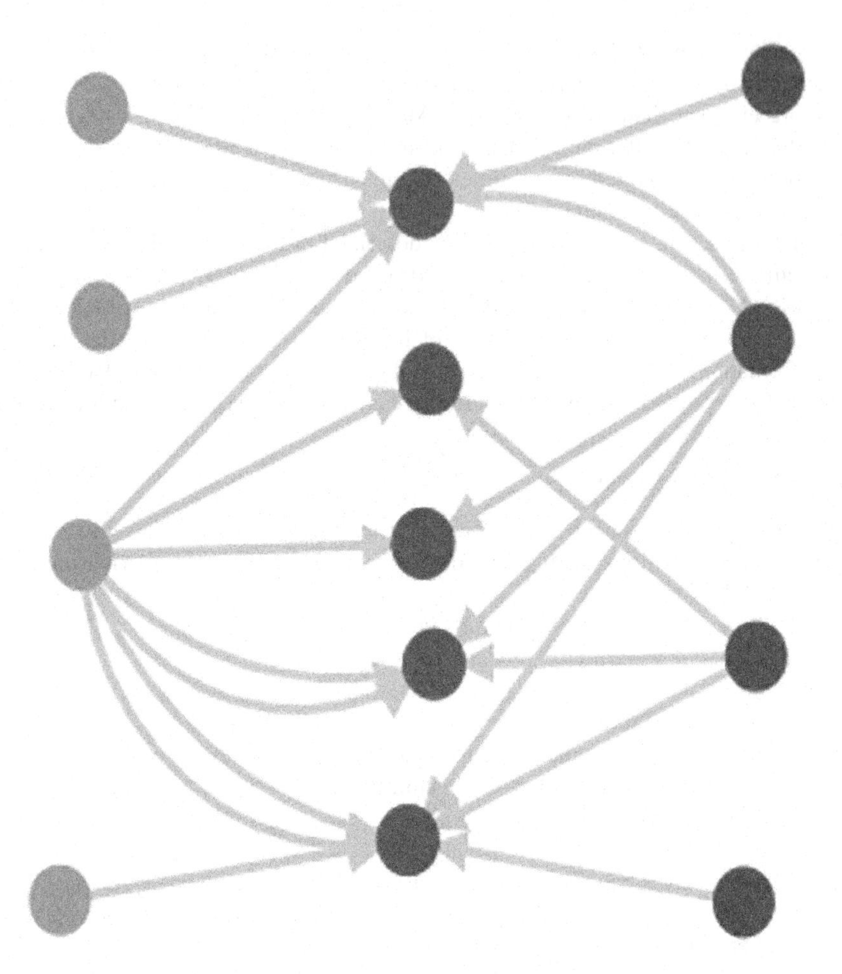

**FIGURE 11.5** Representation of the Tripartite Graph, Which Connects the lncRNAs (green) with the Diseases (blue) Using miRNAs (red) as a Bridge.

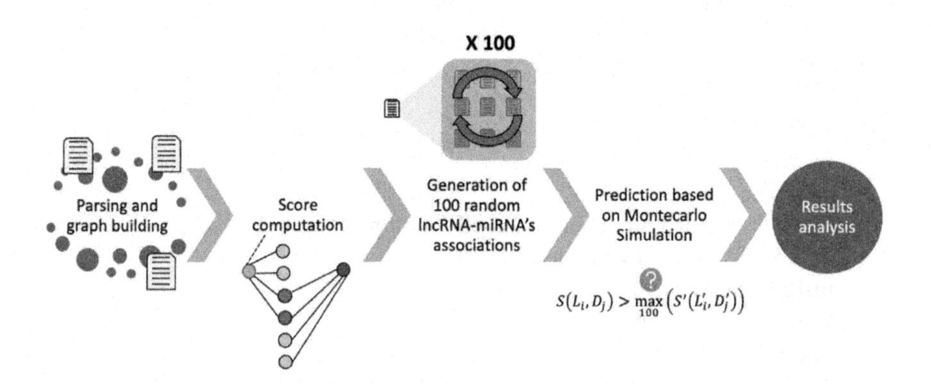


**FIGURE 11.6** Steps of the Implemented System.

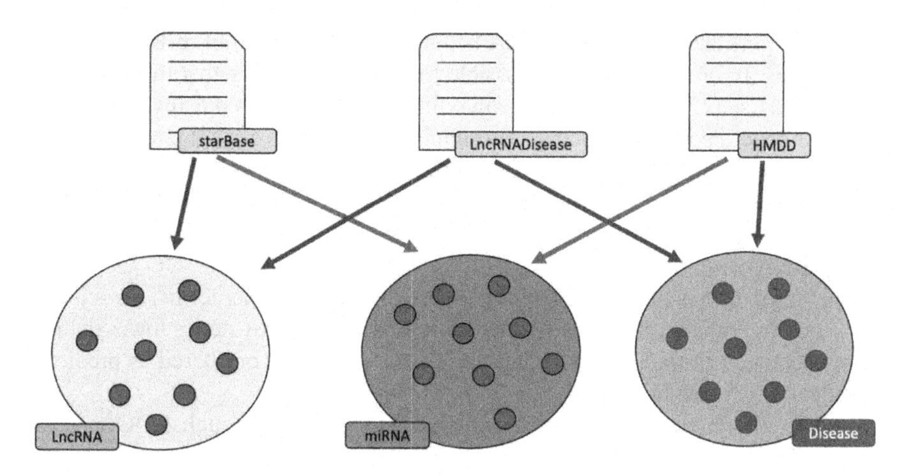


**FIGURE 11.7** Data Extraction from the Datasets.

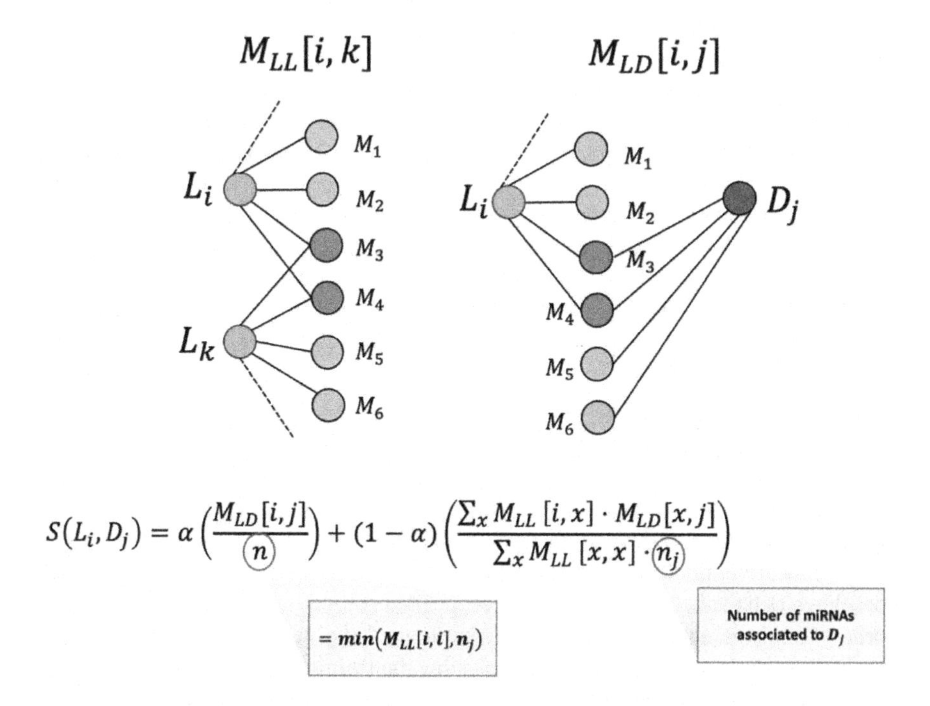


**FIGURE 11.8** Score Computation over lncRNA, *Li*, and Disease, *Dj*.

Only the first three fields are necessary for the following steps, while the others are used only for debugging purposes.

The defined service is reused in the Monte Carlo simulation. To this end, we generate 100 random files of miRNA–lncRNA associations. The random generation is a simple permutation of the miRNA belonging to the dataset of miRNA-lncRNA associations. In order to predict a possible LDA, it is necessary that the score computed during step 2 is higher than the maximum score computed using the 100 random files generated in the Monte Carlo simulation. Since we are not interested in the results of each of the 100 random files, after every iteration we save the maximum score between the iteration $i$ and the iteration $i-1$. At the end of the 100 iterations, we have the maximum score calculated over the 100 random files for each association lncRNA–disease.

The fourth step is the prediction of possible LDAs or, in other terms, the rejection of the null hypothesis. If the maximum score found in the last step is lower than the score calculated in the second step, then the association is considered as predicted. This operation is trivial.

Finally, the last step is the result analysis, performed through ROC analysis. Apache Spark includes MLLib, a library with a set of functionalities to calculate the ROC metrics. In order to calculate the AUC, it is necessary to convert the data into a format (*score, prediction*), in which the score represents the value calculated in Step 2 and the prediction is 1 or 0, depending on the results of Step 4 (respectively True or False).

### 11.5.3 Results

The approach described earlier has been validated on experimental verified data downloaded from starBase (Li et al., 2013) for the LMIs and from HMDD (Li et al., 2014) for the MDAs, resulting in 1,114 lncRNAs, 524 miRNAs, 75 diseases, 10,112 LMIs, and 5,430 MDAs. A golden-standard dataset with 183 LDAs has been obtained from the LncRNADisease database (Chen et al., 2013). The system has run on a cloud instance with 4 core vCPU and 8GB of RAM. The Spark Master has been configured to run as standalone with 12 thread instances.

Before proceeding with the discussion of the results, some considerations are needed. Although a number of approaches for LDAs prediction have been presented recently, including machine learning–based models, only a few of them do not use directly known lncRNA–diseases relationships during the prediction task. However, so far, the experimentally identified known LDAs are still very limited; therefore, using them during prediction could bias the final result. Indeed, when such approaches are applied for de novo LDAs prediction, their performance drastically goes down (Lu et al., 2018). This enforces the idea behind our approach, since neighborhood analysis automatically guides one toward the detection of similar behaviors without the need of positive examples for the training step. With respect to the other approaches which do not use LDAs during prediction (e.g., the $p$-value based approach in Chen, 2015), experimental tests have shown that our approach is able to detect specific situations that are not captured by its competitors. In particular, approaches such as that used by Chen (2015) often fail in detecting true LDAs where the lncRNA and the diseases do not have a large number of shared miRNAs. Instead, our approach is particularly effective in detecting this kind of situation, since neighborhood analysis allows to detect, for example, that there are similar lncRNAs associated with that disease.

The proposed approach has been applied to the known experimentally verified LADs in the lncRNADisease database according to LOOCV. In particular, each known disease–lncRNA association is left out, in turn, as a test sample. If the rank of the test sample exceeds the given threshold, this model is considered in order to provide a successful prediction. When the thresholds are varied, true-positive rate (TPR, sensitivity) and false-positive rate (FPR, specificity) could be obtained. Here, sensitivity refers to the percentage of the test samples whose ranking is higher than the given threshold. Specificity refers to the percentage of samples that are below the threshold. The ROC curve can be drawn by plotting TPR versus FPR at different thresholds. The AUC is further calculated to evaluate the performance of the tested methods. AUC = 1 indicates perfect performance and AUC = 0.5 indicates random performance.

We have implemented the $p$-value based on a hypergeometric distribution for LDAs inference proposed by Chen (2015) and compared our approach against it. As a result, the proposed neighborhoods-based approach achieved an AUC equal to 0.67 (see Figure 11.9), whereas the $p$-value based approach scored AUC = 0.53, showing that the consideration of indirect relationships between lncRNAs and diseases through neighborhood analysis is more effective.

As for data extracted from StarBase and HMDD, our approach has produced 7,941 statistically significant LDAs predictions. Among them, it has been able to detect 66 of the 74 verified integrating multiple heterogeneous networks for novel LDA inference LDAs of the gold-standard dataset that could have been detected in this larger dataset (due to the presence of lncRNAs and diseases in the gold standard), 24 out of which not detected by the $p$-value-based approach. In Figure 11.10,

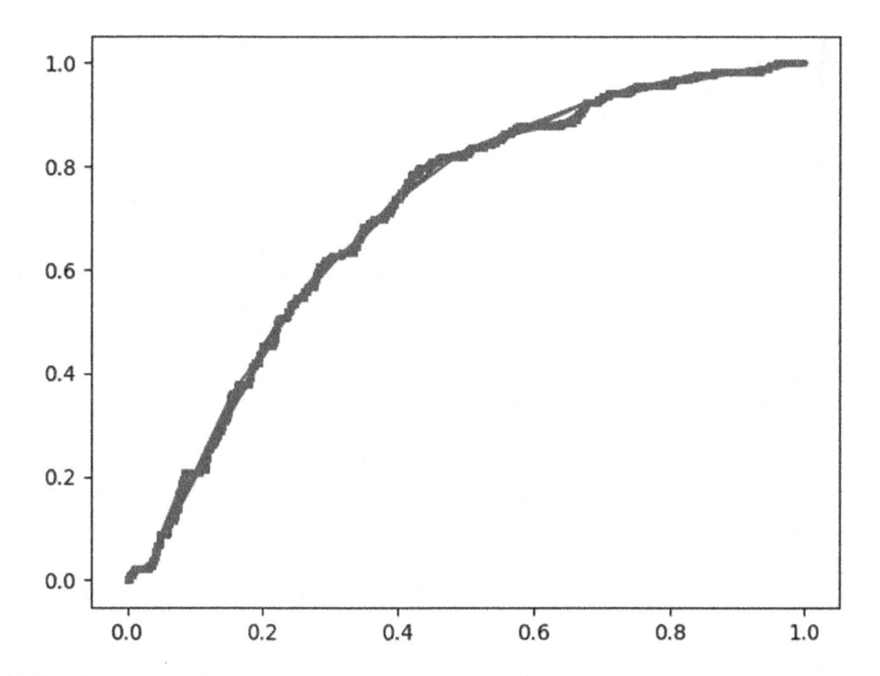


**FIGURE 11.9** ROC Curve Calculated over 85,000 LDAs.

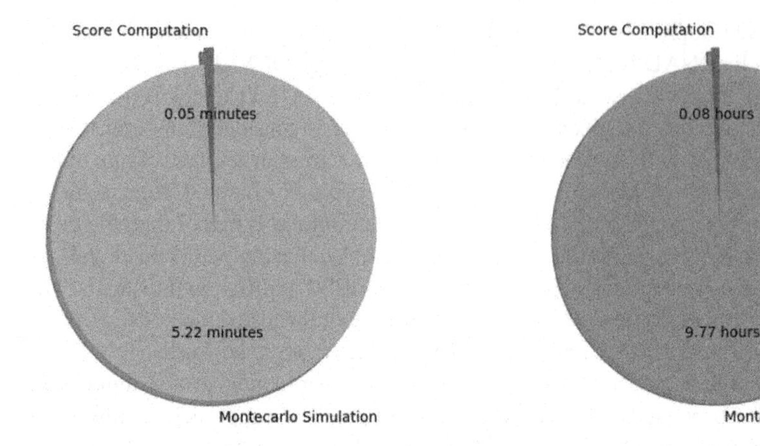


**FIGURE 11.10** On the left, the time spent for the execution of the system over the small dataset. On the right, the time spent for the execution of the system over the complete dataset.

two pie charts show the execution times of the system applied over two dataset configurations. The chart on the left refers to the dataset containing 2,625 LDAs, while on the right the configuration includes 83,550 LDAs. Only two of the five steps are shown, motivated by the fact that only these two steps call an action. As already explained before, in Apache Spark, the transformations are lazily evaluated when an action is called for, so we can only consider the latter one in the time analysis. Since the scale of the two pie charts is significantly different, the first one is minute-based while the other is hour-based. The analysis shows that the system is bounded by the Monte Carlo simulation. This is due to the need of repeating several times the score computation over the random-generated LMIs, in order to get the maximum score association for the null hypothesis rejection phase.

## 11.6 CONCLUDING REMARKS

This chapter has introduced the reader to the problem of predicting, via computational approaches, LDAs. Machine learning approaches and big data technologies have a central role for the design and development of prediction tools with this aim, and we have provided an overview of the main techniques proposed in the last few years on this topic. In order to explain further details on both the problem under consideration and aspects related to its solution, we have focused on a specific approach based on neighborhood analysis through a tripartite graph built on LMIs and MDAs. An important feature of this approach is that it predicts potential LDAs without relying on the information of known disease–lncRNA associations. In particular, neighborhood analysis performs better than other techniques previously presented in the literature and not based on this information, such as $p$-value based on hypergeometric distribution. Moreover, approaches based on integrative networks have indeed been shown to perform better than others. Therefore, we may conclude that a promising future direction for the research in this field may be to consider more complex pipelines, where different types of molecular interactions and associations are taken into account (e.g., gene–lncRNA co-expression relationship, lncRNA–protein interactions, etc.), suitably combined with network neighborhood analysis and network clustering approaches (Pizzuti et al., 2007, 2012, 2014).

## ACKNOWLEDGMENTS

This research has been partially supported by the PRIN research project "Multicriteria Data Structures and Algorithms: from compressed to learned indexes, and beyond", grant n. 2017WR7SHH, funded by the Italian MIUR.

## REFERENCES

Alaimo, S., Giugno, R. and Pulvirenti, A. (2014). ncPred: ncRNA-disease association prediction through tripartite network-based inference. *Frontiers in Bioengineering and Biotechnology*, 2:71.

Bonomo, M., La Placa, A. and Rombo, S. E. (2020). Prediction of lncRNA-disease associations from tripartite graphs. In *Heterogeneous Data Management, Polystores, and Analytics for Healthcare: VLDB Workshops, Poly 2020 and DMAH 2020, Virtual Event, August 31 and September 4, 2020, Revised Selected Papers*. Cham: Springer Nature, p. 205.

Chen, G., et al. (2013). LncRNADisease: a database for long-non-coding RNA-associated diseases. *Nucleic Acids Research*, 41(D1):D983–D986.

Chen, X. (2015). Predicting lncRNA-disease associations and constructing lncRNA functional similarity network based on the information of miRNA. *Scientific Reports*, 5(1):13186.

Chen, X. and Yan, G.-Y. (2013). Novel human lncRNA – disease association inference based on lncRNA expression profiles. *Bioinformatics*, 29(20):2617–2624.

Chen, X., You, Z.-H., Yan, G.-Y. and Gong, D.-W. (2016). IRWRLDA: improved random walk with restart for lncRNA-disease association prediction. *Oncotarget*, 7(36):57919–57931.

Cheng, L., et al. (2016). IntNetLncSim: an integrative network analysis method to infer human lncRNA functional similarity. *Oncotarget*, 7(30):47864–47874.

Chou, C. H. et al. (2016). miRTarBase 2016: updates to the experimentally validated miRNA-target interactions database. *Nucleic Acids Research*, 44(D1):D239–D247.

Cui, Z. et al. (2020). LncRNA-disease associations prediction using bipartite local model with nearest profile-based association inferring. *IEEE Journal of Biomedical and Health Informatics*, 24(5):1519–1527.

Deng, L., Li, W. and Zhang, J. (2020). LDAH2V: exploring meta-paths across multiple networks for lncRNA-disease association prediction. *IEEE Xplore*, 1545–5963.

Fan, X.-N., et al. (2019.) Prediction of lncRNA-disease associations by integrating diverse heterogeneous information sources with RWR algorithm and positive pointwise mutual information. *BMC Bioinformatics*, 20(1):87.

Giancarlo, R., Rombo, S. E. and Utro, F. (2015). Epigenomic k-mer dictionaries: shedding light on how sequence composition influences in vivo nucleosome positioning. *Bioinformatics*, 31(18):2939–2946.

Gibb, E. A., Brown, C. J. and Lam, W. L. (2011). The functional role of long non-coding RNA in human carcinomas. *Molecular Cancer*, 10(1):38.

Gordon, A. D. (1996). *Null Models in Cluster Validation*. Berlin and Heidelberg: Springer, 32–44.

Guo, Z.-H., Yi, H.-C. and You, Z.-H. (2019). Construction and comprehensive analysis of a molecular association network via lncRNA – miRNA – Disease – Drug – Protein graph. *Cells*, 8(8):866.

Hao, Y., et al. (2016). NPInter v3.0: an upgraded database of noncoding RNA-associated interactions. *Database*, 2016:baw057.

Ikematsu, K. and Murata, T. (2013). A fast method for detecting communities from tripartite networks. In *Proceedings of International Conference on Social Informatics*. New York: Springer International Publishing, pp. 192–205.

Jiang, Q., et al. (2009). miR2Disease: a manually curated database for microRNA deregulation in human disease. *Nucleic Acids Research*, 37(suppl_1):D98–D104.

Kibbe, W. A., et al. (2015). Disease Ontology 2015 update: an expanded and updated database of human diseases for linking biomedical knowledge through disease data. *Nucleic Acids Research*, 43(D1):D1071–D1078.

Lan, W., et al. (2016). LDAP: a web server for lncRNA-disease association prediction. *Bioinformatics*, 33(3):458–460.

Li, J.-H., et al. (2013). starBase v2.0: decoding miRNA-ceRNA, miRNA-ncRNA and protein – RNA interaction networks from large-scale CLIP-Seq data. *Nucleic Acids Research*, 42(D1):D92–D97.

Li, J.-H., et al. (2019). A novel approach for potential human lncRNA-disease association prediction based on local random walk. *IEEE/ACM Transactions on Computational Biology and Bioinformatics*, 1.

Li, X., Wu, Z., Fu, X. and Han, W. (2014). lncRNAs: insights into their function and mechanics in underlying disorders. *Mutation Research/Reviews in Mutation Research*, 762:1–21.

Li, Y., et al. (2014). HMDD v2.0: a database for experimentally supported human microRNA and disease associations. *Nucleic Acids Research*, 42(D1):D1070–D1074.

Liao, Q., et al. (2011). Large-scale prediction of long non-coding RNA functions in a coding-non-coding gene co-expression network. *Nucleic Acids Res*, 39(9):3864–3878.

Liu, C.-J., et al. (2017). lncRInter: a database of experimentally validated long non-coding RNA interaction. *Journal of Genetics and Genomics*, 44(5):265–268.

Liu, Y., et al. (2019). A novel network-based computational model for prediction of potential lncRNA – disease association. *International Journal of Molecular Sciences*, 1–12.

Lu, C., et al. (2018). Prediction of lncRNA-disease associations based on inductive matrix completion. *Bioinformatics*, 34(19):3357–3364.

Lu, C., et al. (2020). Predicting human lncRNA-disease associations based on geometric matrix completion. *IEEE Journal of Biomedical and Health Informatics*, 24(8):2420–2429.

Ning, S., et al. (2016). Lnc2Cancer: a manually curated database of experimentally supported lncRNAs associated with various human cancers. *Nucleic Acids Research*, 44(D1):D980–D985.

Ou-Yang, L., et al. (2019). LncRNA-disease association prediction using two-side sparse self-representation. *Frontiers in Genetics*, 10:476.

Park, C., Yu, N., Choi, I., Kim, W. and Lee, S. (2014). lncRNAtor: a comprehensive resource for functional investigation of long non-coding RNAs. *Bioinformatics*, 30(17):2480–2485.

Ping, P., Wang, L., Kuang, L., Ye, S., Iqbal, M. F. B. and Pei, T. (2019). A novel method for lncRNA-disease association prediction based on an lncRNA-disease association network. *IEEE/ACM Transactions on Computational Biology and Bioinformatics*, 16(2):688–693.

Pizzuti, C. and Rombo, S. E. (2007). PINCoC: A co-clustering based approach to analyze protein-protein interaction networks. In *Proceedings of International Conference on Intelligent Data Engineering and Automated Learning*. Birmingham: Springer-Verlag, pp. 821–830.

Pizzuti, C. and Rombo, S. E. (2012). A coclustering approach for mining large protein-protein interaction networks. *IEEE/ACM Transactions on Computational Biology and Bioinformatics*, 9(3):717–730.

Pizzuti, C. and Rombo, S. E. (2014). Algorithms and tools for protein-protein interaction netwwork clustering, with a special focus on population-based stochastic methods. *Bioinformatics*, 30(10):1343–1352.

Schriml, L. M., et al. (2012). Disease ontology: a backbone for disease semantic integration. *Nucleic Acids Research*, 40(D1):D940–D946.

Shi, J.-Y., et al. (2017). Predicting binary, discrete and continued lncRNA-disease associations via a unified framework based on graph regression. *BMC Medical Genomics*, 10(4):65.

Sumathipala, M., Maiorino, E., Weiss, S. T. and Sharma, A. (2019). Network diffusion approach to predict lncRNA disease associations using multi-type biological networks: LION. *Frontiers in Physiology*, 10:1–11.

Taft, R. J., et al. (2010). Non-coding RNAs: regulators of disease. *The Journal of Pathology: A Journal of the Pathological Society of Great Britain and Ireland*, 220(2):126–139.

Tan, H., Sun, Q., Li, G., Xiao, Q., Ding, P., Luo, J. and Liang, C. (2020). Multiview consensus graph learning for lncRNA – disease association prediction. *Frontiers in Genetics*, 11:89.

Vergoulis, T., et al. (2012). TarBase 6.0: capturing the exponential growth of miRNA targets with experimental support. *Nucleic Acids Research*, 40(D1):D222–D229.

Wang, B. and Zhang, J. (2020). Logistic regression analysis for lncRNA-disease association prediction based on random forest and clinical stage data. *IEEE Access*, 8:35004–35017.

Wang, M.-N., et al. (2020). LDGRNMF: lncRNA-disease associations prediction based on graph regularized non-negative matrix factorization. *Neurocomputing*:925–2312.

Wang, Q. and Yan, G. (2019). IDLDA: an improved diffusion model for predicting lncRNA – disease associations. *Frontiers in Genetics*, 10:1259.

Wang, Y., et al. (2019). LncDisAP: a computation model for lncRNA-disease association prediction based on multiple biological datasets. *BMC Bioinformatics*, 20(16):582.

Wei, H., Liao, Q. and Liu, B. (2019). iLncRNAdis-FB: identify lncRNA-disease associations by fusing biological feature blocks through deep neural network. *IEEE/ACM Transactions on Computational Biology and Bioinformatics*:1–1.

Xiao, F., Zuo, Z., Cai, G., Kang, S., Gao, X. and Li, T. (2009). miRecords: an integrated resource for microRNA – target interactions. *Nucleic Acids Research*, 37(suppl_1):D105–D110.

Xie, G., Meng, T., Luo, Y. and Liu, Z. (2019). SKF-LDA: similarity kernel fusion for predicting lncRNA-disease association. *Molecular Therapy-Nucleic Acids*, 18:45–55.

Xuan, Z., et al. (2019). A probabilistic matrix factorization method for identifying lncRNA-disease associations. *Genes*, 10(2):126.

Yu, G., et al. (2017). BRWLDA: bi-random walks for predicting lncRNA-disease associations. *Oncotarget*, 8(36):60429–60446.

Yu, J., Ping, P., Wang, L., Kuang, L., Li, X. and Wu, Z. (2018). A novel probability model forLncRNA – disease association prediction based on the Naïve Bayesian classifier. *Genes*, 9(7):345.

Zaharia, M., et al. (2016). Apache Spark: a unified engine for big data processing. *Communications of the ACM*, 59(11):56–55.

Zhang, H., et al. (2019). Predicting lncRNA-disease associations using network topological similarity. *Mathematical Biosciences*, 315:108229.

Zhang, J., Zhang, Z., Chen, Z. and Deng, L. (2019). Integrating multiple heterogeneous networks for novel lncRNA-disease association inference. *IEEE/ACM Transactions on Computational Biology and Bioinformatics*, 16(2):396–406.

Zhang, Y., et al. (2020). LDAI-ISPS: lncRNA – disease associations inference based on integrated space projection scores. *International Journal of Molecular Sciences*, 21(4):1508.

Zhao, W., Luo, J. and Jiao, S. (2014). Comprehensive characterization of cancer subtype associated long non-coding RNAs and their clinical implications. *Scientific Reports*, 4(1):6591.

Zhao, X., Yang, Y. and Yin, M. (2020). MHRWR: prediction of lncRNA-disease associations based on multiple heterogeneous networks. *IEEE/ACM Transactions on Computational Biology and Bioinformatics*:1–1.

Zhou, B., et al. (2018). EVLncRNAs: a manually curated database for long non-coding RNAs validated by low-throughput experiments. *Nucleic Acids Research*, 46(D1):D100–D105.

Zhu, W., et al. (2019). ALSBMF: predicting lncRNA-disease associations by alternating least squares based on matrix factorization. *IEEE Access*, 8:26190–26198.

# *Section IV*

## *Applications in Clinical Diagnosis*

# 12 Tracking Slow Modulations of Effective Connectivity for Early Epilepsy Detection

## *A Review*

*Vishwambhar Pathak, Praveen Dhyani, Prabhat Mahanti, and Vivek Gaur*

**CONTENTS**



DOI: 10.1201/9781003142751-16

## 12.1 INTRODUCTION

The standard steps for epilepsy classification, diagnosis, and management described by the International League Against Epilepsy use various etiologies. Each etiology carries a set of significant therapeutic implications arising due to various types of seizures, syndromes, and presentations characterizing epilepsy (Sheffer et al., 2017). Among the six etiologies, the structural etiology portends typical ictal associations among cortical regions, visible on structural neuroimaging. An epileptogenic brain can be conceived as belonging to one of the four states, namely, the interictal, the preictal, the ictal, and the postictal. The problem of early detection and the prediction of seizures can be perceived as being a binary classification problem with two classes: preictal and nonictal. Researchers have investigated the patterns of preictal and ictal phase synchronization among local (neighboring) and distant cortical areas. Several such reports, including Dominguez et al. (2005), emphasized the influence of local synchronization at specific frequency bands as neuromarkers of seizures. However, as described in this review, the counterhypothesis that slow (long-time) fluctuations in synchronization of distant cortical regions characterize the seizures. Hence, the ability to identify such subtle patterns is key to effective prognosis and treatment. The testing of such a hypothesis, however, would require investigating exhaustive combinations in spatial and temporal domains. Sparsity preserving algorithmic workflows, on one hand, and deep learning techniques employing computational power of graphics processing units (GPUs), on the other hand, support implementation of such investigations. Barnett et al. (2018) emphasized that the success of any model estimation requires resolving the critical data quality issues, namely non-stationarity, nonlinearity, exogenous influences, noise, sampling rates, and temporal/spatial features related to the neurophysiological recordings. Accordingly, this chapter presents in the following sections a review of existing measures and methods for findings such neuromarkers with critical analysis to underline possibilities of qualitative improvement and speeded computation. Major emphasis has been given to bivariate measures that have been found at the core of required epileptic brain–region–connectivity studies. It further proposes a design of model with a theoretical description convincing that it overcomes the limitations of the traditional methods and enables estimation of the causal–connectivity patterns within the limits of computational tractability.

## 12.2 TRADITIONAL MEASURES AND METHODS OF ICTOGENESIS

Figure 12.1 shows the categorization of the measures applied in ictogenesis studies based on the intended revelations of the measures and methods applied (Carney et al., 2011; Friston et al., 2013; Cohen, 2014). A nonlinear dynamical system like the evolving brain networks is usually characterized by the strength of excitation of individual subsystems and the integration and direction of synchronized excitations of the subsystems. A variety of workflows utilizing either univariate measures, bivariate measures, or a combination of the two have been investigated for quantifying the ictal and preictal patterns to enable epileptic seizure prediction, as surveyed in Carney et al. (2011) and Lehnertz et al. (2007, 2017). However, while the presumptions underlying one or the other model showed success in particular hypotheses tested over selected datasets, attempts to generalize the findings have seen limited success.

### 12.2.1 Measures of Brain Segregation Analyses

The structural (anatomical) connectivity of the specialized cortical and subcortical regions via large-range fiber bundles, known as functional segregation, investigated using diffusion tensor magnetic resonance imaging (MRI), is estimated typically using univariate measures of the cortical signals. Various formulations including Fourier transform, short-term Fourier transform, discrete wavelet transforms, accumulated energy, statistical moments (mean, variance, skewness), correlation dimension, correlation density, autocorrelation and autoregressive model, Kolmogorov entropy, spectral entropy, loss of recurrence, and Lyapunov exponent quantify different properties of the underlying phenomena. For example, the Fourier transform

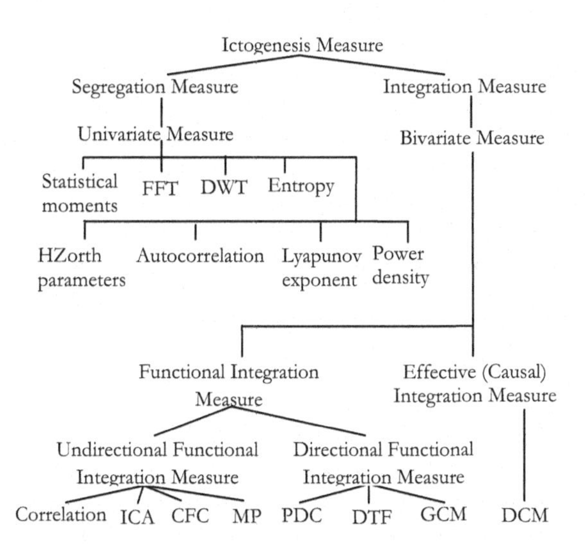


**FIGURE 12.1** Categorization of Common Ictogenesis Measures: Segregation/Integration (functional [directional/undirectional]/effective).

of signals helps identify the major frequency ranges with respective strengths. Extensive descriptions of these methods and the case-wise utility of each can be found in Cohen (2014) and Rasekhi et al. (2015).

### 12.2.2 Measures of Integration

The functional integration of the functionally segregated systems is analyzed in terms of functional connectivity (FC) and effective connectivity (EC) related to resting state, event or task performance mostly using electroencephalogram (EEG), electrocorticography (ECoG), or magnetoencephalography (MEG) signals due to their high temporal resolution as compared to seldom-used functional MRI (fMRI) recording, which provides high spatial resolution but less temporal resolution (Friston et al., 2013). FC is a measure of statistical dependence (or similarity) of excitations of different brain regions, and the contrasting EC describes causal (directed) dependence among the hidden states (cortical and subcortical activities) underlying the generation of the functional synchronizations, typically determined by phase-dependent coherence. Popular bivariate measures of undirected-FC include independent component analysis (ICA), correlation, and coherence and various measures of synchrony, for example, mutual information and cross-frequency coupling (CFC). Common measures of directed FC (temporal precedence) include Granger causality (GC), transfer entropy (TE), partial directed connectivity (PDC), and directed transfer function (DTF). EC (Gerstein & Perkel, 1969) is usually estimated through dynamic causal models (DCMs; Kiebel et al., 2007) expressed in terms of (discrete time) ordinary difference equations (ODEs) or (continuous-time) partial differential equations (PDEs) whose parameters represent various intrinsic and extrinsic neuronal excitations and interactions. The current work focuses on finding tractable and effective workflow for implementation of the effective (causal) connectivity.

#### 12.2.2.1 Estimation of Undirected Functional Connectivity

Bivariate functional connectivity measures have been favored over all-brain multivariate mass-connectivity methods, mostly due to ease of implementation, visualization, and statistical quantification. Even the apparently multivariate connectivity methods are essentially mass-bivariate as they combine the pairwise connectivity values. The availability of various connectivity methods (Cohen, 2014) provides flexibility for custom-tailoring for testing specific research hypotheses. Although these methods, including CFC, PAC, and MP, among others, stand out in revealing (slow) long-range connectivity dynamics with high spatiotemporal resolution, still these measures do not enable directionality analyses.

#### 12.2.2.2 Estimation of Directed Functional Connectivity

Directed FC is commonly estimated using GC, which gives a (bivariate) measure of the directional dependencies among the empirically sampled neurophysical excitations. Multivariate estimations of directed FC are performed using DTF (Kaminski & Blinowska, 1991; Blinowska, 2011) and PDC (Baccalá & Sameshima, 2001) in which non-stationarity is resolved by analyzing over a short time window a signal within which stationarity can be assumed. Time-varying directed FC can be computed using adaptive DTF (Wilke et al., 2007) and adaptive PDC (Astolfi et al., 2008).

#### 12.2.2.3 Estimation of Causal (Effective) Connectivity

Effective connectivity (Gerstein & Perkel, 1969) requires formulation of the underlying generative parameterized dynamic causal model (DCM) (Kiebel et al., 2007) expressed in terms of a difference equation for discrete time representations and differential equations in the case of continuous-time representations. The same can be estimated using model comparison and optimization (Friston, 2011; Freestone et al., 2014; Lehnertz et al., 2017). A noise-invariant DCM employing Kalman smoothing–based expectation maximization was described in Nalatore et al. (2007). But Sommerlade et al. (2012) revealed the model's limitations in handling critical dynamics at the source process. Recently, researchers have observed convergence of the two different views, namely, GC and DCM, and have recommended harnessing the benefits of both (Friston, 2011; Friston et al., 2013; Friston et al., 2014).

### 12.2.3 Concerns and Techniques for Data Preparation

#### 12.2.3.1 Noise, Artifacts, and Volume Conduction

EEG signals mostly contain mix of signal and noise. The spurious oscillations assumed to be caused due to subject movement; device error, volume conductions, and amplifier saturation are considered noise and are often handled by applying suitable filters before using the data. However, what is defined as a signal and what otherwise is defined as noise is largely guided by the goal of the study. The characteristic patterns of epilepsy and other diseases using high-frequency oscillations have been reported in (Gotman, 2010; Modur, 2014; Seo et al., 2020). Artifacts generated by volume conduction arising from a spurious effect of activity of a single source on the recording of oscillations at multiple electrodes with zero or near-zero phase delay can be mitigated by first extracting orthogonal components by orthogonalizing their spectrum power envelopes before performing the connectivity analysis. Parametric methods like Morlet wavelets and Hilbert transforms or nonparametric transformations like fast Fourier transform (FFT) are commonly applied to achieve orthogonalization. However, functional connectivity models based on the nonparametric direct data-decomposition techniques like FFT and wavelets do not represent the generative process of the EEG, which is essentially correlated across frequency bands, thereby obscuring the exploitation of the underlying time–frequency structure. The resulting decomposition therefore must yield multiple noncorrelated components in different frequency bands. Moreover, in this process, the data are decomposed into components within a series of fixed frequent bands corresponding to basic functions. The time–frequency structure of these resultant components is intractable and necessitates ad hoc assumptions to define the frequency bands in a way to further derive interpretations. Model-based approaches allow for the separation of complex signals into possibly correlated components on the basis of their time-frequency behavior. Parametric spectral estimation by an autoregressive (AR) model provides a flexible framework for dynamic modelling of the signals. Multivariate autoregressive (MVAR) models represent interactions between multiple EEG signals in the form of linear difference equations in state space using which the directional influences, as well as the direction of the information flow between channels, can be estimated. Deep learning methods applying hyperparameter tuning for artifact removal have been reviewed in Val-Calvo et al. (2019), who emphasized invariant feature encoding for fast real-time estimation. Koopman embedding employed in the proposed workflow in Section 12.4 is a promising scheme.

#### 12.2.3.2 Handling Non-Stationarity

To handle the non-stationarity present in the signals, the characteristic piecewise stationarity of the signals is exploited. Such a strategy is to compute short-term Fourier transforms over suitable segments of signals; however, the chosen resolution must ensure balance considering the time–frequency trade-off. A schematic quantifying the degree of non-stationarity was proposed in Mormann et al. (2005), which first computed the loss of recurrence, thence determining the predictability of a seizure onset. Autocorrelations among signals at different time points along the seizure onset zone have been estimated using autoregressive models in vast research to characterize the preictal patterns (Mormann, 2003; Mormann et al., 2007). These models exploit the piecewise stationarity of the underlying signal by applying variety of strategies, for example, computing analytic measures over sliding windows of suitable length. Window length is a crucial parameter of such models. The regime-switching vector autoregression (VAR) model described in Ting et al. (2017) employed windows of varying length. The DCM described in Papadopoulou et al. (2015) highlighted the different timescales of the neuronal bursting (fast) and the variations in synaptic excitations (slow) and hence assumed local stationarity within successive epochs.

## 12.3 ESTIMATING NEUROMARKERS OF EPILEPSY BASED ON SLOW MODULATIONS IN CONNECTIVITY

### 12.3.1 Evidence of Patterns of Long-Range Slowing Connectivity in Preictal Period and Related Measures and Models

Preictal patterns of the shifts of phase distribution (0.5–0 rad) of the phase–amplitude coupling, specifically, between slow-wave phases (delta band) and lower gamma amplitude have been found (Steriade et al., 1996; Steriade, 2003; Labyt et al., 2007; Le Van Quyen et al., 2010; Valderrama et al., 2012; Alvarado-Rojas et al., 2014; Friston et al., 2014). To achieve high resolution for estimating slow modulations occurring over long-range relations, the power spectral densities were computed over windows of 5 sec. In the method devised by Mormann et al. (2005), the effects of slow changes in neural dynamics were investigated by comparing the connectivity among the channels during the preictal period of a seizure against the preceding interictal period. However, clustering of seizures in some patients caused difficulty in generalizing this scheme. For the viability of comparison of distributions, the authors recommended interictal interval(s) to be taken longer than 35 min before and after the seizure point. This constraint, in turn, imposed rejection of seizures showing up at a lesser interval. Their investigation discovered that the seizures caused a considerable reduction of power in lower frequencies (delta band) and the resultant increase in values of Hjorth parameters (Hjorth Mobility [HM] and Hjorth Complexity [HC]) and a decrease in decorrelation time. On the other hand, a simultaneous relative increase in other bands was observed. The largest Lyapunov exponent and the correlation dimension were observed to increase simultaneously, in contradiction to their inverse correlativity reported in earlier studies (Pereda et al., 1998). The authors argued that the increase in the previously mentioned measures and the decrease in the local flow were caused due to the decrease in delta power. Studies reported by Labyt et al. (2007), Richardson (2012), and Alvarado-Rojas et al. (2014) revealed patterns of slow preictal changes in neuro-cortical excitations evolving over long-term time scale. Alvarado-Rojas et al. (2014) formulated triggering of preictal alarm based on the power of the delta (0.1–4 Hz), theta (4–8 Hz) and gamma (30–140 Hz) bands, relative to the optimal threshold over the power of the whole frequency range. High resolution was achieved by computing power spectral density over windows of 5 sec. They assessed the sensitivity and specificity of cross-frequency coupling changes from long-term interictal data and discovered preictal patterns of strong modulations between slow wave phases (delta band) and lower gamma amplitude, a result in coherence with earlier similar recommendations by Steriade et al. (1996), Steriade (2003), Le Van Quyen et al. (2010), and Valderrama et al. (2012). They also discovered slow modulation of power of higher gamma band (40–140 Hz) within intracranial regions. They observed a shift of phase distribution (0.5–0 rad) of the coupling during the preictal period along 10 minutes before a seizure. These preictal changes were not typically described by conventional power spectral analyses. Table 12.1 presents a summary of the neuromarkers of epilepsy reported in the surveyed research.

### 12.3.2 Time-Varying Connectivity Inference Methods

Time-varying connectivity has traditionally been measured using the model-free descriptive approaches and the model-based generative approaches. But of larger interest of the ictogenesis research community are the generative model-based approaches. We reviewed in detail the two popular generative model-based methods namely the Granger causal model (GCM), the DCM, and recent deep learning models. A summary of the measures and models for estimating connectivity among brain regions, as reported in the surveyed research, has been presented in Table 12.2 and is followed by Table 12.3, which presents a summary of how different models address the issues critical to the success of estimation.

#### 12.3.2.1 GC

Although being a popular measure of directional connectivity, the GC carries the challenge of determining the optimal autoregressive order to ensure convergence (Friston et al., 2014; Driver, 1977; Jirsa & Ding, 2004). Robust methods for nonparametric estimates of GC employing spectral matrix factorization have been proposed, for example, the Wilson–Burg estimates described by Nedungadi et al. (2009).

**TABLE 12.1**
**Preictal EEG Patterns as Neuromarkers for Epilepsy Prediction**

| Paper | Neuromarkers | Measures Applied |
|---|---|---|
| Steriade et al. (1996), Steriade (2003), Labyt et al. (2007), Le Van Quyen et al. (2010), Valderrama et al. (2012) | Strong modulations between slow-wave phases (delta band) and lower gamma amplitude | CFC, PSD |
| Mormann et al. (2005) | Reduction of power in lower frequencies (delta band), hence increase in HM and HC, and decreased decorrelation time but opposite effect in other bands; increase in the largest Lyapunov exponent and correlation dimension | Hjorth parameters, PSD, Lyapunov exponent; correlation dimension |
| Alvarado-Rojas et al. (2014) | Strong modulations between slow wave phases (delta band) and lower gamma amplitude; slow modulation of power in higher gamma band (40–140 Hz); shift of phase distribution (0.5 rad to 0 rad) of the phase–amplitude coupling in delta and lower gamma bands | CFC, PSD, over 5-sec window using Burg method |

**TABLE 12.2**
**Measures and Models for Estimation of Connectivity among Brain Regions**

| Paper | Model | Measure/Method | Advantage | Limitation |
|---|---|---|---|---|
| Granger and Joyeux (1980), Silvennoinen and Terasvirta (2009), Friston et al. (2014) | GC | VAR, VARFIMA, GARCH | Models colinearity, non-stationarity, long-term memory, heteroscedasticity | Non-convergence due to decaying autocorrelations |
| Nedungadi et al. (2009), Stokes and Purdon (2017), Dhamala (2018) | GGC | Spectral matrix factorization (SMF) | Better nonparametric estimates of GC in the presence of minimum phase linear dependencies | Bias and high variance |
| Barnett et al. (2018), Quiroga et al. (2002) | GGC | SMF | Bias and high variance of GGC resolved by single full model regression | Still higher sensitivity to noise |

| Paper | Model | Measure/Method | Advantage | Limitation |
|---|---|---|---|---|
| Carney et al. (2011), Martinez-Vargas et al. (2017) | GC | DTF, PDC (linear measures of coupling) | Better estimates of connectivity than the nonlinear methods | DTF detects indirect connectivities too; PDC's weak dependence on frequency |
| Ting et al. (2017) | GC | RSVAR | Smooth detection of slow and abrupt effective connectivity dynamics | Difficulty in determining optimal lag level (p-order) |
| He and Parida (2014), Ren et al. (2020) | MVGC | Transductive-HSIC-LASSO; HSIC-Lasso-GC | Causal connectivity for exhaustive combinations of brain regions, minimum redundancy | Explicit traditional handling of non-stationarity |
| Papadopoulou et al. (2015), Friston et al. (2016), Rosch et al. (2018) | DCM | Transfer function measure; visualized using a correlation matrix of mean Fourier amplitude for each region, time point. | Patterns of variations along pre- and ictal states and slow changes along seizure onset; long-range effective connectivity | Assumes homogeneity of functional architecture across all participants |
| Barnett et al. (2018), Friston and Penny (2011), Friston et al. (2013) | GC + DCM | GC learns brain's dynamics, DCM estimates models of connectivity | Exhaustive and double sure connectivity estimation | Optimal lag-level (p-order) for GC and the window-length for DCM |
| Noé and Nuske (2013), Brunton et al. (2016b), Hernandez et al. (2017), Friston et al. (2017), Omidvarnia et al. (2017), Wehmeyer and Noé (2018), Zhang et al. (2018) | Deep learning | VAEs, KL divergence, auxiliary network | Nonparametric estimation; handles sparsity, non-stationarity, overfitting | Interpretability of the determining measures is difficult |
| Lusch et al. (2016), Li et al. (2017), Otto and Rowley (2017), Takeishi et al. (2017), Yeung et al. (2017), Mardt (2018), Zhang et al. (2020) | Deep learning | Koopman eigenfunctions; phase coherence | Compact embedding and interpretable minimal description of system dynamics; implicit handling of non-stationarity | (Scope for research) No literature found on its application for epileptic-seizure characterization |

**TABLE 12.3**
**Approaches to Address Issues Critical to the Success of Connectivity Analysis**

| Model<br>Issue | GC (VAR/ DTF/PDC/ GGC/ Partial GC) | DCM | Deep Learning |
|---|---|---|---|
| **Noise, and Exogenous influence** | Separation of complex signals into correlated components by AR-based parametric estimation | The generative model includes spontaneous, stochastic fluctuations | Hyperparameter tuning for invariant feature encoding (e.g., Koopman encoding) required for fast estimation |
| **Non-Stationarity** | Piecewise stationarity by computing connectivity measures over sliding windows; choice of window length is critical; long-term memory effects not well handled | Generative model assumes local stationarity within successive epochs | Using features in time-frequency domain; apply time-frequency analysis method, e.g., empirical mode decomposition |
| **Nonlinearity** | Linear GC models, e.g., DTF, PDC found better than nonlinear estimation models like GGC | Essentially nonlinear formulation of the generative model of the system's dynamics | Essentially nonlinear learning |
| **Temporal/ Spatial Aggregation** | Time-varying spatiotemporal representations in DTF, PDC | Separation of temporal scales highlight role of slow fluctuations in seizure excitations | Convolution absorbs spatial relations |

Heteroscedasticity, if present, even in a stationary signal, may affect statistical significance tests. Models and tools reported by Ding et al. (2000), Seth (2010), Seth et al. (2015), Barnett and Seth (2014, 2015), and Barnett et al. (2018) have addressed the major issues in time-varying multivariate GC estimation using state–space VAR models, namely colinearity, non-stationarity, long-term memory, and heteroscedasticity. Stronger methods like vector autoregressive fractionally integrated moving average (VARFIMA) (Granger & Joyeux, 1980), generalized autoregressive conditional heteroscedastic (GARCH) Silvennoinen & Terasvirta, 2009), and other formulations, for example, those devised by Ozaki (2012), enable better handling of the signals carrying slow-moving averages and nonexponentially decaying autocorrelations. The VAR models underlying the GCMs provide a balance between the less expensive but limited utility model-free methods, for example, mutual information and transfer entropy, on one hand, and on being the model-based approach of DCMs that are promising but make explicit assumptions (state equations) about the underlying generative process to fit into the observed data (observation equation), on the other hand.

Friston et al. (2014) found that *long-range correlations* exist and display *slowing* near (transcritical) bifurcations because of slowly decaying fast (e.g., gamma) activity, due to which the corresponding eigenvalues approach zero from below. This causes the nonconvergence of the autoregressive measures computed from the eigenvalues, for example, the Lyapunov exponent. Such studies revealed unreliability of spectral GC computed from finite-order autoregressive processes. Solutions employing even higher order autoregressive processes may also turn infinite in the case of a (nearly) singular cross-covariance matrix. Limitations of autoregressive formulations have also been argued highlighting the essential delays occurring in dynamics of biological systems, which induce an infinite number of hidden states in a state-space model (Driver, 1977; Jirsa & Ding, 2004). Several schematics have been devised to solve this difficulty. Applicability of spectral matrix factorization for producing nonparametric estimates of GC was recommended by Nedungadi et al. (2009). Granger–Geweke causal (GGC) analysis has also used spectral matrix factorization to compute a reduced model parameters from full model (Dhamala, 2018). As argued by Stokes and Purdon (2017), the bias and high variance arise as limitations of the GGC estimation due to the use of "separate, independent, full, and reduced regressions". However, as reviewed by Barnett et al. (2018), such limitations have been resolved in further schematics employing spectral matrix factorization–based models and deriving GGC estimates from single full-model regression, for example, Wilson's frequency-domain algorithm (Dhamala et al., 2008), Whittle's time-domain algorithm (Barnett & Seth, 2014), and a state-space method involving a solution of Riccati equation (Barnett & Seth, 2015). The nonparametric Wilson–Burg estimates have a greater degree of freedom as compared to the number of coefficients of an AR model, and thus, they offer better estimates of GC in the presence of minimum phase linear dependencies. Thus, nonparametric GC measures have been recommended for characterizing the neurophysiological time series representing coupled and delayed dynamics having slow modes. However, these nonlinear methods have been claimed as being sensitive to noise by some authors, for example, Quiroga et al. (2002). Carney et al. (2011) argued the lack of evidence of the nonlinearity of brain (dys)functioning at meso- or macro-scale in the case of epilepsy, albeit the underlying subsystems' functions, being essentially nonlinear and nonstationary. Considering the context of estimating event-related time-varying connectivities in the presence of noise, linear measures, namely, DTF as well as PDC, give better results than a nonlinear one (Blinowska, 2008; Martinez-Vargas et al., 2017). Winterhalder et al. (2005) revealed the drawback of DTF is that it also detects indirect flows. The disadvantage of PDC is the obscurity of resultant patterns of transitions as the measure emphasizes less the sinks than the sources. Another unfavorable feature of PDC is its inability to distinguish the roles of frequency bands due to weak relation to frequency. Thus, DTF and PDC do not stand out as convincing tools for the said purpose of estimation of dynamic directional connectivity across various rhythms and brain regions. DTF and GGC have been found to be sensitive to white noise input and latent variables. This issue was resolved by nonparametric inference of partial GC estimation (Guo et al., 2008; Roelstraete & Rosseel, 2012).

The regime-switching VAR model for effective connectivity estimation from fMRI data, as described by Ting et al. (2017), has been shown to be robust enough for smooth detection of slow and abrupt effective connectivity dynamics as well as offering scalability to inexpensively investigate sufficiently large numbers of brain regions. The implementation of VAR-based models, however, requires grounded mechanism for determining optimal lag level (p-order). Ren et al. (2020) exposed the bias and variance-related limitations of the traditional VAR-based bivariate GC in computing nonlinear connectivity estimates and described improved workflow for generalizing the traditional bivariate GC to facilitate recognition of causal relationships among multivariate time series. They first computed stationarity test and reconstructed the state space to compute historical information. Then a "Hilbert–Schmidt independence criterion Lasso Granger causality" (HSIC-Lasso-GC) model was estimated using generalized information criterion. Employing further significance tests thus enabled the estimation of nonlinear multivariate causal connectivity analysis for exhaustive combinations of brain regions. The HSIC-Lasso regularization method selects features with minimum redundancy and associates only the most probable label with the selected features. Further improvement of predictive accuracy of such nonlinear sparse regression can be achieved by employing transductive-HSIC-Lasso (He & Parida, 2014), which overcomes the errors/limitations arising in traditional models due to use of only training data by including unlabeled (test) data during redundancy computation.

#### 12.3.2.2 DCM

In a key work on employing DCM for cross-spectral density to track the slow modulations in ictogenetic synaptic gains and their coupling reported by Papadopoulou et al. (2015) provided a balance for the huge number of the hidden states and time bins. They obviated the computation of a huge number of hidden states across numerous time bins by estimating piecewise linear approximations to slow-changing unknown second-order cross-spectral parameters, for example, the transfer functions from the time–frequency responses. The model estimation involved a chain of Bayesian updating by pooling of posterior of one model of seizure as prior for the next. Veridical evidence obtained with these empirical analyses well defined the fluctuations within preictal and ictal states and the slow changes (seconds to minutes) along seizure onset. Variations among the competing models were based on varying the parameters of the assumed neural mass model, namely, the intrinsic (local) excitations and inhibitions as well as the extrinsic (between long-range regions) forward and backward couplings. However, as a way to reduce model parameters, the model validation involved only the selected frequency bands in the range 8 Hz to 48 Hz and limited time span along the pre- and postictal period. A generalization of the DCM-based workflow therefore ought to be difficult owing to the growing model parameters and difficulties associated with a decision for the length of the sliding windows. A Bayesian model inversion method (Rosch et al., 2018) produced a hierarchical DCM model for which the cross-spectral densities were estimated for each time window and each subject using an MVAR process. The process was iterated over short segments of windows of varying lengths. The Bayesian model evidence was evaluated to determine the winning models via an exhaustive search across specific effective connectivity network architectures formed by combinations of <{with, without} × {forward-backward connections, abnormal intrinsic connectivity, extrinsic connectivity change} × {ROI}> for each individual subject.

Then Bayesian model reduction and selection would produce the underlying shared effective connectivity network architectures. The process was shown to effectively discover the slow changes in long-range directed connectivities. Slow fluctuations in distributed activity were visualized using a correlation matrix of mean Fourier amplitude for each region and time point, following the methods described in Friston et al. (2016). This process, however, essentially assumes homogeneity of the functional architecture across all participants (Omidvarnia et al., 2017).

#### 12.3.2.3 Blending GC and DCM

Contemporary research has inspired models exploiting the complementarity of GC and DCM in a way to develop comprehensive models of the brain's dynamics (Valdes-Sosa et al., 2005; Friston & Penny, 2011; Friston et al., 2013; Barnett, 2018). Accordingly, a typical workflow would first apply GC to learn the brain's dynamic responses in 'resting' states or in different spontaneous conditions. Then in the next step DCM based model estimation would be exploited to investigate several hypotheses about connectivities involving the regions learned by GC in the first step. This combined method will help understand characteristic brain regions associated with the seizures. Theoretically promising, however, these methods too are fraught with the challenges related to the decision about GC's optimal autoregressive p-order, on one hand, and difficulty of determining the length of sliding windows in DCM, on the other hand.

#### 12.3.2.4 Deep Learning and Regularization

Exploiting deep learning models with regularization enables handling huge combinations and sparse connectivity. Recent advancements in deep learning and parallel computing technologies have inspired novel methods for scalable and tractable estimation of the time-varying long-range connectivity relations. The ergodic dynamics of the brain have been linked in classical literature (Omidvarnia et al., 2017) to variational Bayesian learning, giving rise to the notion of the "Bayesian brain", and the underlying Kullback–Leibler (KL) divergence estimation has been assimilated to the natural phenomena of minimization of entropy or "free energy" according to the principle of free energy (Friston et al., 2017) of dynamical systems. Efficient implementations of deep learning variational autoencoders (VAEs) have been shown to be capable of identifying slow-moving system parameters (Noé & Nuske, 2013; Brunton et al., 2016b; Hernandez et al., 2017; Wehmeyer & Noé, 2018; Zhang et al., 2018). Deep learning methods for data-driven discovery of Koopman eigenfunctions (Li et al., 2017; Otto & Rowley, 2017; Takeishi et al., 2017; Yeung et al., 2017; Mardt, 2018; Wehmeyer & Noé, 2018) have been developed to produce an interpretable minimal description of the system dynamics and at same time avoid overfitting. Using deep learning–based autoencoders to produce compact embedding of a continuous spectrum generated by nonlinear dynamics was described by Lusch et al. (2016), who employed an auxiliary network, adapting the methods from classical asymptotic and perturbation handling in dynamics systems.

### 12.3.3 Performance Issues and Resolutions

#### 12.3.3.1 Handling High Dimensionality with Low-Dimensional Embedding

The dimensionality reduction is required to improve computation time of multivariate GC model estimation like Markov models and state-space models. Dimensionality reduction has been, in general, performed based on factor analysis or principal component analysis (PCA). Although PCA determines the modes and orders them as per respective representation of variance in the data. However, it does not derive temporal dynamics inherent in neural recording explicitly. As an alternative to FFT and PCA, which could only derive static modal decomposition, empirical mode decomposition (EMD) methods fetched Hilbert spectrum of the oscillatory modes from time-varying nonlinear and non-stationary data (Huang, 1998). However, being single-channel time-domain transforms, EMDs do not derive spatial structures. Singular value decomposition (SVD) has been indicated as a valid approximation of the eigendecompostion. With an increase of size, this method also becomes computationally inhibitive. Hence, there has been a focus on employing modified dynamic mode decomposition (DMD) methods (Kutz et al., 2016) capable of performing spatiotemporal derivations. The DMD achieves tractable computation by determining eigen-components of an augmented matrix of input data with a truncated SVD basis (Brunton et al., 2016a). Other widely used alternatives of PCA include temporal independent component analysis (tICA), sparse tICA, soft-max Markov state models, and diffusion maps. However, VAEs have been largely considered better substitutes for the traditional computationally demanding dimensionality reduction methods. An implementation of VAEs harnessing the deep learning technology has been adapted in the workflow proposed in Section 12.4.

#### 12.3.3.2 Handling Sparsity

To reduce computation cost caused by sparsity, Brunton et al. (2016b) adapted the schematics for handling sparsity in a dynamical system with noise by employing sparse regression (Tibshirani, 1996; James et al., 2013) and compressed sensing (Donoho, 2006; Baraniuk, 2007; Tropp, 2007; Candès & Wakin, 2008). To efficiently capture model evolution and to ensure convergence regularization techniques compensating for sparsity have been recommended. The ℓ1-regularized regression–based least absolute shrinkage and selection operator (LASSO; Tibshirani, 1996) was recommended for data with zero-mean i.i.d. Gaussian noise but is poor in scalability. Brunton et al. (2016b) recommended a sequential thresholded least-squares-based method. Low-rank approximation was performed by Brunton et al. (2014) using the faster proper orthogonal decomposition (POD) (Berkooz et al., 1993) to handle high dimensions. The data-driven Galerkin models were shown to be capable of identifying the characteristic bifurcation regimes within the underlying nonlinear dynamic system. Although being scalable, these methods were found to generate error in the absence of sparsity. For effective reconstruction of sparse attractor dynamics using time-series data, Brunton et al. (2016b) recommended time-delay coordinates (Daniels & Nemenman, 2015)–based modelling of underlying nonlinear generative process best described as chaotic Lorenz systems represented with nonlinear ODEs. These schematics facilitate determining the governing equations from data and hence are highly significant for the context of our said problem as the directional connectivity states are essentially sparse when processed from multichannel EEG over long recordings consisting of ictal, as well as nonictal, periods.

#### 12.3.3.3 Recognizing Parsimonious Network

A robust method for establishing the network of affective synchronized brain regions would be the one that establishes a parsimonious network that is sparsely connected and hence simple enough to keep only an optimal topological ensemble of interactions that best encodes the underlying functional dynamics. Such robust methods for analyzing spatiotemporal dynamics of parsimonious networks have been developed in recent research which exploited the Koopman operator (Koopman, 1931) for data-driven linear embedding of nonlinear dynamics of complex systems by performing eigendecomposition of signals into orthogonal, coherent components (Gaspard, 1998; Budišić et al., 2012; Salova et al., 2019). Such a workflow presented, by Marrouch (2019), applied the Koopman operator to decompose the complex system dynamics into a low-dimensional manifold represented as a hierarchy of periodic and quasi-periodic patterns of connectivity. More important, it was done without performing bandpass filtering. Applying Koopman operator (eigenfunction decomposition) in place of PCA/ICA/Fourier transforms for dimensionality reduction prior to estimating directional connectivity measures will improve computation time. Efficient methods are, however, required to deal with the huge computation involved, more so when decomposing a system with continuous spectra. Recent methods based on deep learning instill optimism.

#### 12.3.3.4 Deriving Causal Connectivity from Phase-Dependent Coherence

Borrowing from Zhang et al. (2020), Yang et al. (2018), and Galilei (1960), causal connectivity can be derived from the instantaneous phase dependency between cause and effect.

#### 12.3.3.5 DMD

The adaptation of the data-driven DMD method of fluid dynamics has been described by Brunton et al. (2016a). Consider the multichannel time series $X$ containing measurements at $m - 1$ time points separated at $\Delta t$ and another time-series $X'$ overlapping $X$ by one $\Delta t$ as in Equation 12.1:

$$X = \begin{bmatrix} | & \cdots & | \\ x_1 & \cdots & x_{m-1} \\ | & \cdots & | \end{bmatrix}, X' = \begin{bmatrix} | & \cdots & | \\ x_2 & \cdots & x_m \\ | & \cdots & | \end{bmatrix} \quad (12.1)$$

Here, each column vector $x_k$, having $|x_k| = n$ would represent measurements at $n$ channels at $k$th time point.

Let us consider a linear operator $A$ of dimension $n \times n$ such that

$$X' = AX \quad (12.2)$$

Then the dynamic mode decomposition of $X$ and $X'$ is given by the eigendecompositon of $A$ defined as in Equation 12.2. Thence, $A$ would represent a linear regression of the nonlinear dynamics relating $X$ and $X'$. However, for a large $n$, considering the expense of computing eigendecompositon of $n \times n$ matrix A, the approach in DMD is to find approximation instead, beginning with an approximation of the observations $X$ by $\hat{X}_t$ given by Equation 12.3:

$$\hat{X}_t = \phi \Lambda^{t} z_0 \tag{12.3}$$

where $z_0$ is the set of weights such that $x_1 = \phi z_0$ and can be solved as a pseudoinverse problem.

The DMD approximation of $X$, denoted $\hat{X}$, is approximation of $X$ with the same dimensionality (Brunton et al., 2016a). A DMD mode $\phi_i$ is a vector that has the same dimension as $x_i$. The magnitude $|\phi_i|$ represents the spatial correlations between the $n$ observable locations. The eigenvalue $\lambda_i$ encodes the temporal dynamics of the spatial mode $\phi_i$. The magnitude and the phase components of $\lambda_i$ reflect the rate of growth or decay and frequency of oscillations.

To harness the advantages of Koopman embedding of complex dynamical systems, DMD methods have been exploited in recent research to approximate the Koopman operator with finite-dimensional representation (Rowley et al., 2009; Schmid, 2010; Kutz et al., 2016). However, the DMD methods have been discredited for an inability to capture nonlinearity due to their inherent linear system measurement. Nonlinearity-absorbing models, namely, extended DMD (eDMD) and VAC, initially developed to simulate molecular dynamics have been later adapted for the purpose (Noé & Nuske, 2013). The cross-validation of the variational score used in VAC enables objective assessment of Koopman embedding. The correspondence of eDMD, VAC, and tICA was reviewed by Klus et al. (2018).

## 12.4 PROPOSED WORKFLOW

Based on the critical review of existing methods, a computationally plausible workflow is proposed herewith for computing the effective connectivity estimates and their slow modulations underlying epileptic cortical activities. The schematic is shown in Figure 12.2.

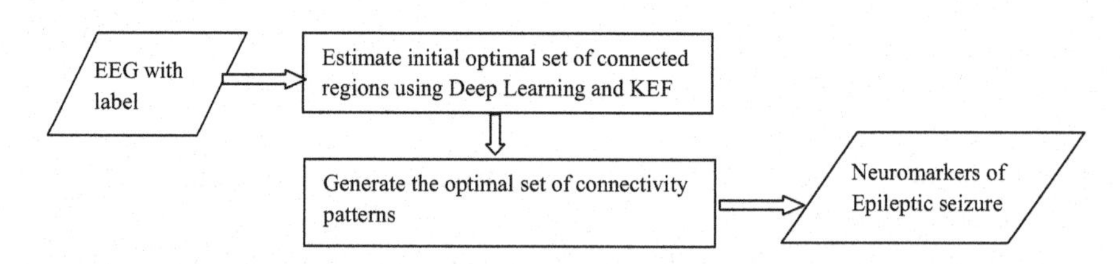


**FIGURE 12.2** The Proposed Workflow.

### 12.4.1 Steps of the Workflow

Step 1. Estimate initial optimal set of connected regions: Apply a deep learning–based method (Lusch et al., 2016) to approximate KEF in a way to produce linear embedding of nonlinear dynamics.

Step 2. Generate the optimal set of connectivity patterns: Compute the power spectrum of the eigenvectors to reflect the phase-dependent coherence, hence the causal connectivity (Brunton et al., 2016a).

#### 12.4.1.1 Low-Dimensional Scalable and Interpretable Embedding of a Parsimonious Network (Step 1)

The model adapts the design of the deep learning method described by Lusch et al. (2016) with a modification in the system's definition to incorporate phase parameter taking inspiration from Mezić (2017) for the treatment of time-periodic systems to find KEF, the eigenfunctions of the Koopman operator. This formulation enables estimation of causal connectivity in terms of phase-dependence of coherent patterns following the interpretation provided by Zhang et al. (2020). This step obviates computationally expensive functions like PCA/ICA to fetch the characteristic modes from non-stationary signals. Accordingly, the evolution of state $\{X_i \in R^n | i = 0..\infty\}$ of a discrete time-periodic dynamical system can be defined by Equation 12.4:

$$\left.\begin{aligned} X &= F(X) \\ X_{k+1} &= F\left(X_k\right) \end{aligned}\right\} \tag{12.4}$$

where F(.) is unknown, typically a nonlinear function that maps the successive progression of the state of the system with time. The preceding process models the system evolution at different phases $s_1, s_2, \ldots, s_M \in [0, T]$, where $T$ is the period. The evolution of F(.) can be formulated in terms of differential equations in Equation 12.4. However, solutions to these dynamics involving high-dimensional $X$ are complicated and often obscure. Following Koopman's formalism, the evolution of states $Y$ of a dynamical system can be observed as evolution of functions in the Hilbert space given by Equation 12.5:

$$Y_s = g\left(X_s\right) \tag{12.5}$$

Hence, the progression of the state measurement function is determined by the Koopman operator, which maps in an infinite dimensional linear space guided by Equation 12.6:

$$\left.\begin{aligned} Kg &\triangleq g \cdot F \\ K\,g\left(x_{k,s}\right) &= g\left(x_{k+1,s}\right) \end{aligned}\right\} \tag{12.6}$$

Thus, we obtain a linear representation of nonlinear dynamics, enabling an advanced nonlinear prediction, an estimation through respective methods applicable for linear systems.

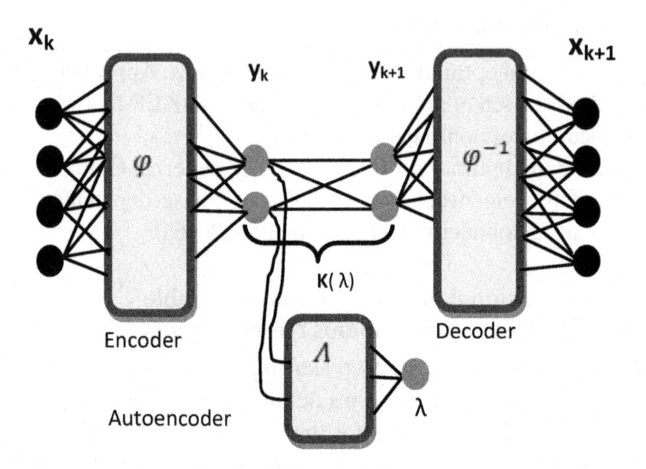


**FIGURE 12.3** Autoencoder with an Auxiliary Network.

The Koopman operator's eigenfunctions spanning an invariant subspace can be generated using Equation 12.7; the resultant matrix produced on the application of a Koopman operator is bounded by this subspace.

$$\varphi\left(x_{k+1,s}\right) = \mathrm{K}\varphi\left(x_{k,s}\right) = \lambda\varphi\left(x_{k,s}\right) \tag{12.7}$$

The identification of the eigenfunction of the Koopman operator is, however, computationally expensive. Deep learning autoencoder network in Figure 12.3, and as described in Lusch et al. (2016), used an auxiliary network to identify the reduced set of parameters to define the continuous eigenvalue spectrum λ.

The learning of the network shown in Figure 12.3 is regularized by specific loss function at each functional step:

1. Identification of intrinsic coordinates: The autoencoder learns the encoder $\varphi$ to identify the optimal set of key intrinsic coordinates $y = \varphi(x)$ spanned by a set of eigenfunctions $\left\{\varphi : \Re^n \rightarrow \Re^p\right\}$ of the Koopman operator K, enabling the description of a dynamical system as $y_{k+1} = \mathrm{K}y_k$. Here $n = 1, \ldots, \infty$ and $p$ is the reduced dimension of the autoencoder subspace defined using prior knowledge of the system. It simultaneously learns the decoder $\varphi^{-1}$ so that the states $x$ be reconstructed using $x_s = \varphi^{-1}(y)$. The evolution of learning of the autoencoder is regularized using the loss function $\|x_s - \varphi^{-1}(\varphi(x))\|$. The norm $\|.\|$ is mean-squared error, obtained by averaging over dimensions followed by averaging over then for the samples and then applying $\ell_2$ regularization.
2. Learn linear dynamics K: Learning is further regularized to ensure optimal predictability of the linear dynamics $y_{k+1,s} = \mathrm{K}y_{k,s}$ using loss function $\|\varphi(x_{k+1,s}) - \varphi(x_{k,s})\|$.

3. Future prediction: The learning process tunes the network for m steps predictability using the generalized loss function $\|\varphi(x_{k+m,s}) - K^m\varphi(x_{k,s})\|$.

To absorb the continuous spectrum, the auxiliary network learns the function $\lambda = \Lambda(y_s)$, which guides the variation of the eigenvalues of the matrix K. The block diagonals $K(\mu, \omega)$ are thence identified using the eigenvalues $\lambda_{\pm} = \mu \pm \omega$. The discrete time K is a Jordan block, defined as in Equation 12.8 for each pair of complex eigenvalues:

$$B(\mu,\omega) = \exp(\mu\Delta t)\begin{bmatrix} \cos(\omega\Delta t) & -\sin(\omega\Delta t) \\ \sin(\omega\Delta t) & \cos(\omega\Delta t) \end{bmatrix}. \tag{12.8}$$

This network structure facilitates a small number of eigenfunctions as it allows eigenvalue computation across the phase space. Network training begins with the generation of trajectories from random initial conditions and then using separate sets for training, validation, and test to avoid overfitting and ensuring early convergence.

#### 12.4.1.2 Determining Effective (Causal) Connectivity Estimates from the Phase-Coherence Structure (Step 2)

The eigenfunctions produced through the preceding procedure embeds sufficient information about the structure and connectivity among the sources underlying the multichannel data. The formulation of phase-dependence-based causal connectivity measurement can be adapted for the eigenfunctions derived by the deep learning-based method in Step 1. The interpretation of the eigenfunctions as per the DMD method serves the purpose—the magnitude of eigenfunction $|\emptyset_i|$ represents the spatial correlations between the $n$ observable locations. The eigenvalue $\lambda_i$ encodes the temporal dynamics of the spatial mode $\emptyset_i$. The magnitude and the phase components of $\lambda i$ reflect the rate of growth or decay and frequency of oscillations. Let $\psi_i$ be the phase information of $\lambda_i$. The causality of channel pair $\{x_1, x_2\}$ is represented by the causality of $\{\emptyset_1, \emptyset_2\}$.

The phase coherence between eigenfunction pair $\{\emptyset_1, \emptyset_2\}$ is given by Equation 12.9:

$$\eta(\varnothing_1, \varnothing_2) = \frac{1}{T}\left| \int_0^T exp\left(j\left(\psi_1(t) - \psi_2(t)\right)\right) \cdot dt \right|, \tag{12.9}$$

where $\psi_1$ and $\psi_2$ are the respective instantaneous phases. Then absolute causal strength is given by Equation 12.10:

$$\sigma(x_2 \rightarrow x_1) = \left|\eta(x_1, x_2) - \eta((x_1 - \varnothing_1), x_2)\right|$$
$$\sigma(x_1 \rightarrow x_2) = \left|\eta(x_1, x_2) - \eta(x_1, (x_2 - \varnothing_1))\right| \cdot \tag{12.10}$$

The relative causal strength $C$ is defined as the relative ratio of the two absolute causal strengths as in Equation 12.11, where $C \in [0,1]$:

$$\left.\begin{aligned} C(x_1 \rightarrow x_2) &= \frac{\sigma(x_1 \rightarrow x_2)}{\sigma(x_1 \rightarrow x_2) + \sigma(x_2 \rightarrow x_1)} \\ C(x_2 \rightarrow x_1) &= \frac{\sigma(x_2 \rightarrow x_1)}{\sigma(x_1 \rightarrow x_2) + \sigma(x_2 \rightarrow x_1)} \end{aligned}\right]. \tag{12.11}$$

Accordingly, a ratio of 0 or 1 indicates a strong differential causal relation between $x_1$ and $x_2$, whereas a ratio of 0.5 would indicate a reciprocal causality or absence of causality.

### 12.4.2 Strength of the Model

In the proposed workflow, Step 1 avoids the steps of segmentation (windows) to exploit piecewise stationarity and that of performing PCA or ICA for identifying influential components and removing artifacts, essentially required in the traditional workflow of learning the directional functional connectivity or effective connectivity, including GCM and DCM. The Koopman operator provides linear embedding of nonlinear dynamic system. However, it is computationally challenging task to obtain eigenfunctions of the Koopman operator for a system having continuous spectrum of eigenvalues. Deep learning autoencoder trained with an auxiliary network learns reduced set of parameters of the intrinsic coordinates and the multiple loss functions ensure the improved accuracy of reconstruction. This mechanism combining the strengths of deep learning and the Koopman embedding thus produces a parsimonious, as well as interpretable, network. The selection of the dimension $p$ of the autoencoder coordinates $y$, however, requires prior knowledge of the system.

In Step 2 of the proposed workflow, the instantaneous phase-dependence among the approximate Koopman eigenfunctions of each pair of input signals describe the causal connectivity dynamics over a long timescale. The underlying algorithm remarkably controls against the inherent exponential computation of products. Deep learning methods have been implemented using programming frameworks like TensorFlow (Abadi et al., 2015), which parallelize computation over GPUs and achieve scalable solutions (Cui et al., 2016).

## 12.5 CONCLUSION

The chapter presented a critical survey of methods for estimation of patterns of slow changes in seizure-related connectivity among various regions of the brain and underlying changes in the brain states. The exploration has been limited to the estimation of bivariate measures. The same contributes with prescription of effective workflow, borrowing from the disruptive models of complex dynamic systems exploited in fluid dynamics (Koopman operator) and molecular dynamics (variational tensors) and the strategies based on deep learning for their tractable computation. A critical account of au courant measures and workflows aimed at estimating the patterns of slow modulations of sparse effective connectivity from epileptiforms has been presented. Traditional measures and methods with corresponding utility and limitations were briefly described in Section 12.2. The major challenge of non-stationarity of an EEG, caused by several reasons, including volume conduction, has been addressed by exploiting the piecewise stationarity of segments of suitable lengths. However, identifying the segment length has been a tricky task as it depended on prior knowledge and often required trial and error process. Parametric and nonparametric methods for minimizing the covariability induced due to volume conduction have been suggested. But their lack of correspondence to the underlying time–frequency structure, on one hand, and the capability of the model-based methods like the MVAR to discover the time-varying directional connectivity structures, on the other, has incited research for wider adaptation of the latter for brain connectivity analyses. Section 12.3 presented a detailed discussion of models for estimation of slow modulations in connectivity, highlighting the performance issues, as well as exploring viable solutions by adapting relevant strategies from fluid dynamics and deep learning technologies. Non-stationarity of the EEG signals was handled in traditional models, including VAR, DTF, and PDC, exploiting piecewise stationarity by applying ICA and performing analyses over windows of suitable length. The implementation of the traditional methods of dynamic connectivity estimation methods, namely, GC and DCM, have raised difficulties in decisions about the optimal system parameters and, hence, sanctioned further model advancements. The selection of nonredundant low-dimensional features is but a crucial issue for fair model performance. The literature review introduced impressive methods for feature encoding and model regularization. The survey of methods for the linear encoding of nonlinear complex dynamic systems, largely applied in fluid dynamics, inducted non-stationarity and sparsity-absorbing noise-invariant Koopman eigendecomposition for data-driven low-dimensional approximation. The definition of the time-varying connectivity properties of dynamic systems in terms of phase-coherence and encoding using with KEFs obviates the need for explicit handling of non-stationarity and a better handling of noise and exogenous inputs. Considering the underlying huge computational requirements, recent deep learning methods employing VAEs capable of data-driven discovery of Koopman eigenfunctions have been incorporated. After generating the sparsity preserving reduced representation of the system dynamics, the phase-dependent power spectrum would reflect the desired characteristic effective (causal) connectivity patterns.

The preceding review thus inspires a series of investigations into case-based empirical effectiveness of the promising mechanisms, for example, the transductive-HSIC LASSO, variational tensors-based autoencoders, the Koopman operator, and others, which enable data-driven discovery of governing equations, and have not been explored in surveyed published research to harness and improve the strength of GC- and DCM-based sparse causal connectivity estimation. Based on the findings in the survey, an atypical comprehensive model has been conceptualized in Section 12.4, which reaps the benefits of robust theoretical formulation of DMD as well as exploits the computational strength of deep learning technologies, as highlighted previously. The proof of concept with synthetic data and real datasets is undertaken as ongoing research and will be the scope of future reports.

## 12.6 DATA STATEMENT

No data have been used, as the scope of this chapter is to present a critical review of the existing methods for the stated problem. Based on the critical analysis, a theoretically robust and tractable model has been proposed.

## ACKNOWLEDGMENT

*This study has been conducted as part of the ongoing project funded by the Department of Science and Technology (DST) of Govt. of India under Cognitive Science Research Initiative. The fund supported procurement of reference material for survey [Grant no. DST/CSRI/2018/276].*

## REFERENCES

Abadi, M., et al. (2015). TensorFlow: Large-scale machine learning on heterogeneous systems. *arXiv:1603.04467.*

Alvarado-Rojas, C., et al. (2014). Slow modulations of high-frequency activity (40–140 Hz) discriminate preictal changes in human focal epilepsy. *Scientific Reports, Nature*, 4, 4545. https://doi.org/10.1038/srep04545

Astolfi, L., et al. (2008). Tracking the time-varying cortical connectivity patterns by adaptive multivariate estimators. *IEEE Transactions on Biomedical Engineering*, 55(3), 902–913. https://doi.org/10.1109/TBME.2007.905419

Baccalá, L. A., & Sameshima, K. J. B. C. (2001). Partial directed coherence: A new concept in neural structure determination. *Biological Cybernetics*, 84(6), 463–474. https://doi.org/10.1007/PL00007990

Baraniuk, R. G. (2007). Compressive sensing [Lecture Notes]. *IEEE Signal Processing Magazine*, 24(4), 118–121. https://doi.org/10.1109/MSP.2007.4286571

Barnett, L., Barrett, A. B., & Seth, A. K. (2018). Misunderstandings regarding the application of Granger causality in neuroscience. *PNAS*, 115(29), E6676–E6677, First published July 10, 2018. https://doi.org/10.1073/pnas.1714497115

Barnett, L., & Seth, A. K. (2014). The MVGC multivariate Granger causality toolbox: A new approach to Granger-causal inference. *Journal of Neuroscience Methods*, 223, 50–68. https://doi.org/10.1016/j.jneumeth.2013.10.018

Barnett, L., & Seth, A. K. (2015). Granger causality for state-space models. *Physical Review E, Statistical, Nonlinear, and Soft Matter Physics*, 91(4), 040101. https://doi.org/10.1103/PhysRevE.91.040101

Berkooz, G., Holmes, P., & Lumley, J. L. (1993). The proper orthogonal decomposition in the analysis of turbulent flows. *Annual Review of Fluid Mechanics*, 23, 539–575.

Blinowska, K. J. (2008). Methods for localization of time-frequency specific activity and estimation of information transfer in brain. *International Journal of Bioelectromagnetism*, 10(1), 2–16.

Blinowska, K. J. (2011). Review of the methods of determination of directed connectivity from multichannel data. *Medical & Biological Engineering & Computing*, 49(5), 521–529. https://doi.org/10.1007/s11517-011-0739-x

Brunton, B. W., et al. (2016a). Extracting spatial – temporal coherent patterns in large-scale neural recordings using dynamic mode decomposition. *Journal of Neuroscience Methods*, 258, 1–15.

Brunton, S. L., et al. (2014). Compressive sensing and low-rank libraries for classification of bifurcation regimes in nonlinear dynamical systems. *SIAM Journal on Applied Dynamical Systems*, 13(4), 1716–1732.

Brunton, S. L., Proctor, J. L., & Kutz, J. N. (2016b). Discovering governing equations from data by sparse identification of nonlinear dynamical systems. *Proceedings of the National Academy of Sciences of the United States of America*, 113(15), 3932–3937. https://doi.org/10.1073/pnas.1517384113

Budišić, M., Ryan, M., & Igor, M. (2012). Applied koopmanism. *Chaos: An Interdisciplinary Journal of Nonlinear Science*, 22(4). https://arxiv.org/abs/1206.3164v3

Candès, E. J., & Wakin, M. B. (2008). An introduction to compressive sampling. *IEEE Signal Processing Magazine*, 25(2), 21–30.

Carney, P. R., Myers, S., & Geyer, J. D. (2011). Seizure prediction: Methods. *Epilepsy & Behavior*, 22(Suppl 1), S94–S101. https://doi.org/doi:10.1016/j.yebeh.2011.09.001

Cohen, M. X. (2014). *Analyzing Neural Time Series Data: Theory and Practice*. Cambridge, MA: The MIT Press.

Cui, H., et al. (2016). Geeps: Scalable deep learning on distributed GPUs with a GPU-specialized parameter server. In *Proc. 11th European Conference on Computer Systems*, 4. https://dl.acm.org/citation.cfm?id=2901318

Daniels, B. C., & Nemenman, I. (2015). Automated adaptive inference of phenomenological dynamical models. *Nat Commun*, 6, 8133. https://doi.org/doi: 10.1038/ncomms9133

Dhamala, M. (2018). Granger-Geweke causality: Estimation and interpretation. *NeuroImage*, 175, 460–463. https://doi.org/10.1016/j.neuroimage.2018.04.043

Dhamala, M., Rangarajan, G., & Ding, M. (2008). Estimating Granger causality from fourier and wavelet transforms of time series data. *Physical Review Letters*, 100, 018701.

Ding, M., et al. (2000). Short-window spectral analysis of cortical event-related potentials by adaptive multivariate autoregressive modelling: Data preprocessing, model validation, and variability assessment. *Biological Cybernetics*, 83(1), 35–45.

Dominguez, L. G., et al. (2005). Enhanced synchrony in epileptiform activity? Local versus distant phase synchronization in generalized seizures. *Journal of Neuroscience*, 25(35), 8077–8084. https://doi.org/10.1523/JNEUROSCI.1046-05.2005

Donoho, D. L. (2006). Compressed sensing. *IEEE Transactions on Information Theory*, 52(4), 1289–1306.

Driver, R. D. (1977). *Ordinary and Delay Differential Equations*. Berlin-Heidelberg-New York: Springer-Verlag.

Freestone, D. R., et al. (2014). Estimation of effective connectivity via data-driven neural modelling. *Frontiers in Neuroscience*, 8, 383. https://doi.org/10.3389/fnins.2014.00383

Friston, K. J. (2011). Functional and effective connectivity: A review. *Brain Connectivity*, 1(1). https://doi.org/10.1089/brain.2011.0008

Friston, K. J., & Penny, W. (2011). Post hoc Bayesian model selection. *Neuroimage*, 56, 2089–2099.

Friston, K. J., et al. (2013). Analysing connectivity with Granger causality and dynamic causal modelling. *Current Opinion in Neurobiology*, 23(2), 172–178. https://doi.org/10.1016/j.conb.2012.11.010

Friston, K. J., et al. (2014). Granger causality revisited. *NeuroImage*, 101, 796–808.

Friston, K. J., et al. (2016). Bayesian model reduction and empirical Bayes for group (DCM) studies. *Neuroimage*, 128, 413–431. PMID:26569570

Friston, K. J., et al. (2017). Active inference: A process theory. *Neural Computation*, 29, 1–49. https://doi.org/doi:10.1162/NECO_a_00912

Galilei, G. (1960). *On Motion, and on Mechanics: Comprising De Motu (ca. 1590)*, 5. Madison: University of Wisconsin Press.

Gaspard, P. (1998). *Chaos, Scattering and Statistical Mechanics*. Cambridge Nonlinear Science Series. Cambridge: Cambridge University Press. https://doi.org/10.1017/CBO9780511628856

Gerstein, G. L., & Perkel, D. H. (1969). Simultaneously recorded trains of action potentials: Analysis and functional interpretation. *Science*, 164, 828–830.

Gotman, J. (2010). High frequency oscillations: The new EEG frontier? *Epilepsia*, 51(Suppl 1), 63–65. https://doi.org/doi:10.1111/j.1528-1167.2009.02449.x

Granger, C. W. J., & Joyeux, R. (1980). An introduction to long-memory time series models and fractional differencing. *Journal of Time Series Analysis*, 1, 15–30.

Guo, S., et al. (2008). Partial Granger causality – eliminating exogenous inputs and latent variables. *Journal of Neuroscience Methods*, 172, 79–93.

He, D., & Parida, L. (2014). Transductive HSIC lasso. *Proceedings of 2014 SIAM International Conference on Data Mining*. Philadelphia, PA: SIAM, pp. 154–162. eISBN: 978-1-61197-344-0. https://doi.org/10.1137/1.9781611973440.18

Hernandez, C. X., et al. (2017). Variational encoding of complex dynamics. *arXiv:1711.08576v2*.

Huang, N. E. (1998). The empirical mode decomposition and the Hilbert spectrum for nonlinearand non-stationary time series analysis. *Proceedings of the Royal Society of London. Series A: Mathematical, Physical and Engineering Sciences*, 454(1971), 903–995.

James, G., et al. (2013). *An Introduction to Statistical Learning*. New York, NY: Springer. ISBN-13: 978–1461471370.

Jirsa, V. K., & Ding, M. (2004). Will a large complex system with time delays be stable? *Physical Review Letters*, 93(7), 070602.

Kaminski, M. J., & Blinowska, K. J. (1991). A new method of the description of the information flow in the brain structures. *Biol Cybern*, 65, 203–210.

Kiebel, S. J., Garrido, M. I., & Friston, K. J. (2007). Dynamic causal modelling of evoked responses: The role of intrinsic connections. *Neuroimage*, 36, 332–345.

Klus, S., Nüske, F., Koltai, P., et al. (2018). Data-driven model reduction and transfer operator approximation. *Journal of Nonlinear Science,* 28, 985–1010. https://doi.org/10.1007/s00332-017-9437-7

Koopman, B. O. (1931). Hamiltonian systems and transformation in Hilbert space. *Proceedings of the National Academy of Sciences*, 17(5), 315–318. https://doi.org/doi:10.1073/pnas.17.5.315. PMC:1076052. PMID:16577368

Kutz, J. N., et al. (2016). *Dynamic Mode Decomposition: Data-Driven Modelling of Complex Systems*. Philadelphia, PA: Society for Industrial and Applied Mathematics.

Labyt, E., et al. (2007). Modelling of entorhinal cortex and simulation of epileptic activity: Insights into the role of inhibition-related parameters. *IEEE Transactions on Information Technology in Biomedicine*, 11(4), 450–461. https://doi.org/doi:10.1109/titb.2006.889680

Lehnertz, K., et al. (2007). State-of-the-art of seizure prediction. *Journal of Clinical Neurophysiology*, 24(2), 147–153.

Lehnertz, K., et al. (2017). Capturing time-varying brain dynamics. *EPJ Nonlinear Biomedical Physics*, 5, 2. https://doi.org/10.1051/epjnbp/2017001

Le Van Quyen, M., et al. (2010). Large-scale microelectrode recordings of high frequency gamma oscillations in human cortex during sleep. *The Journal of Neuroscience*, 30, 7770–7782.

Li, Q., et al. (2017). Extended dynamic mode decomposition with dictionary learning: A data-driven adaptive spectral decomposition of the koopman operator. *Chaos*, 27, 103111.

Lusch, B., et al. (2016). Deep learning for universal linear embeddings of nonlinear dynamics. *Nature Communications*, 9, 4950 (2018). https://doi.org/10.1038/s41467-018-07210-0
Mardt, A. (2018). VAMPnets: Deep learning of molecular kinetics. *Nature Communications*, 9(1), 5. https://doi.org/10.1038/s41467-017-02388-1
Marrouch, N. (2019). Data-driven Koopman operator approach for computational neuroscience. *Annals of Mathematics and Artificial Intelligence*, 88, 1155–1173. https://doi.org/10.1007/s10472-019-09666-2
Martinez-Vargas, J. D., et al. (2017). Improved localization of seizure onset zones using spatiotemporal constraints and time-varying source connectivity. *Frontiers in Neuroscience*, 11, 156. https://doi.org/10.3389/fnins.2017.00156
Mezić, I. (2017). *Spectral Operator Methods in Dynamical Systems: Theory and Applications*. New York, NY: Springer.
Modur, P. N. (2014). High frequency oscillations and infraslow activity in epilepsy. *Annals of Indian Academy of Neurology*, 17(Suppl 1), S99–S106. https://doi.org/10.4103/0972-2327.128674
Mormann, F. (2003). Automated detection of a preseizure state based on a decrease in synchronization in intracranial electroencephalogram recordings from epilepsy patients. *From Physical Review. E, Statistical, Nonlinear, and Soft Matter Physics*, 67, 021912.
Mormann, F., et al. (2005). On the predictability of epileptic seizures. *Clinical Neurophysiology*, 116, 569–587. https://doi.org/10.1016/j.clinph.2004.08.025
Mormann, F., et al. (2007). Seizure prediction: The long and winding road. *Brain*, 130, 314–333.
Nalatore, H., Ding, M., & Rangarajan, G. (2007). Mitigating the effects of measurement noise on Granger causality. *Physical Review. E, Statistical, Nonlinear, and Soft Matter Physics*, 75(3 Pt 1), 031123.
Nedungadi, A. G., et al. (2009). Analyzing multiple spike trains with nonparametric Granger causality. *Journal of Computational Neuroscience*, 27(1), 55–64.
Noé, F., & Nuske, F. A. (2013). Variational approach to modelling slow processes in stochastic dynamical systems. *Multiscale Modelling and Simulation*, 11, 635–655. https://github.com/sklus/d3s/
Omidvarnia, A., Pedersen, M., Rosch, R. E., Friston, K. J., & Jackson, G. D. (2017). Hierarchical disruption in the Bayesian brain: Focal epilepsy and brain networks. *NeuroImage: Clinical*, 15, 682–688. https://doi.org/10.1016/j.nicl.2017.05.019
Otto, S. E., & Rowley, C. W. (2017). Linearly-recurrent autoencoder networks for learning dynamics. Preprint at http://arxiv.org/abs/1712.01378
Ozaki, T. (2012). *Time Series Modelling of Neuroscience Data*. Boca Raton, FL: CRC Press.
Papadopoulou, M., et al. (2015). Tracking slow modulations in synaptic gain using dynamic causal modelling: Validation in epilepsy. *NeuroImage*, 107, 117–126.
Pereda, E., et al. (1998). Non-linear behaviour of human EEG: Fractal exponent versus correlation dimension in awake and sleep stages. *Neuroscience Letters*, 250(2), 91–94.
Quiroga, Q. R., Kreuz, T., & Grassberger, P. (2002). Event synchronization: A simple and fast method to measure synchronicity and time delay patterns. *Physical Review. E, Statistical, Nonlinear, and Soft Matter Physics*, 66(4 Pt 1), 041904.
Rasekhi, J., et al. (2015). Epileptic seizure prediction based on ratio and differential linear univariate features. *Journal of Medical Signals and Sensors*, 5(1), 1–11.
Ren, W., Li, B., & Han, M. (2020). A novel Granger causality method based on HSIC-Lasso for revealing nonlinear relationship between multivariate time series. *Physica A: Statistical Mechanics and Its Applications*, 541(1), 123245.
Richardson, M. P. (2012). Large scale brain models of epilepsy: Dynamics meets connectomics. *Journal of Neurology, Neurosurgery & Psychiatry*, 83, 1238–1248.

Roelstraete, B., & Rosseel, Y. (2012). Does partial Granger causality really eliminate the influence of exogenous inputs and latent variables? *Journal of Neuroscience Methods*, 206(1), 73–77. https://doi.org/10.1016/j.jneumeth.2012.01.010. Epub 2012 Feb 11. PMID:22330817.

Rosch, R. E., et al. (2018). Calcium imaging and dynamic causal modelling reveal brain-wide changes in effective connectivity and synaptic dynamics during epileptic seizures. *PLOS Computational Biology*, 14(8), e1006375. https://doi.org/10.1371/ journal.pcbi.1006375

Rowley, C. W., et al. (2009). Spectral analysis of nonlinear flows. *Journal of Fluid Mechanics*, 645, 115–127.

Salova, A., et al. (2019). Koopman operator and its approximations for systems with symmetries. *Chaos*, 29, 093128. https://doi.org/10.1063/1.5099091

Schmid, P. J. (2010). Dynamic mode decomposition of numerical and experimental data. *Journal of Fluid Mechanics*, 656, 5–28.

Seo, J.-H., et al. (2020). Pattern recognition in epileptic EEG signals via Dynamic mode decomposition. *Mathematics*, 8, 481.

Seth, A. K. (2010). A MATLAB toolbox for Granger causal connectivity analysis. *Journal of Neuroscience Methods*, 186, 262–273.

Seth, A. K., Barrett, A. B., & Barnett, L. (2015). Granger causality analysis in neuroscience and neuroimaging. *Journal of Neuroscience*, 35(8), 3293–3297. https://doi.org/10.1523/JNEUROSCI.4399-14.2015

Sheffer, I. E., et al. (2017). Latest survey on classification of epilepsy and clues for early detection and prediction. *ILAE Position Paper, Epilepsia*, 58(4), 512–521. https://doi.org/10.1111/epi.13709

Silvennoinen, A., & Terasvirta, T. (2009). Multivariate GARCH models. In Mikosch, T., et al. (Eds.), *Handbook of Financial Time Series*. Berlin: Springer-Verlag, pp. 201–229.

Sommerlade, L., et al. (2012). Inference of Granger causal time-dependent influences in noisy multivariate time series. *Journal of Neuroscience Methods*, 203(1), 173–185.

Steriade, M. (2003). *Neuronal Substrates of Sleep and Epilepsy*. Cambridge: Cambridge University Press.

Steriade, M., Amzica, F., & Contreras, D. (1996). Synchronization of fast (30–40 Hz) spontaneous cortical rhythms during brain activation. *The Journal of Neuroscience*, 16, 392–417.

Stokes, P. A., & Purdon, P. L. (2017). A study of problems encountered in Granger causality analysis from a neuroscience perspective. *Proceedings of the National Academy of Sciences of the United States of America*, 114, E7063–E7072. https://doi.org/10.1073/pnas.1704663114

Takeishi, N., Kawahara, Y., & Yairi, T. (2017). Learning koopman invariant subspaces for dynamic mode decomposition. In Guyon, I., et al. (Eds.), *Advances in Neural Information Processing Systems*. Curran Associates, Inc., pp. 1130–1140. https://papers.nips.cc/paper/6713-learningkoopman-invariant-subspaces-for-dynamic-mode-decomposition

Tibshirani, R. (1996). Regression shrinkage and selection via the lasso. *Journal of the Royal Statistical Society, B*, 58(1), 267–288.

Ting, C.-M., et al. (2017). Estimating time-varying effective connectivity in high-dimensional fMRI data using regime-switching factor models. *arXiv:1701.06754*. https://arxiv.org/pdf/1701.06754.pdf

Tropp, J. A. (2007). Gilbert AC Signal recovery from random measurements via orthogonal matching pursuit. *IEEE Transactions on Information Theory*, 53(12), 4655–4666.

Val-Calvo, M., Álvarez-Sánchez, J. R., Ferrández-Vicente, J. M., & Fernández, E. (2019). Optimization of real-time EEG artifact removal and emotion estimation for human-robot interaction applications. *Frontiers in Computational Neuroscience*, 13. https://doi.org/10.3389/fncom.2019.00080. ISSN=1662–5188

Valderrama, M., et al. (2012). Cortical mapping of gamma oscillations during human slow wave sleep. *PLoS ONE*, 7, e33477.

Valdes-Sosa, P. A., et al. (2005). Estimating brain functional connectivity with sparse multivariate regression. *Philosophical Transactions of the Royal Society B*, 360(2005), 969–981.

Wehmeyer, C., & Noé, F. (2018). Time-lagged autoencoders: Deep learning of slow collective variables for molecular kinetics. *The Journal of Chemical Physics*, 148, 241703.

Wilke, C., Ding, L., & He, B. (2007). An adaptive directed transfer function approach for detecting dynamic causal interactions. *Conference Proceedings – IEEE Engineering in Medicine and Biology Society*, pp. 4949–4952, EMBC'07, Leon, France.

Winterhalder, M., et al. (2005). Comparison of linear signal processing techniques to infer directed interactions in multivariate neural systems. *Signal Processing*, 85(11), 2137–2160. https://doi.org/10.1016/j.sigpro.2005.07.011

Yang, A. C., Peng, C.-K., & Huang, N. E. (2018). Causal decomposition in the mutual causation system. *Nature Communications*, 9(1), 1–10.

Yeung, E., Kundu, S., & Hodas, N. (2017). Learning deep neural network representations for Koopman operators of nonlinear dynamical systems. Preprint at http://arxiv.org/abs/1708.06850

Zhang, C., Bütepage, J., & Kjellstr, H. (2018). Advances in variational inference. *arXiv:1711.05597v3.*

Zhang, H., Zhao, M., Wei, C., Mantini, D., Li, Z., &Liu, Q. (2020). EEGdenoiseNet: A benchmark dataset for deep learning solutions of EEG denoising. *arXiv:2009.11662v2*. https://arxiv.org/abs/2009.11662v2

# 13 An Application for Predicting Type II Diabetes Using an IoT Healthcare Monitoring System

*Dalvin Vinoth Kumar A, Cecil Donald A, and Margaret Mary T*

**CONTENTS**



DOI: 10.1201/9781003142751-17

## 13.1 INTRODUCTION

The main attempt of the chapter is to exemplify how the Internet of Things (IoT) is changing healthcare. Monitoring and recording numerous medical parameters of the patient outside the hospital have become widespread phenomena. Health actually plays a significant role in one's life, but people are very busy in their routine lives and neglect their healthcare. In this context, the IoT monitoring system gives alerts and awareness about their health. IoT monitoring has taken a huge position in patients' lives also. For all these reasons, this chapter explains why IoT healthcare monitoring is important in human life and how it plays a significant role and is needed in this current scenario. An IoT design is a system for watching a patient's body at any time by Internet connection. The use of this system is to assess some biological parameters of a patient's body like temperature, heartbeat, blood pressure, and so on. By using sensors to sense and send the values to an IoT cloud platform through a Wi-Fi module (Ashwini, Akshay, Annapurna, & Nagesh, 2020). All health information for a patient will be stockpiled in the cloud. This permits the general physician to monitor a patient's health continuously using their smartphone. Based on the patient's health condition, the doctor may suggest/give immediate treatment during an emergency. Hence, IoT monitoring in one's life is part of parcel in our healthcare.

According to Gartner, 26 billion devices will be used for IoT by 2022 (Gartner, 2013). These devices seamlessly gather and share information directly with each other and the cloud, making it possible to collect records and analyze data. This information provides insight into one's health and supplements actions to advance health, without the hindrance of a person's daily routine. In this chapter, there's a greater role of the IoT and IT in managing the patient's medical data with its security. Knowledge abducted in this chapter is current trends, challenges, a case study, and a real-world application. This chapter also depicts the IoT's application in healthcare and ways to improve the primary health needs of those in developing nations. The IoT may be a vision that remains in its terribly early stages, wherever everybody interprets the visions of IoT as supporting the items and digital and linguistics views (Zhong, & Ge, 2018).

All these three perspectives on the IoT should be integrated with each other seamlessly, as shown in Figure 13.1, for extracting the full benefits of IoT architecture:

- Things-oriented vision
- Internet-oriented vision
- Semantic-oriented vision

**Things-oriented vision:** This vision gives the viewpoint that all the genuine physical objects can have sensors joined to the IoT to get the constant data from them. This can be cultivated by the sensors-based organization of inserted gadgets utilizing radio-frequency ID (RFID), near-field communication, and other remote innovations (Mezaall, Yousif, Abdulkareem, Hussein, & Khaleel, 2018). This vision gives the base for integration, all "things" considered utilizing diverse sensor-based organizations to work together and coincide.

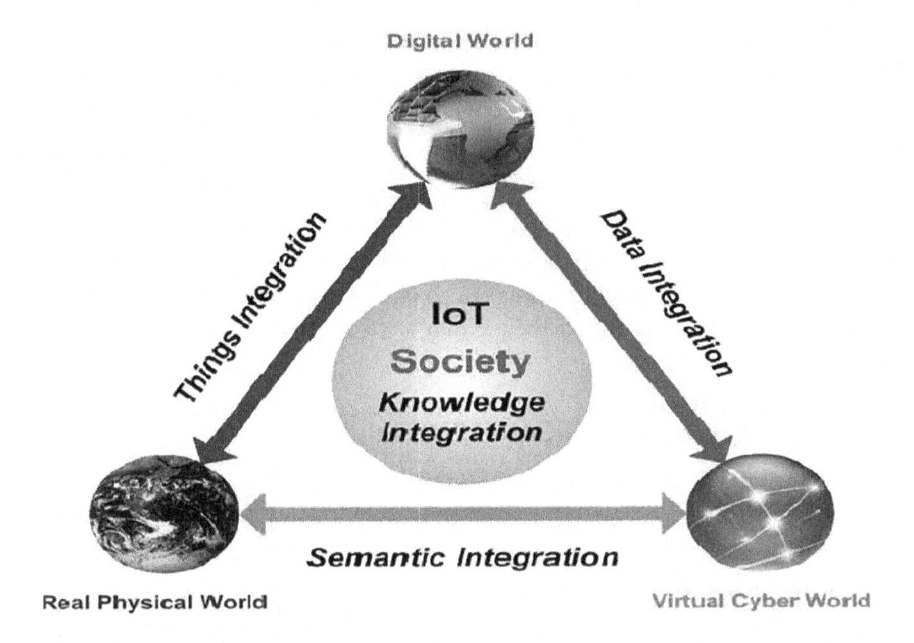


**FIGURE 13.1** Perspectives of the IoT.

**Internet-oriented vision:** This vision gives the point of view that all the gadgets can be associated through the Internet and can be portrayed as brilliant objects. This can be refined by utilizing one of a unique Internet protocol for each associated object. This vision gives the base to the data integration of the relative multitude of the entire smart object, which can be persistently monitored (Singh, Tripathi, & Jara, 2014).

**Semantic-oriented vision:** This vision gives the viewpoint that all the information gathered from various sensors requires to be analyzed for important understanding. This can be cultivated with semantic strategies, which isolate raw information from the significant data and their understanding. This vision gives the base to the semantic incorporation using semantic middleware (Shen & Liu, 2011).

## 13.2 THE NEED OF AN IOT HEALTH MONITORING SYSTEM

The worldwide IoT in medical services are expected to reach USD 534.3 billion by 2025, growing at a compound annual growth rate of 19.9% over the period, as per another report by Grand View Research, Inc. A rise in ventures for the execution of IoT arrangements in the medical service area is a key component driving the market.

Expanding the entrance of associated gadgets in different medical care foundations and the selections of IoT frameworks and programming arrangements in medical care and operational exercises are additionally among the key components enlarging market development. As its popularity among the healthcare practitioners to manage chronic diseases, real-time data monitoring is expected to boost the demand for an IoT network of health institutions. Amalgamations of remote monitoring, analytics, and mobile platforms have reduced the rate of readmissions of patients suffering from congestive heart failure, diabetes, and blood pressure. This dawning of technologies, nurtured a little further, shows the result of positivity in managing a patient's chronic diseases and medication dosage, which has further fostered the market growth (Grand View Research is registered in the State of California at Grand View Research, 2019), such as the connected wheelchair that allows people with disabilities to interconnect their health alerts to care teams and involve care suppliers.

## 13.3 INTERNET OF MEDICAL THINGS (IoMT)

Despite the growth of IoT in the healthcare industry, it also introduced the brand-new IoMT, which uploads positive developments in treating and managing various diseases, such as insomnia, cardiopathy, autism, diabetes, and asthma. The IoMT will influence 136.8 billion US Dollars by 2021. Be that as it may, what is significant are the advantages for doctors and patients (Katherine, 2020) of an IoT in healthcare? According to Andrew Keller, "IoMT simply collects and exchange data one or the other with users or the devices via Internet, and keeps doctors to be well aware of a patient's condition on timely basis" (Katherine O, IoT in Healthcare: Benefits, 2020). The IoT architecture is shown in Figure 13.2.

This architecture is common to the IoT process. The sensors help to transform the physical world data (e.g., temperature, pressure, humidity, etc.) and count human health data (heart rate, oxygen saturation, blood pressure, blood glucose, etc.) to the digital world and the actuators transform the digital data to physical actions (e.g., Infusion pumps, dialysis system, etc.). These IoT devices consist of sensors for receipt signals from the environment for analysis, or actuators for controlling the environment based on the inputs or both sensors and actuators (Atzori, Iera, & Morabito, 2010). These devices are connected through an Internet transfer and to cloud storage for communication with similar devices and people, as shown in

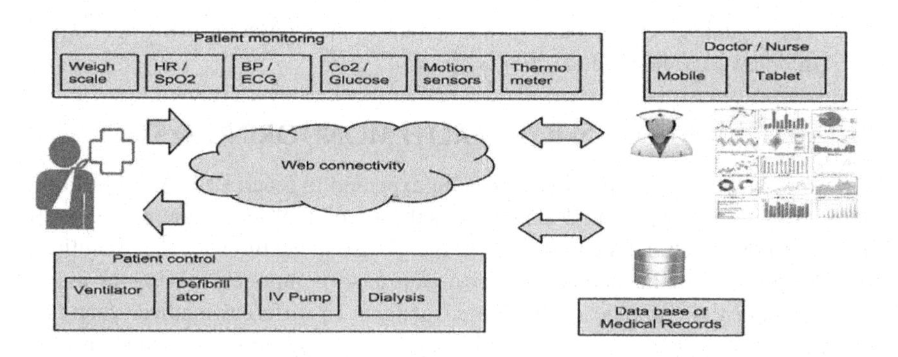


**FIGURE 13.2** IoT Architecture.

Figure 13.2. There are multiple studies from several research companies for the projected figures of these IoT devices extending from 26 to 212 billion IoT devices in 2020 (Gartner, Press Release, 2013).

## 13.4 IoT DIABETES DATA ANALYSIS USING BIGML

In this modern era, human beings encounter different health issues. Most of the health issues are due to the food habits of the individuals (Moreno, Kersting, De Henauw, González-Gross, Sichert-Hellert, Matthys, & Ross, 2005). In this chapter, an analytical approach is projected to pretreat diabetic mellitus, which has three phases: data collection, data storage, and analytics, as shown in Figure 13.3.

This methodology assumes a significant part in foreseeing diabetes and pretreating diabetic patients. In the first phase, data collection is done through IoT gadgets and different sources. The gathered information is scrubbed utilizing pre-preparing strategies. Phase II deals with data storage stockpiling. The pre-prepared information is put away in stockrooms. To store the monstrous amount of information, distributed storage is utilized. The information put away in the cloud is dissected to build up a relationship between the different boundaries, for example, body mass index (BMI), air pollution level, blood pressure, and so forth, with diabetic mellitus. The third phase of the proposed method manages the predictive analytics, whereby the choices made depend on an affiliation among diet pattern, physical fitness, current medicine intake, and other factors.

### 13.4.1 Phase I: Data Collection

#### 13.4.1.1 Collection of Clinical Data from Medical Laboratories and Storing It in the Raw Data Store

The dataset selection is one of the important processes in data mining. For this, the most relevant data should be collected from a particular domain to attain values. The derived values should be more flexible and informative in that domain. In this study, the data collection is done through questionnaires, sensors, and clinical experiments.

#### 13.4.1.2 Collection of E-Questionnaire Data from Users through an Online Survey

A standardized questionnaire, including items of common risk factors of diabetes, has been sent to the participants to obtain info on demographic characteristics, family diabetes history, and lifestyle risk features. It contains a set of questions under three categories:

**Category A (Demographic):** It contains **personal information** including, namely, name, age, sex, and address.
**Category B:** It encompasses facts about the **family**, such as diabetic history of parents, financial status, and educational status.
**Category C:** It includes the **physical data** such as height, weight, smoking habit, consumption of alcohol, work type, and physical activity.

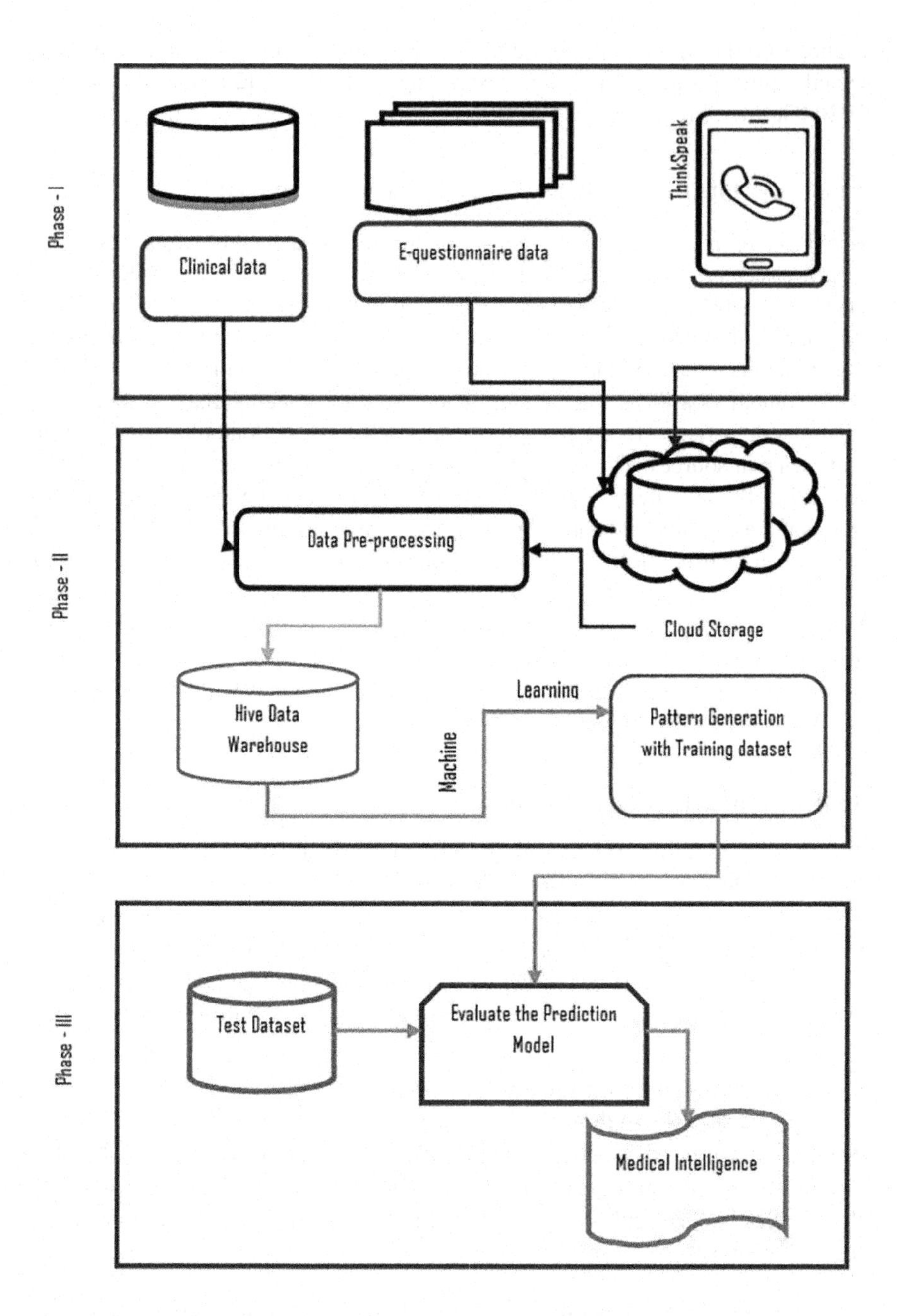


**FIGURE 13.3** Methodology Diagram for Predicting Type II Diabetes.

Anthropometric measurements have been taken from participants. BMI has been calculated (weight in kilograms divided by the square of height in meters ($kg/m^2$)). BMI ≥ 25 is defined as overweight. In category C, the data related to the following physical activities have been collected: cigarette smoking at least 500 cigarettes in one's life, alcohol consumption of at least 100 g per week for one year or longer, and participation in modest or vigorous physical activity for 30 minutes or more per day for at least three days in a week.

#### 13.4.1.3 Collection of IoT Data from Users through the ThingSpeak IoT Cloud

IoT data are collected from the devices connected to the Internet. MQ-7, MQ-3, and Honeywell HPm Particle Sensor are utilized for air class data. The readings of the sensors are stored in this microcontroller temporarily and sent to the cloud. In this data collection process, the ThingSpeak web API is used for storing data in the cloud. The ThingSpeak IoT analytics platform service allows aggregating, visualizing, and analyzing live statistics streams in the cloud.

The data related to blood pressure (diastolic and systolic) and blood sugar, along with their screening data; BMI; dietary history; physical activity; and air pollution level (Pm2.5 and Pm10) are collected using this system. The selected populations are also included in the blood tests for the collection of random blood sugar levels. Blood pressure is screened using an arm blood pressure digital monitor. The data relating dietary history and physical activity are collected through the questionnaire.

### 13.4.2 Phase II: Data Storage and Analytics

#### 13.4.2.1 Preprocessing and Diabetic Data Warehouse Creation

The preprocessing of the large volume of clinical data becomes essential. Data preprocessing is related to the data mining process as an important step. Observing data that have not been carefully examined for such issues can produce misleading outcomes. If there is an ample amount of incorrect and redundant information or noisy and unreliable data, then knowledge discovery becomes challenging. Data preprocessing includes various steps such as cleaning, normalization, transformation, feature extraction, and selection. The outcome of data preprocessing is the complete dataset with reduced attributes. The major drawback with the clinical dataset is the existence of redundant records. These redundant records cause the learning algorithm to be biased. So eliminating redundant records is essential to enhance detection accuracy. During preprocessing, the raw data are supplied as input, and several suitable data preprocessing methods are applied, thereby decreasing the invalid instances in the dataset. In our diabetes dataset, data transformation and data validation are considered as the important preprocessing techniques for eliminating impure and invalid data. The preprocessed data with zero missing values and zero errors are shown in Figure 13.4.

The cleaned dataset is named the Diabetics Dataset to form a data warehouse in a JSON format as shown in Figure 13.5.

#### 13.4.2.2 Model Construction and Validation

The preprocessed data are now deployed in the analysis to build prediction model. In this project work, three type II diabetes predicative models are developed using the BigML tool (Zainudin & Shamsuddin, 2016). BigML provides machine learning algorithms as SaaS. It can be retrieved through three models, namely, web interface,

Sources
Datasets
Supervised
Unsupervised
Predictions
Tasks
WhizzML
TNSet926
Diabetes
Name
Type
Count
Missing
Errors
Histogram
Age
Air Quality
BMI
Blood pressure
Diabetes
Diabetes pedigree
Glucose
Insulin
Pregnancies
18,432

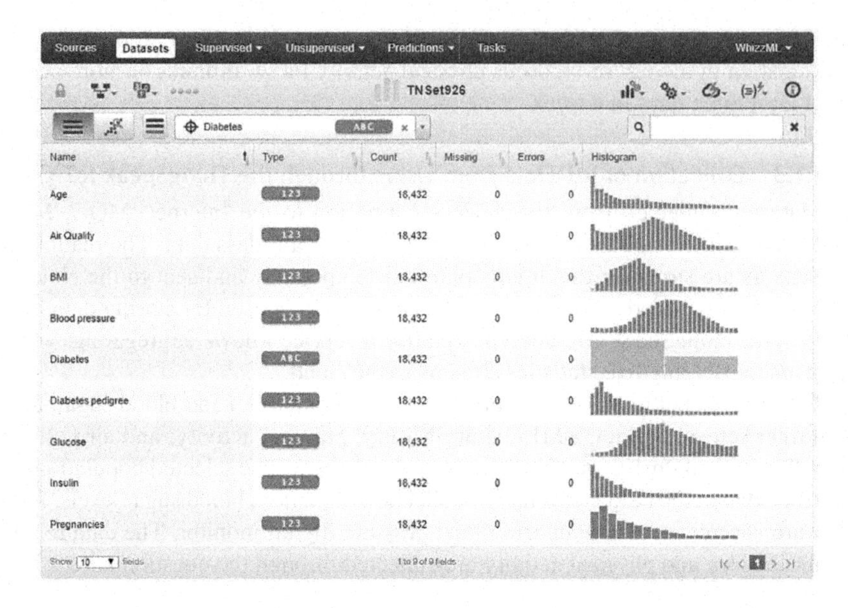

FIGURE 13.4 Preprocessed Data Sorted in Alphabetical Order.

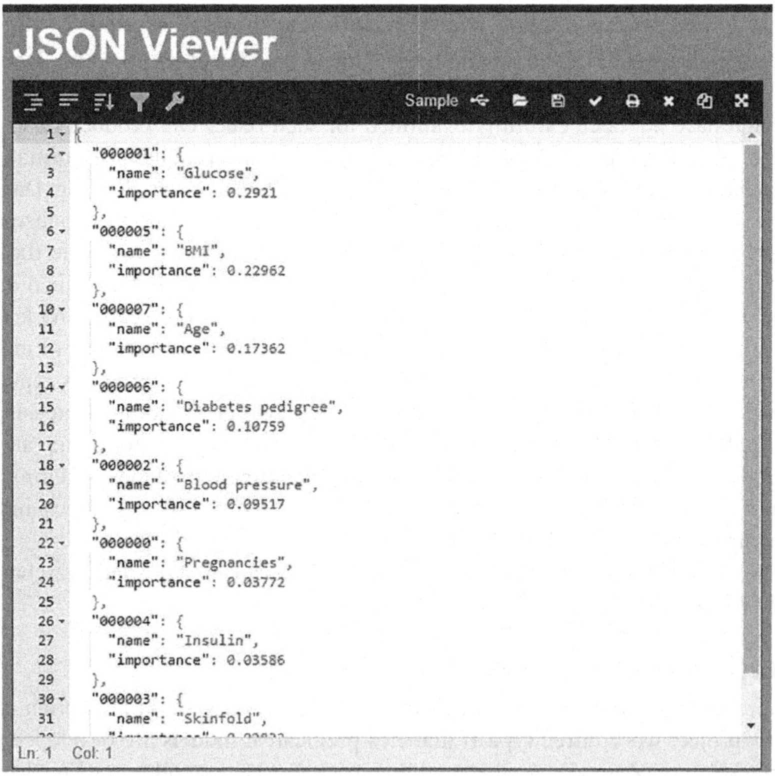

JSON Viewer
Sample
{
"000001": {
"name": "Glucose",
"importance": 0.2921
},
"000005": {
"name": "BMI",
"importance": 0.22962
},
"000007": {
"name": "Age",
"importance": 0.17362
},
"000006": {
"name": "Diabetes pedigree",
"importance": 0.10759
},
"000002": {
"name": "Blood pressure",
"importance": 0.09517
},
"000000": {
"name": "Pregnancies",
"importance": 0.03772
},
"000004": {
"name": "Insulin",
"importance": 0.03586
},
"000003": {
"name": "Skinfold",
Ln: 1 Col: 1

FIGURE 13.5 Diabetics Dataset Data Warehouse.

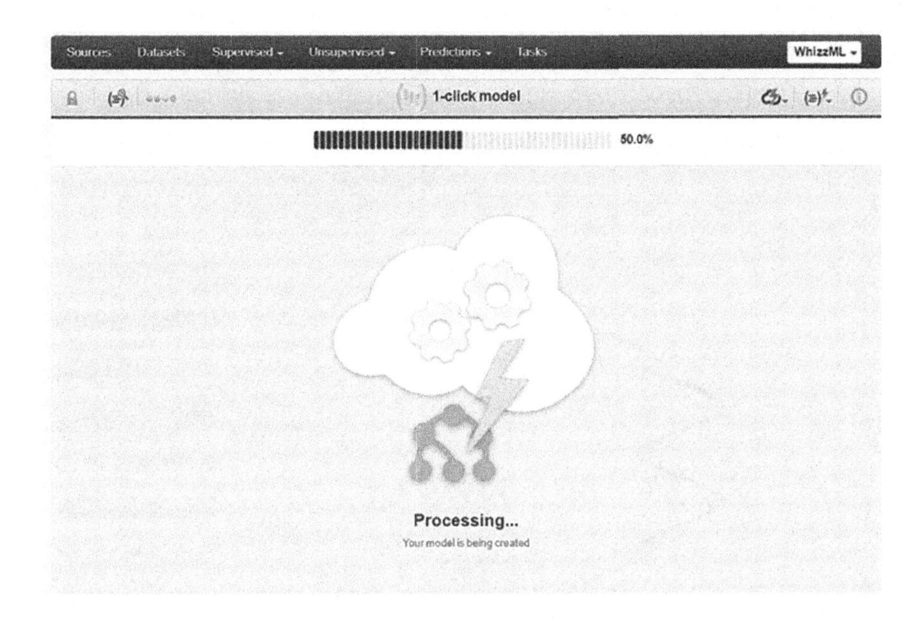


**FIGURE 13.6** Model Creation Process in BigML.

command line, and restful API. The model is developed using the available web interface. The model creation process using BigML is shown in Figure 13.6.

In standard statistical analysis, log_lenier is complex due to the exponential number of variables. Association discovery approaches are very much suitable for high-dimensional data. The relationship between the values is discovered using association discovery rather than the relation between the variables. The FilteredTop_K Technique is used for association discovery. The created model is named as Tamil Nadu Diabetes Diagnosis Model 1, which is shown in Figure 13.6. The association discovered using the developed model is shown in Figures 13.7, 13.8, and 13.9.

The maximum number of associations that can be obtained from variables in BigML is 500. It ranges between 0 and 500. The obtained associations are sorted from maximum to minimum confidence value. It is observed that the potential association increases exponentially when there are large numbers of fields, values, and instances. The associations are discovered for the item in the antecedent ranges from 0 to 10. Leverage and lift are the two commonly preferred search strategies. Leverage produces the result of the most relevant case, whereas lift produces the least relevant case. The association graph shown in Figure 13.10 is for Diabetics Dataset data.

The number of instances is 14,745, the number of item fields is 27, the number of associations discovered is 100, the search strategy chosen is leveraged, the percentage set for the minimum support is 4.2970%, and the percentage set for the minimum confidence is 23.8000%. Actionable Association for Dataset Diabetics Dataset and JavaScript Object Notation Predictive Markup Language (JSON- PML) are given in Figure 13.11.

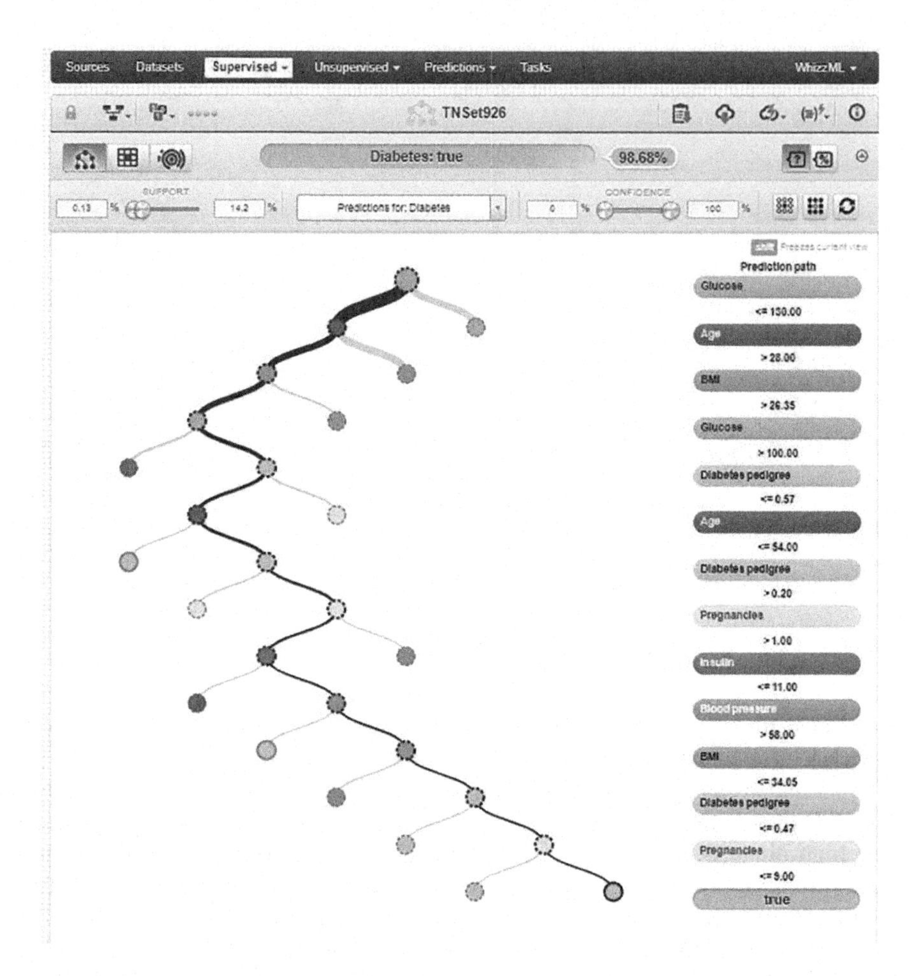


**FIGURE 13.7** Association between Variables with Confidence of 98.68%.

## 13.5 IoT-BASED PATIENT HEALTH MONITORING SYSTEM DEVELOPMENT USING ESP8266 AND ARDUINO

ThingSpeak is the IoT storage platform used in this application. ThingSpeak is an open-source IoT application and API for storing and retrieving data from things using the HTTP protocol over the Internet or through the network. This IoT device could sense the blood pressure rate and measure the surrounding environmental temperature. It continuously monitors the pulse rate and surrounding environmental temperature and updates them to an IoT platform. The Arduino Sketch outfits the various functionalities of the project such as converting them into strings, passing them to the IoT platform, reading sensor data, and presenting measured pulse rate and temperature on the character LCD is shown as a block diagram in Figure 13.11.

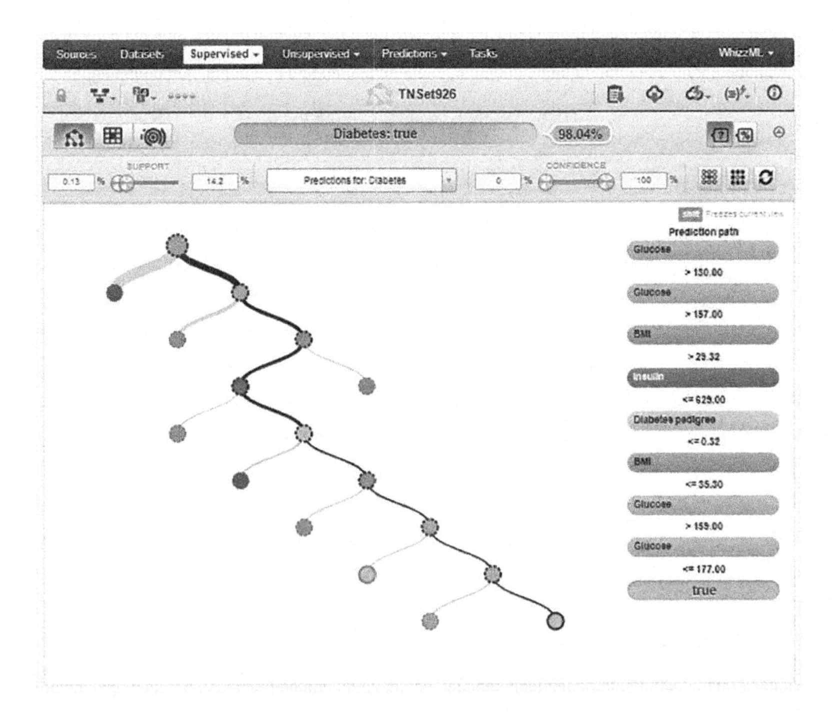


**FIGURE 13.8** Association between Variables with Confidence of 98.04%.

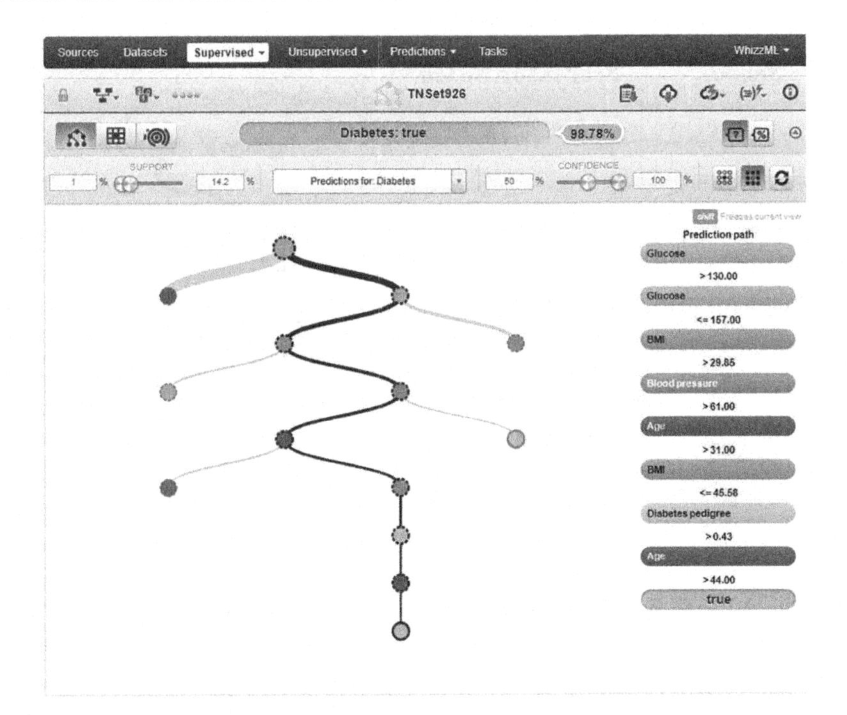


**FIGURE 13.9** Association between Variables with Confidence of 98.78%.

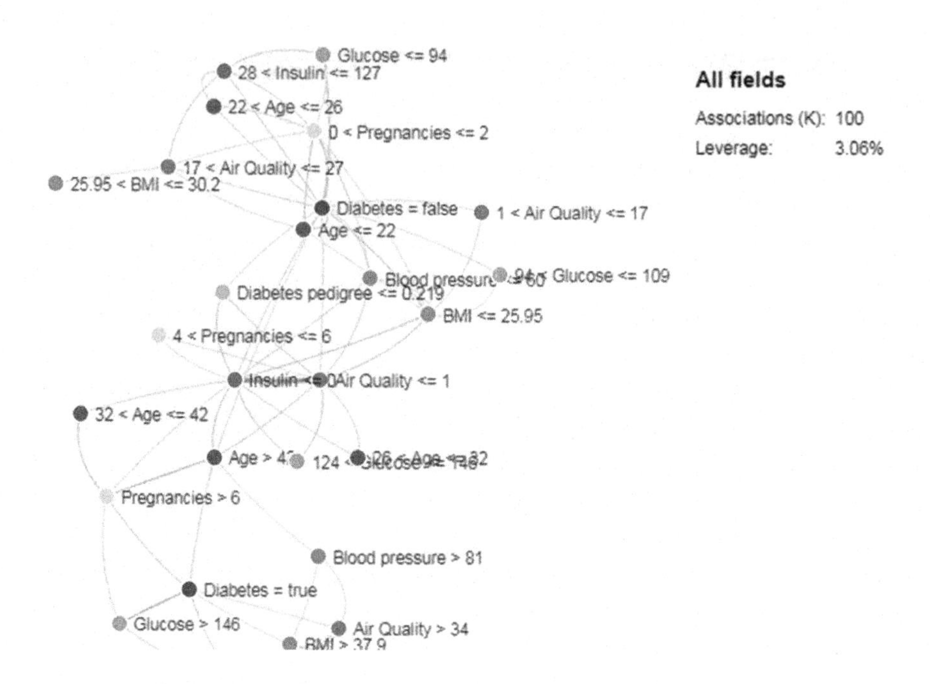


**FIGURE 13.10** Association Graph for the Diabetics Dataset.

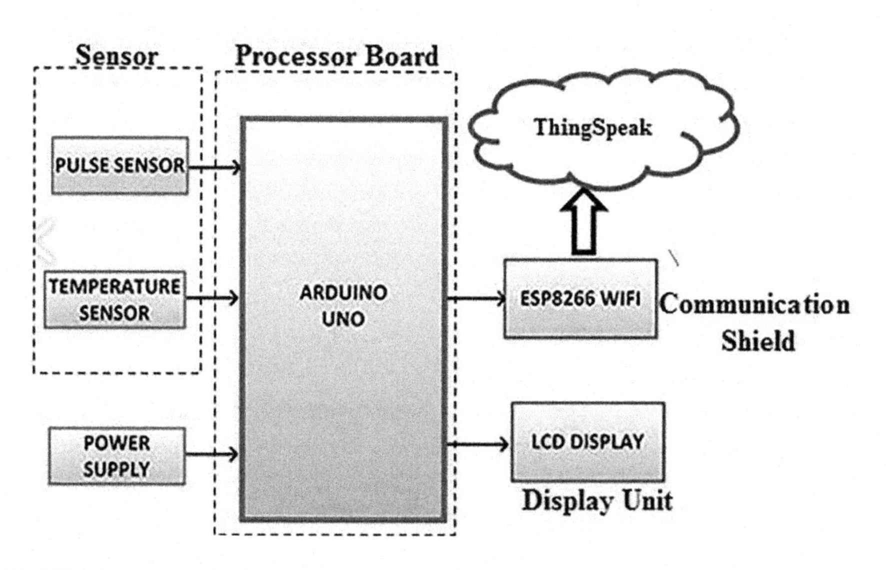


**FIGURE 13.11** Block Diagram of the IoT Healthcare Monitoring Application.

**TABLE 13.1**
**Components Required**

| S. No. | Component Name |
|---|---|
| 1 | Arduino Board |
| 2 | ESP8266–01 |
| 3 | LCD Display |
| 4 | Potentiometer |
| 5 | Pulse Sensor |
| 6 | Temperature Sensor |
| 7 | Resistor 2k, 1k |
| 8 | LED |
| 9 | Connecting Wires |
| 10 | Breadboard |

The components required for developing a health-monitoring IoT application using Arduino are given in Table 13.1.

### 13.5.1 Pulse Sensor

The pulse sensor, an attachment module to the heart rate sensor for the Arduino, is shown in Figure 13.12. It tends to be utilized by athletes, students, game and mobile developers, and artists who need to effortlessly incorporate live pulse information into their projects/devices. The embodiment is an incorporated optical enhancing circuit and garbage value elimination (noise) circuit sensor. Affix the pulse sensor to your ear cartilage or fingertip and attach it to the processor board (Arduino). The heartbeat sensor has three pins: VCC, GND, and Analog Pin. VCC and GND are the power pins to the sensor; the signal pin is the output pin connected to the processor board.

### 13.5.2 Temperature Sensor

The LM35 arrangement is precision-coordinated circuit temperature equipment with an O/P voltage directly proportional to the centigrade temperature, as shown in Figure 13.13. The LM35 has some benefits over direct temperature sensors devised in Kelvin, as it is not required to deduct a huge consistent voltage from the o/p to get helpful centigrade scaling. It has VCC, GND and signal pins. The VCC and GND are the power pins to the sensor; the signal pin is the output pin connected to the processor board.

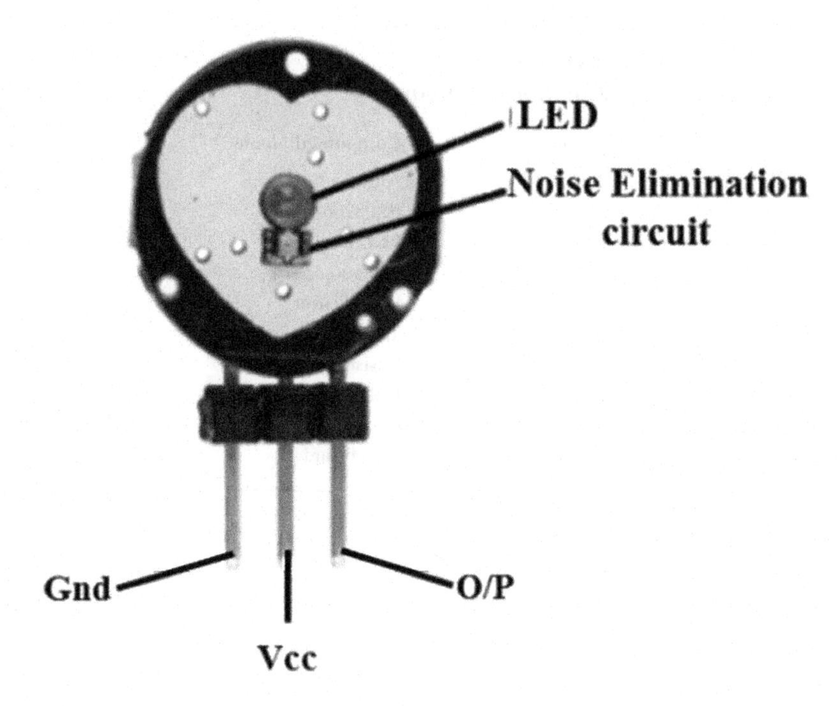


**FIGURE 13.12** Pulse Sensor.

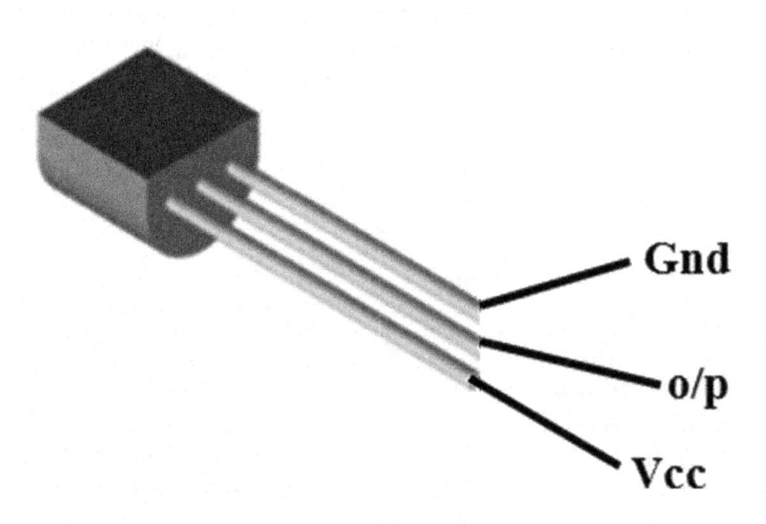


**FIGURE 13.13** Temperature Sensor.

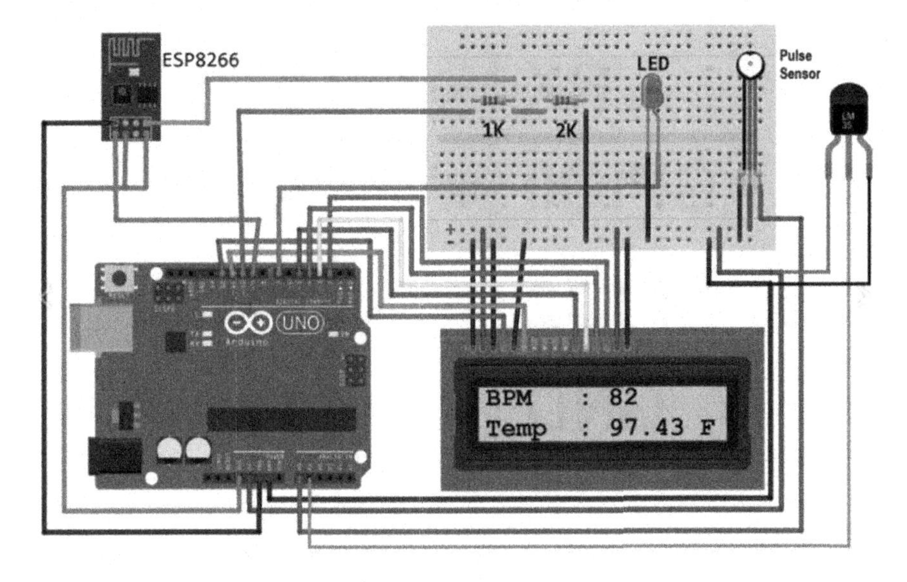


**FIGURE 13.14** Circuit Diagram for the IoT Healthcare Monitoring System.

### 13.5.3 Circuit Diagram

The circuit diagram for the IoT Healthcare Monitoring System is shown in Figure 13.14.

**Application Development Procedure**

***Step 1:*** Connect heartbeat Sensor output pin to A0 of the processor board (Arduino) and other two pins to power (VCC and GND).

***Step 2:*** Connect LM35 temperature sensor output pin to A1 of Arduino and other two pins to VCC and GND.

***Step 3:*** Connect red LED to digital pin 7 of Arduino with a 220-ohm resistor.

***Step 4:*** LCD Pins 1, 3, 5, 16 should be connected to GND.

***Step 5:*** Pins 2, 15 of LCD should be connected to VCC.

***Step 6:*** Connect Pins 4, 6, 11, 12, 13, 14 of LCD to digital pin 12, 11, 5, 4, 3, 2 of Arduino, respectively.

***Step 7:*** The RX pin of the Wi-Fi shield (ESP8266) works on 3.3 V, and it will not interconnect with the Arduino when connected directly to the Arduino. So a voltage divider is needed, for it will turn the 5V into 3.3V, which was achieved by connecting the 2.2K and 1K resistor. Thus, the RX pin of the Wi-Fi shield is connected to the Arduino pin 10 through the resistors. The block diagram of Wi-Fi shield is shown in Figure 13.15.

***Step 8:*** Connect the TX pin of the ESP8266 to pin 9 of the Arduino.

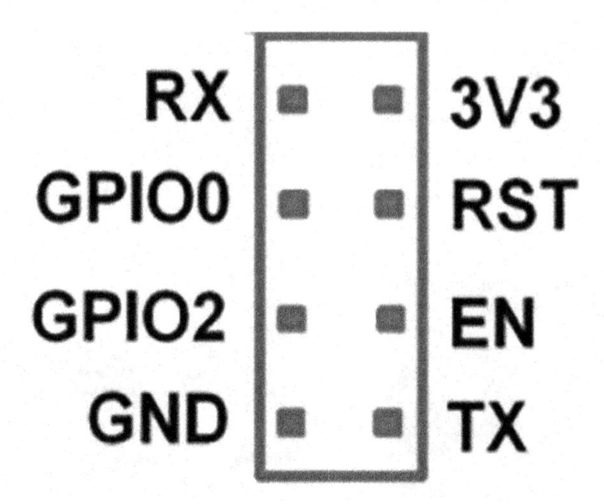


**FIGURE 13.15** Block Diagram of the Wi-Fi Shield.

## 13.6 CODE FOR IoT HEALTHCARE MONITORING SYSTEM USING ARDUINO

The code running on this device is Embedded C. It is developed using Arduino IDE, the project developed in Arduino IDE is called a sketch. This code has three parts: variable declaration part, setup method, and loop method. The detailed code for the application development is displayed in Appendix to this chapter.

## 13.7 BENEFITS OF IoT IN HEALTHCARE

The IoT in healthcare gains more popularity every year. It means that such a solution has several vital benefits. Let's discuss them more precisely.

**Health Monitoring:** Smart devices can track health issues. IoT applications for medical care can impart signs of a crisis if the patient has an asthma attack, cardiovascular failure, or another medical issue. For instance, Apple has coordinated a Fall Detection System in the Apple Watch. It distinguishes if the user falls and shows an alert. The person needs to tap 'I'm OK' for one minute. For another situation, the Apple Watch calls emergency services. These applications additionally make an impression on emergency contacts.

**Better Patient Experience:** The IoT in clinics improves the medical healthcare system and gives patients a more agreeable approach for connecting with specialists. It prompts improving patient experience and picking up client reliability. A clinic from the United States—Mount Sinai Medical Center—reduced the wait time of their emergency patients by 50% with their AutoBed software. The medical clinic has about 1,100 beds but admits at least 59,000 patients every year. The essential task of AutoBed is coordinating accessible beds with new patients.

**Drug Management:** One of the IoT medical care benefits is better drug management. IoT innovation permits controlling the measure of taking medication. Specialists can screen the portion and track the adequacy of treatment. Additionally, the IoT permits sending suggestions to patients about when they should ingest their medications. Now and again, it's conceivable that a relative could be advised when a patient hasn't taken medication on schedule.

**Healthcare Automation:** IoT gadgets can help computerize authoritative, manual, and routine errands. IoT clinical applications can analyze a significant amount of information and make various measurements to perceive any adjustments in the patients' ailments. Robotized cycles of gathering information reduce the number of errors in making a diagnosis.

**Preventive Healthcare:** People die in mass numbers due to chronic diseases such as cardiopathy, diabetes, and more, and that's just the beginning. Utilizing the logical abilities of IoT, giving patients more customized types of treatment and care is conceivable. Smart, connected, or wearable gadgets can screen medical issues of older or patients with ongoing illnesses. Having such information, specialists will have the option of giving a doctor better treatment options and seeing the first symptoms of the disease (Katherine, 2020).

## 13.8 CHAPTER SUMMARY

The IoT plays a vital role in healthcare, and a lot of technology is involved in monitoring the health conditions of patients. The objective of this technical chapter was to demonstrate how IoT monitoring is transforming healthcare and the role of IT in healthcare. Monitoring and recording of numerous medical parameters of the patient outside the hospital have become widespread phenomena. A growing preference among the healthcare practitioners for real-time data monitoring to manage chronic diseases is expected to propel the demand for an IoT network in healthcare institutions. Datasets collected from various reliable sources were collected using a questionnaire. The collected dataset was analyzed using the BigML tool. It was observed that air quality has a significant impact on type II diabetes. An IoT application was developed based on the analyzed results. The developed model can be used for real-time diabetic predication.

# Appendix

```
#include <LiquidCrystal.h> // Declaration
LiquidCrystal lcdisp(12, 11, 5, 4, 3, 2);
#include <SoftwareSerial.h>
float PulseSignal= 0;
float temp = 0;
SoftwareSerial ser(9,10);
String apiKey = "OO707TGA1BLUNN12";
// Variables
int SensorpulsePin = A0; // PulseSignalSensor purple wire connected to analog pin 0
int ledblinkPin = 7; // pin to blink led at each beat
int ledfadePin = 13; // pin to do fancy classy fading blink at each beat
int fadeRate = 0; // used to fade LED on with PWM on fadePin
// Variables, used in the interrupt service routine!
int BODYBPM; // int that holds raw Analog in 0. updated every 2mS
int Temp; // holds the incoming raw data
int IBI = 600; // int that holds the time interval between beats! Must be seeded!
boolean PulseSignal= false; // "True" when User's live heartbeat is detected. "False"
when nota "live beat".
boolean QS = false; // becomes true when Arduoino finds a beat.

// Regards Serial OutPut – Set This Up to your needs

static boolean serialVisual = true; // Set to 'false' by Default. Re-set to 'true' to see
Arduino Serial Monitor ASCII Visual Pulse
int rate[10]; // array to hold last ten IBI values
unsigned long sampleCounter = 0; // used to determine PulseSignaltiming
unsigned long lastBeatTime = 0; // used to find IBI
int P = 512; // used to find peak in PulseSignalwave, seeded
int T = 512; // used to find trough in PulseSignalwave, seeded
int thresh = 525; // used to find instant moment of heart beat, seeded
int amp = 100; // used to hold amplitude of PulseSignalwaveform, seeded
boolean firstBeat = true; // used to seed rate array so we startup with reasonable
BODYBPM
boolean secondBeat = false; // used to seed rate array so we startup with reasonable
BODYBPM

void setup()
{
lcdisp.begin(16, 2);
pinMode(blinkPin,OUTPUT); // pin that will blink to your heartbeat!
pinMode(fadePin,OUTPUT); // pin that will fade to your heartbeat!
Serial.begin(115200); // we agree to talk fast!
```

```
interruptSetup(); // sets up to read PulseSignalSensor temp every 2mS

// LCD Display

lcdisp.clear();
lcdisp.setCursor(0,0);
lcdisp.print("Patient Health");
lcdisp.setCursor(0,1);
lcdisp.print("Monitoring ");
delay(4000);
lcdisp.clear();
lcdisp.setCursor(0,0);
lcdisp.print("Initializing. . . .");
delay(5000);
lcdisp.clear();
lcdisp.setCursor(0,0);
lcdisp.print("Getting Data. . . .");
ser.begin(9600);
ser.println("AT");
delay(1000);
ser.println("AT+GMR");
delay(1000);
ser.println("AT+CWMODE=3");
delay(1000);
ser.println("AT+RST");
delay(5000);
ser.println("AT+CIPMUX=1");
delay(1000);

String cmd="AT+CWJAP=\"Alexahome\",\"98765432\"";
ser.println(cmd);
delay(1000);
ser.println("AT+CIFSR");
delay(1000);
}

// Main Function for Execution
void loop()
{
serialOutput();
if (QS == true) // A Heartbeat Was Found
{

// BODYBPM and IBI have been Determined
// Quantified Self "QS" true when arduino finds a heartbeat
```

```
fadeRate = 255; // Makes the LED Fade Effect Happen, Set 'fadeRate' Variable to
255 to fade LED with pulse
serialOutputWhenBeatHappens(); // A Beat Happened, Output that to serial.
QS = false; // reset the Quantified Self flag for next time
}
ledFadeToBeat(); // Makes the LED Fade Effect Happen
delay(20); // take a break
read_temp();
esp_8266();
}
void ledFadeToBeat()
{
fadeRate -= 15; // set LED fade value
fadeRate = constrain(fadeRate,0,255);          // keep LED fade value from going
into negative numbers!
analogWrite(fadePin,fadeRate); // fade LED
}
void interruptSetup()
{
// Initializes Timer2 to throw an interrupt every 2mS.
TCCR2A = 0x02;
TCCR2B = 0x06;
OCR2A = 0X7C;
TIMSK2 = 0x02;
sei();
}
void serialOutput()
{
if (serialVisual == true)
{
arduinoSerialMonitorVisual('-', Temp);
}
else
{
sendDataToSerial('S', Temp); // goes to sendDataToSerial function
}
}
void serialOutputWhenBeatHappens()
{
if (serialVisual == true) // Code to Make the Serial Monitor Visualizer Work
{
Serial.print("*** Heart-Beat Happened *** "); //ASCII Art Madness
Serial.print("BODYBPM: ");
Serial.println(BODYBPM);
}
else
```

```
{
sendDataToSerial('B',BODYBPM);
sendDataToSerial('Q',IBI);
}
}
void arduinoSerialMonitorVisual (char symbol, int data)
{
const int sensorMinimum = 0;
const int sensorMaximum = 1024;
int sensorRead = data;
int range = map(sensorReading, sensorMinimum, sensorMaximum, 0, 11);
// do something different depending on the
// range value:
switch (range)
{
case 0:
Serial.println("“"); /////ASCII Art Madness
break;
case 1:
Serial.println("---");
break;
case 2:
Serial.println("------");
break;
case 3:
Serial.println("---------");
break;
case 4:
Serial.println("------------");
break;
case 5:
Serial.println("--------------|-");
break;
case 6:
Serial.println("--------------|---");
break;
case 7:
Serial.println("--------------|-------");
break;
case 8:
Serial.println("--------------|----------");
break;
case 9:
Serial.println("--------------|----------------");
break;
case 10:
```

```
Serial.println("--------------|-------------------");
break;
case 11:
Serial.println("--------------|-----------------------");
break;
}
}

void sendDataToSerial(char symbol, int data)
{
Serial.print(symbol);
Serial.println(data);
}
ISR(TIMER2_COMPA_vect) //triggered when Timer2 counts to 124
{
cli(); // disable interrupts while we do this
Temp = analogRead(SensorpulsePin); // read the PulseSignalSensor
sampleCounter += 2; // keep track of the time in mS with this variable
int N = sampleCounter – lastBeatTime; // monitor the time since the last beat to avoid
noise
// find the peak and trough of the PulseSignalwave

if(Temp < thresh && N > (IBI/5)*3) // avoid dichrotic noise by waiting 3/5 of last IBI
{
if (Temp < T) // T is the trough
{
T = Temp; // keep track of lowest point in PulseSignalwave
}
}
if(Temp > thresh && Temp > P)
{// thresh condition helps avoid noise
P = Temp; // P is the peak
} // keep track of highest point in PulseSignalwave
// NOW IT'S TIME TO LOOK FOR THE HEART BEAT
// temp surges up in value every time there is a pulse
if (N > 250)
{// avoid high frequency noise
if ((Temp > thresh) && (PulseSignal== false) && (N > (IBI/5)*3))
{
PulseSignal= true; // set the PulseSignalflag when we think there is a pulse
digitalWrite(blinkPin,HIGH); // turn on pin 13 LED
IBI = sampleCounter – lastBeatTime; // measure time between beats in mS
lastBeatTime = sampleCounter; // keep track of time for next pulse

if(secondBeat)
{
```

```
secondBeat = false; // clear secondBeat flag
for(int i=0; i<=9; i++) // seed the running total to get a realisitic BODYBPM at startup
{
rate[i] = IBI;
}
}
if(firstBeat) // if it's the first time we found a beat, if firstBeat == TRUE
{
firstBeat = false;
secondBeat = true;
sei();
return;
}

word runningTotal = 0; // clear the runningTotal variable
for(int i=0; i<=8; i++)
{// shift data in the rate array
rate[i] = rate[i+1]; // and drop the oldest IBI value
runningTotal += rate[i]; // add up the 9 oldest IBI values
}
rate[9] = IBI; // add the latest IBI to the rate array
runningTotal += rate[9];
runningTotal /= 10;
BODYBPM = 60000/runningTotal;
QS = true; // set Quantified Self flag
PulseSignal= BODYBPM;
}
}
if (Temp < thresh && PulseSignal== true)
{
digitalWrite(blinkPin,LOW);
PulseSignal= false;
amp = P – T;
thresh = amp/2 + T;
P = thresh;
T = thresh;
}
if (N > 2500)
{
thresh = 512;
P = 512;
T = 512;
lastBeatTime = sampleCounter;
firstBeat = true;
secondBeat = false;
}
```

```
sei();
}
void esp_8266()
{
String cmd = "AT+CIPSTART=4,\"TCP\",\"";
cmd += "184.106.153.149"; // api.thingspeak.com
cmd += "\",80";
ser.println(cmd);
Serial.println(cmd);
if(ser.find("Error"))
{
Serial.println("AT+CIPSTART error");
return;
}
String getStr = "GET /update?api_key=";
getStr += apiKey;
getStr +="&field1=";
getStr +=String(temp);
getStr +="&field2=";
getStr +=String(pulse);
getStr += "\r\n\r\n";
// send data length
cmd = "AT+CIPSEND=4,";
cmd += String(getStr.length());
ser.println(cmd);
Serial.println(cmd);
delay(1000);
ser.print(getStr);
Serial.println(getStr);                    //thingspeak (Cloud storage)
delay(3000);
}
void read_temp()
{
int temp_val = analogRead(A1);
float mv = (temp_val/1024.0)*5000;
float cel = mv/10;
temp = (cel*9)/5 + 32;
Serial.print("Temperature:");
Serial.println(temp);
lcdisp.clear();
lcdisp.setCursor(0,0);
lcdisp.print("BODYBPM:");
lcdisp.setCursor(7,0);
lcdisp.print(BODYBPM);
lcdisp.setCursor(0,1);
lcdisp.print("Temp.:");
```

```
lcdisp.setCursor(7,1);
lcdisp.print(temp);
lcdisp.setCursor(13,1);
lcdisp.print("F");
}
```

## REFERENCES


Ashwini, J., Akshay, Annapurna, & Nagesh. (2020, 3 September). *Patient Health Monitoring System Using IOT Devices: Electronics Project*. Retrieved from www.seminarsonly.com/Engineering-Projects/Electronics/patient-health-monitoring-system.php.

Atzori, L., Iera, A., & Morabito, G. (2010). The Internet of things: A survey. *Computer Networks*, *54*(15), 2787–2805.

Gartner. (2013). *Press Release*. Retrieved from www.gartner.com/newsroom/id/2636073.

Grand View Research. (2019, March). *IoT in Healthcare Market Growth & Trends*. Retrieved from www.grandviewresearch.com/press-release/global-iot-in-healthcare-market.

Katherine Orekhova. (2020, January). *The Role of IoT in Healthcare Industry: Benefits and Use Cases*. Retrieved from www.cleveroad.com/blog/iot-in-healthcare.

Mezaall, Y. S., Yousif, L. N., Abdulkareem, Z. J., Hussein, H. A., & Khaleel, S. K. (2018). Review about effects of IOT and Nano-technology techniques in the development of IONT in wireless systems. *International Journal of Engineering & Technology*, *7*(4), 3602–3606.

Moreno, L. A., Kersting, M., De Henauw, S., González-Gross, M., Sichert-Hellert, W., Matthys, C., . . . & Ross, N. (2005). How to measure dietary intake and food habits in adolescence: the European perspective. *International Journal of Obesity*, *29*(2), S66–S77.

Shen, G., & Liu, B. (2011, May). The visions, technologies, applications and security issues of Internet of things. In *2011 International Conference on E-Business and E-Government (ICEE),* Shanghai, China (pp. 1–4). IEEE.

Singh, D., Tripathi, G., & Jara, A. J. (2014, March). A survey of Internet-of-things: Future vision, architecture, challenges and services. In *2014 IEEE World Forum on Internet of Things (WF-IoT),* Seoul, Korea (South) (pp. 287–292). IEEE.

Zainudin, Z., & Shamsuddin, S. M. (2016). Predictive analytics in Malaysian dengue data from 2010 until 2015 using BigML. *International Journal of Advances in Soft Computing and Its Applications*, *8*(3).

Zhong, R. Y., & Ge, W. (2018). Internet of things enabled manufacturing: A review. *International Journal of Agile Systems and Management*, *11*(2), 126–154.

# *Section V*

## *Issues in Security and Informatics in Healthcare*

# 14 A Conceptual Model for Assessing Security and Privacy Risks in Healthcare Information Infrastructures

## *The CUREX Approach*

*Georgia Kougka, Anastasios Gounaris, Apostolos Papadopoulos, Athena Vakali, Diana Navarro Llobet, Jos Dumortier, Eleni Veroni, Christos Xenakis, and Gustavo Gonzalez-Granadillo*

**CONTENTS**



## 14.1 INTRODUCTION

Nowadays, a uniform and widely adopted model of the healthcare processes and medical records constitutes an emerging need with a view to standardizing health data processing and facilitating the sharing of medical records, for example, to allow patients to receive informed treatment even when they are away from the location where they have received treatment in the past. To this end, several new data models and standards have been proposed. In this chapter, after providing an overview of these efforts, an important gap in both the literature and practice is identified, namely, to adopt a formalized procedure to assess the healthcare infrastructure in terms of privacy and cybersecurity risks, which is an essential step toward establishing trust in sharing and processing medical records across health institutions. To fill this gap, a conceptual model is introduced, which is materialized in the context of

DOI: 10.1201/9781003142751-19

a modular, configurable, and context-awareness-oriented platform, named CUREX. CUREX[1] stands for "*seCUre and pRivate hEalth data eXchange*" and is an EU H2020–funded project that is presented in Diaz-Honrubia et al. (2019).

In summary, the main contribution of this chapter is the proposal of a platform-independent high-level conceptual model to perform cybersecurity and risk assessment. This model covers all the steps in the risk assessment process, including the detection of assets along with their vulnerabilities, identification of the associated threats, and derivation of multifaceted risk assessment reports and guidelines. In addition, the accompanying reference architecture is presented, followed by a discussion of legal and ethical issues, and how the CUREX solution manages to become General Data Protection Regulation (GDPR)–compliant by design.

## 14.2 DATA MODELLING PROPOSALS AND STANDARDS FOR HEALTHCARE SYSTEMS

To date, several standards and modelling examples for the healthcare systems have been introduced. Nevertheless, the emerging privacy and cybersecurity demands raise new challenges in modern healthcare infrastructures, which are under continuous attacks and threats of data leakage. There is a great need to adopt policies and technical measures that, while allowing unprohibited data access to legitimate end users, such as healthcare employees and patients, take privacy and security concerns into consideration, especially during health data sharing across health platforms. To date, the proposed data models focus on aspects such as secure interoperable data exchange, support of standards of data sharing between existing health information systems (HISs), and the establishment of means to allow the patients to consciously distribute processing actions between a cloud infrastructure and the healthcare infrastructure. An overview of the most relevant models and standards in the literature, for example, security and privacy models, is presented in the following.

In general, a data model in the healthcare domain is typically adopted to provide syntactic, semantic specifications and metadata about the shared data according to Demski et al. (2016). For example, the authors in openEHR (2021) propose modelling tools, such as the ADL Workbench, that provide the data schema definitions for a data exchange interface to guide implementation of the corresponding software artifacts. Then, the data models can be translated into formats that can be handled by non-experts, such as XML Schema Definitions (XSD) and JavaScript Object Notation (JSON) templates. Another proposed model is the data model presented by He et al. (2010) that supports uniform standards of data sharing between existing HISs. This model employs a Virtual Private Network–based model, in which public networks are exploited to establish a private cloud in order to allow different healthcare infrastructures to share HIS information. More specifically, the storage and management of sensitive patient records are assigned to a private cloud, while additional data, such as business operations, are stored into a commercial cloud server. In addition, a cloud-hosted control center is responsible for coordinating infrastructure resource sharing.

There are also patient-centric data models that deal with the risks of the privacy leaks in the cases of adopting cloud infrastructures as Li et al. (2010) has presented, while Yu et al. (2010) shift the responsibility for managing medical records from the data owner to the cloud or the healthcare infrastructure. The data model that Li et al. (2010) proposed is a patient-centric model in the sense that it shifts the privacy responsibility to the patients: The patients have full control over their electronic health record data in terms of how these are distributed and the period of granted permission to the third parties to access them. Additionally, any processing or computation actions in the context of the data model presented by Yu et al. (2010) take place without providing access to the full patient data; this can be achieved through using security mechanisms, such as Key-Policy Attribute-Based Encryption, Proxy Re-Encryption, and lazy re-encryption.

Another security model has been proposed by Kaletsch and Sunyaev (2011), which is referred to as Online Referral and Appointment Planner (ORAP). ORAP is employed as part of a privacy framework, whereby the healthcare employees transfer healthcare information to other experts through the ORAP model. This model allows the secure transfer of the health records, but it does not guarantee the security of the medical history or action plan. Also, this model does not provide permission to patients to update or process the health record entries. Other privacy models for healthcare systems are also reviewed by Rahim et al. (2014), in which privacy models were considered regarding several privacy preferences and dimensions, such as awareness (Schwaig et al., 2013; Tesema, 2010; Samy et al., 2009), trust (Tesema, 2010; Samsuri et al., 2011), sharing needs (Tesema, 2010; Damschroder et al., 2007), and access control (Mont et al., 2011; Ardagna & Capitani, 2011), are discussed. For example, according to Schwaig et al. (2013), Tesema (2010), and Samy et al. (2009), there have been proposed data models that focus on security awareness, that is the patients are aware of the policies related to their data-handling actions and are informed in the case the health records are being disclosed without their consent, as typically requested by modern data protection legislation. An additional example is the data models that can be applied to a healthcare infrastructure for monitoring security policies through a framework in which a user is able to manage the health records' access control as Mont et al. (2011) and Ardagna and Capitani (2011) presented. Tesema (2010) highlight the relationship that may exist between security awareness and a patient's concern through a patient's data, while the proposal that Samsuri et al. (2011) introduced focuses on the trust aspect and prompts to include free accessibility for patients, transparency in medical record management, and patient consent in a privacy policy. The sharing-needs aspect is also discussed by Damschroder et al. (2007), where the patients are informed about how their records are going to be used by a specific infrastructure, as it is observed that the major threats of patient privacy are derived from inside the healthcare infrastructure, from which there is legitimate access to medical records.

Furthermore, there are several clinical information models in the literature that have been proposed as open data models and can be exploited by healthcare systems. A modern HIS may adopt a model-driven software development methodology that applies related, established data models. Consequently, the standardization of these data models plays a fundamental role in the quality, security, and privacy of modern HISs. A variety of related standards has been nowadays proposed, in order to ensure the proper communication and interoperability across the HISs. For example, the Clinical Data Interchange Standards Consortium (CDISC, 2021) and ODM (2021) provides an operational data model that defines the exact data that should be shared among the healthcare platforms divided into three categories: data shared during study setup, operation, and analysis (CDISC, 2021; ODM 2021). In addition, there have been proposed standards for diseases, diagnoses, health management, and clinical purposes, such as the International Classification of Diseases.[2] Finally, there are standards for exchanging clinical and administrative healthcare data, such as the Fast Healthcare Interoperability Resource,[3] and standards that consider the representation of quality measures as electronic documents, for example, the Quality Reporting Document Architecture.[4]

Overall, although there are several data models that have been proposed for dealing with the risks of privacy leaks in the case of adopting cloud infrastructures, none of the existing model proposals or standards considers the assessment of privacy and cybersecurity risks. In other words, the existing models and standards do not deal with aspects such "*Infrastructure X has a vulnerable asset Y because of the list of reasons Z*". This is addressed by CUREX, as explained in the next sections.

## 14.3 THE CUREX CONCEPTUAL DATA MODEL FOR RISK ASSESSMENT

In Figure 14.1, the conceptual model of the CUREX platform is represented by an Entity-Relationship (ER) diagram of the data elements and the associated relationships that are processed and stored in the CUREX platform but can be generalized to any HIS.

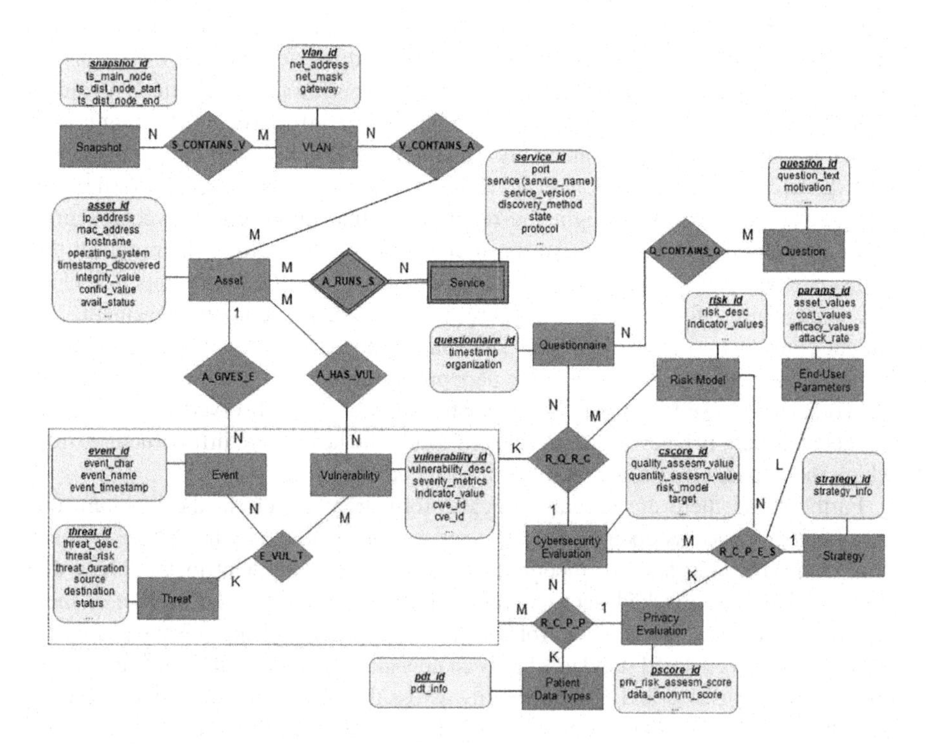


**FIGURE 14.1** The ER Diagram of CUREX Platform.

The set of entities that participate in the diagram are the following:

> **Snapshot**, **VLAN**, **Asset**, **Service**, **Event**, **Vulnerability**, **Threat**, **Questionnaire**, **Question**, **Risk Model**, **Privacy Evaluation**, **Cybersecurity Evaluation**, **End-User Parameters**, **Strategy,** and **Patient Data Types**.

In a nutshell, HIS infrastructure is monitored in a systematic manner. Within a monitored **VLAN**, several **assets** are identified, each running a set of **services**. The monitoring activity is repeated periodically; therefore, the monitored data refer to specific **snapshots**. Furthermore, each asset is annotated with several **vulnerabilities**, for example, as these are defined in well-established repositories like common vulnerabilities and exposures (CVE).[5] For example, the entries contain information like *"The asset with MAC address A7:4F:D1:2E:AA:2F suffers from the vulnerability with the name 'Drupal Core SQL Injection Vulnerability,' which is associated with the CVE identifier CVE-2014–3704."* Moreover, the network traffic is monitored in an online manner using standard tools, and the post-processed traffic data are encapsulated into **events**, which are associated with specific assets. Several events combined with specific vulnerabilities give rise and/or associated to security or privacy **threats**. This continuous monitoring is coupled with additional information, such as **questionnaire** data regarding the procedures and security measures, the policies enforced, and the organization business profile; in summary, the questionnaire includes several types of **questions** to acquire a complete view of the security- and privacy-related status within the organization running the HIS. Each questionnaire corresponds to an organization for which data processing activities are identified and a **risk model** is defined. A risk model is a set of threat scenarios and inputs related to different vulnerabilities, incidents, and/or infrastructure contexts, which may affect the risk level and cause economic losses in terms of confidentiality, integrity, and availability based on qualitative and quantitative risk scores. Additionally, **privacy and cybersecurity evaluation** take place in order to quantify the privacy and security risks imposed by identified vulnerabilities on the assets. Then, based on the defined risks, critical Internet security controls, and **end-user parameters**, such as system performance usability cost and direct costs, the best **safeguard strategy** is proposed. These strategies indicate the proper management of the **patient data types**.

These entities participate in different relationships that are identified and presented in Table 14.1 along with the associated entities that participate in those relationships.

Then, the logical representation of the entities, as well as the relational representation of the relationships, is described in Tables 14.2 and 14.3, while the primary keys are shown underlined. Following the standard database methodology when creating the relation schema from the ER diagram, the ER diagram is mapped onto the corresponding set of relations, assigning both keys from a many-to-many relationship as the primary key of the resulting relation. Moreover, each entity corresponds to a relation with the same primary key as the one used in the entity. The extracted relations of the CUREX conceptual model are shown in Tables 14.2 and 14.3.

**TABLE 14.1**
**Relationships in the ER Diagram**

| Relationships | Associated Entities |
|---|---|
| S_CONTAINS_V | Snapshot, VLAN |
| V_CONTAINS_A | VLAN, Asset |
| A_RUNS_S | Asset, Service |
| A_GIVES_E | Asset, Event |
| A_HAS_VUL | Asset, Vulnerability |
| E_VUL_T | Event, Vulnerability, Threat |
| Q_CONTAINS_Q: | Questionnaire, Question |
| R_Q_R_C | EVT_Report, Questionnaire, Risk Model, Cybersecurity Evaluation |
| R_C_P_P | EVT_Report, Cybersecurity Evaluation, Privacy Evaluation, Patient Data Types |
| R_C_P_E_S | Risk Model, Cybersecurity Evaluation, End-User Parameters, Strategy |

**TABLE 14.2**
**Relations Extracted by the Entities in the CUREX Conceptual Model**

| Relations | Relation Attributes |
|---|---|
| Snapshot | snapshot_id, timestamp_main_node, timestamp_dist_node_start, timestamp_dist_node_end |
| VLAN | vlan_id, network_address, network_mask, gateway |
| Asset | asset_id, ip_address, mac_address, hostname, operating_system, timestamp_discovered, integrity_value, confid_value, availability_status |
| Service | service_id, port, service_name, service_version, discovery_method, state, protocol |
| Event | event_id, event_char, event_name, event_timestamp |
| Vulnerability | vulnerability_id, vulnerability_description, severity_metrics, indicator_value, cwe_id, cve_id (CWE and CVE are public repositories) |
| Threat | threat_id, threat_description, threat_risk, threat_duration, source, destination, status |
| Questionnaire | questionnaire_id, timestamp, organization |
| Question | question_id, question_text, motivation |
| Risk Model | risk_id, risk_description, indicator_values |
| Privacy Evaluation | pscore_id, priv_risk_assesment_scores |
| Cybersecurity Evaluation | cscore_id, quality_assesment_value, quantity_assesment_value, risk_model, target |
| End-User Parameters | parameters_id, asset_values, cost_values, efficacy_values, attack_rate |
| Strategy | strategy_id, strategy_info |
| Patient_Data_Types | patient_data_type, patient_data_info |

**TABLE 14.3**
**Relations Extracted by the Relationships in the CUREX Conceptual Model**

| Relations | Relation Attributes |
|---|---|
| S_CONTAINS_V | snapshot_id, vlan_id |
| V_CONTAINS_A | vlan_id, asset_id |
| A_RUNS_S | asset_id, service_id |
| A_GIVES_E | asset_id, event_id |
| A_HAS_VUL | asset_id, vulnerability_id |
| E_VUL_T | event_id, vulnerability_id, threat_id |
| Q_CONTAINS_Q | questionnaire_id, question_id |
| R_Q_R_C | event_id, vulnerability_id, threat_id, risk_id, questionnaire_id, cscore_id |
| R_C_P_P | event_id, vulnerability_id, threat_id, cscore_id, patient_data_type_id, patient_score_id |
| R_C_P_E_S | risk_id, cybersecurity_core_id, privacy_score_id, parameters_id, strategy_id |
| S_CONTAINS_V | snapshot_id, vlan_id |
| V_CONTAINS_A | vlan_id, asset_id |
| A_RUNS_S | asset_id, service_id |
| A_GIVES_E | asset_id, event_id |
| A_HAS_V | asset_id, vulnerability_id |

## 14.4 ARCHITECTURE OF CUREX PLATFORM

The aforementioned model is developed in a HIS-agnostic manner and is largely independent of the exact implementation details followed by CUREX. However, the important point is that it is a practical model that, when materialized, can yield tangible outcomes.

The details of the architecture implemented by CUREX are as follows. CUREX aims to produce a novel, flexible, and scalable situational awareness-oriented platform that can address comprehensively the protection of the confidentiality and integrity of health data focusing on health data exchange cases according to Diaz-Honrubia et al. (2019). This platform consists of several loosely coupled components. The CUREX platform is GDPR-compliant by design as we explain later. It allows the healthcare providers to be able to assess the possible cybersecurity and privacy risks of the patient's data protection and finally, suggest optimal safeguards for protecting by numerous threats. The CUREX architecture is decentralized and is enhanced with a private blockchain infrastructure in order to ensure the integrity of the risk assessment process; for example, assessment reports cannot be modified, as presented in CUREX Consortium: Deliverable 2.2 (2019). More specifically, the outcome of the assessment of all HISs is stored in a blockchain to increase reliability and data integrity.

As shown in Figure 14.2, the CUREX architecture consists of different layers including a variety of subcomponents. The figure shows the interaction between these components and health-related infrastructure (e.g., systems of a hospital) in order to ensure the proper and safe communication among them, as the authors in

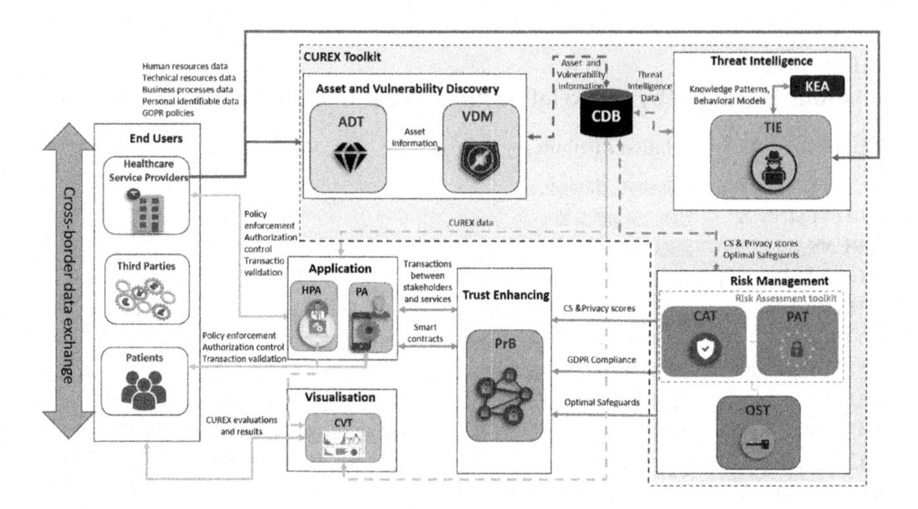


**FIGURE 14.2** The CUREX Platform Architecture.

CUREX Consortium: Deliverable 2.2 (2019) described. The following are the main CUREX components:

- Asset Discovery Tool (ADT), which performs an analysis of the network of a healthcare institution in order to discover the healthcare assets, such as not only the hospital's devices, of both servers and other types, connected to the network of a healthcare institution but also the necessary information from these assets, which are in high-risk of attack (vulnerable assets)
- Vulnerability Discovery Manager (VDM), which identifies, analyzes, and reports vulnerabilities related to the assets discovered by the ADT. The vulnerabilities are further enriched by the output of Knowledge Extraction Analytics (KEA), described next
- KEA, which detects threat patterns by exploiting the knowledge extracted by other components of CUREX or the network communication interfaces for constructing models and designing methods; it also detects anomalies in the network traffic at runtime as Bellas et al. (2020) presented. Advanced analytics and both supervised and unsupervised machine learning are employed to this end. The KEA output is combined with TIE and is also fed to VDM.
- Threat Intelligence Engine (TIE), which encapsulates KEA, processes, and analyses data, such as vulnerabilities, past incidents, network traffic, and system logs with a view to identifying suspicious behaviors constituting threats and produce alerts
- Cybersecurity Assessment Tool (CAT), which is a part of the Risk Management toolkit and conducts cybersecurity risk assessment by collecting and analyzing cybersecurity events in real time and correlating them in order to assess the possible risks. It leverages the results of TIE.
- Private Assessment Tool (PAT) is another component of the risk assessment toolkit that conducts the privacy assessment by providing privacy levels

regarding the level of conformance to the GDPR directives in order to protect the sensitive health data and personally identifiable information (PII) of the patients.

- Optimal Safeguard Tool (OST), which constitutes the third component of the Risk Management toolkit and complements the cybersecurity and privacy risk assessments with the recommendation of optimal safeguards
- Private Blockchain (PrB), which operates as a decentralized database that not only stores risk assessment and privacy reports and scores, as they derived by the CAT and PAT CUREX components, but also stores a subset of the OST output considering cyber strategies and tool recommendations. Each partner running the CUREX platform forms an individual node in the private blockchain infrastructure.
- Health Professional Application (HPA) and Patient Application (PA) that act as interaction points for the health professionals or patients with the CUREX platform, respectively

In Figure 14.2, it is shown that there is another component, denoted as CDB, standing for the CUREX database. This is a conceptual database that conforms to the model presented earlier and allows for the information flow between components and sharing data.

Finally, apart from the HPA and PA, which provide the interface to healthcare professionals and patients, respectively, for example, to retrieve an assessment report, the CUREX platform is deployed on an infrastructure running a HIS and is managed by the same administrators running the corresponding HIS.

## 14.5 LEGAL AND ETHICAL ISSUES

The proposed approach allows a processing methodology, where no personal data on individual behavior or other personal data that could be in conflict with the right of self-determination are collected, stored, or analyzed. Within the CUREX platform and its associated conceptual and more detailed logical schemata, no piece of data can be treated as a direct identifier of a data subject.

However, CUREX heavily relies on the monitoring and assessment of an HIS and, in general, of the IT infrastructure of a health organization. This monitoring includes metadata regarding IPs and MAC addresses of hardware devices, mentioned as assets. Given the fact that these devices are used by an individual or a small set of hospital staff members cannot be excluded; such data are legally qualified as personal data under the scope of the GDPR and similar legislation acts.

The CUREX approach is heavily influenced by the requirement to comply with Regulation (EU) 2016/679 ("the General Data Protection Regulation" or "the GDPR"). At the level of data modelling and management, the following technical and policy measures have to be taken to achieve data protection by design and all of them have been adopted by the CUREX approach:

- **Data minimization:** Data processing must be lawful, fair, and transparent. **It should involve only data that are necessary and proportionate** to achieve the specific task or purpose for which they were collected, that is, data that are directly relevant and necessary for deriving risk assessment reports. **Therefore, the CUREX risk assessment methodology needs only to collect the data that are needed for its objectives**, since collecting unnecessary/unrelated data may be deemed unethical and unlawful. To this end, the CUREX solution monitors the minimum amount required to identify IT assets, such as MAC addresses, given that exact risk mitigation policies may need to be applied to these specific assets (therefore, they need to remain identifiable).

- **Data confidentiality:** CUREX components can be implemented in such a manner that they keep internally data in databases that are not exposed to healthcare employees but only to the CUREX platform admins. For example, past logs may be kept within the KEA component to allow machine learning training. Apart from enforcing strict access control policies leveraging the corresponding mechanisms of database tools, it is straightforward to enforce suitable retention policies; for example, log data should be kept only for a predefined period.
- **The pseudonymization or anonymization of personal data/applied cryptography through hashing:** This applies to IP and MAC addresses when they are still needed for training machine learning models but without the need to be possible to be linked to real devices. The IP and MAC addresses kept internally in KEA and TIE for future machine learning training are anonymized through hashing.
- Through using asset identifiers as a key attribute in the data model, all information stored/kept regarding a specific asset can be easily retrieved and checked.

These measures complement the broader CUREX policy regarding GDPR compliance, such as hospital staff to be informed via a data protection notice applying to the personal data processing in the context of the CUREX platform as a whole.

In summary, the basic data protection principles (ensuring GDPR compliance by design) that apply in the context of the CUREX modelling are as follow (see Article 5 of the GDPR):

1. *Lawfulness, fairness, and transparency:* CUREX provides a generic data model that it is straightforward to check that IP and MAC addresses of hardware assets; that is, only data attributes that may deemed personal are processed for the project purposes.
2. *Purpose limitation:* The personal data are processed for the risk assessment purposes only, and there is no interface so that someone can extract them and process them for other purposes in the reference architecture presented.
3. *Data minimization:* CUREX monitors only the minimal asset data (IPs and Mac addresses to identify hardware assets).

4. *Accuracy:* All data regarding identifiable assets are accurate; that is, the CUREX model does not include any data that can only be collected in a fuzzy or approximate manner.
5. *Storage limitation:* Any retention policy can be enforced by the CUREX platform admins.
6. *Security:* Data are processed in a manner that ensures protection against unauthorized or unlawful processing and against accidental loss, destruction, or damage, using appropriate technical or organizational measures ('integrity and confidentiality').

## 14.6 CONCLUSION

In this chapter, first, an overview of several proposed data models and standards for the healthcare domain was presented; based on this overview, the need to develop and adopt a model-based formalized procedure to assess the healthcare infrastructure in terms of privacy and cybersecurity risks, which is an overlooked aspect so far, is identified. Therefore, a conceptual model is proposed in order to ensure trust in sharing and processing medical records across institutions. The presented model has been adopted by the CUREX healthcare platform. Finally, the legal and ethical aspects were discussed along with the relevant technical and policy measures.

## ACKNOWLEDGMENTS

The research work that has been presented in this chapter has been supported by the European Commission under the Horizon 2020 Programme, through funding of the CUREX project (G.A. n826404). The authors would also like to thank all the CUREX consortium members for their work: the University of Piraeus Research Center, Atos Spain SA, Almerys SAS, Cyberlens BV, INTRASOFT International SA, Suite5 LTD, Timelex, Eight Bells LTD (8BELLS), Ubitech Limited, the University of Greenwich, the Universidad Politécnica de Madrid, the University of Cyprus, Aristotle University of Thessaloniki, Servicio Madrilenño de Salud, Fundació Privada Hospital Asil de Granollers, and Karolinska Institutet.

## NOTES

1 https://curex-project.eu/
2 www.who.int/classifications/icd/en/
3 www.hl7.org/fhir/overview.html
4 https://ecqi.healthit.gov/qrda
5 https://cve.mitre.org/

## REFERENCES

Ardagna, C. A., & Capitani, S. D. (2011). Privacy models and languages: Access control and data handling policies. In *Digital Privacy*, part III. Berlin: Springer, pp 309–329.

Bellas, C., Naskos, A., Kougka, G., Vlahavas, G., Gounaris, A., Vakali, A., Papadopoulos, A., Biliri, E., Bountouni, N., & Granadillo, G. G. (2020). A methodology for runtime detection and extraction of threat patterns. *SN Computer Science*, 1(5):238.

CDISC. (2021). *Mission and Principles*. Accessed February 2021 from www.cdisc.org/about/mission.

CUREX Consortium. Deliverable 2.2. (2019). *Overall Architecture Design*. Project Deliverable.

Damschroder, L. J., Pritts, J. L., Neblo, M. A., et al. (2007). Patients, privacy and trust: Patients' willingness to allow researchers to access their medical records. *Social Science & Medicine*, 64:223–235.

Demski, H., Garde, S., & Hildebrand, C. (2016). Open data models for smart health interconnected applications: The example of openEHR. *BMC Medical Informatics and Decision Making*, 16(1):137.

Diaz-Honrubia, A. J., Rodriguez Gonzalez, A., Mora Zamorano, J., Rey Jiménez, J., Gonzalez-Granadillo, G., Diaz, R., Konidi, M., Papachristou, P., Nifakos, S., Kougka, G., & Gounaris, A. (2019). An overview of the CUREX platform. In *2019 IEEE 32nd International Symposium on Computer-Based Medical Systems (CBMS)*, pp. 162–167. https://ieeexplore.ieee.org/abstract/document/8787402?casa_token=g3NuU_4SFV8AAAAA:vtilgbkwLS2vIsg8h_78WryNqzieVaMJ7afZGsVmL6wd9TazrtcF3RVw9R8GnUJOkjNz1vSl2w

"General Data Protection Regulation" Regulation (EU) 2016/679 of the European Parliament and of the Council of 17 April 2016 on the protection of natural persons with regard to the processin of personal data and on the free movement of such data, and repealing Directive 95/46/EC. https://ec.europa.eu/info/law/law-topic/data-protection/data-protection-eu_en

He, C., Jin, X., Zhao, Z., & Xiang, T. (2010). A cloud computing solution for hospital information system. In *2010 IEEE International Conference on Intelligent Computing and Intelligent Systems*, Vol. 2, pp. 517–520. https://ieeexplore.ieee.org/abstract/document/5658278?casa_token=1XPNUyb7aLIAAAAA:ahoGc8CIBtimAFqwtwjaBlqYE4aeUW5qKBIb2jKcaIDfj1OFvfu-4mJdQpC6X8hsV1YnEVcRaA

Kaletsch, A., & Sunyaev, A. (2011). Privacy engineering: Personal health records in cloud computing environments. In *ICIS 2011 Proceedings, Article* 2.

Li, M., Yu, S., Ren, K., & Lou, W. (2010). Securing personal health records in cloud computing: Patient-centric fine-grained data access control in multi-owner settings. In *Security and Privacy in Communication Networks*, Vol. 50, pp. 89–106. Berlin and Heidelberg: Springer Berlin Heidelberg.

Mont, M. C., Pearson, S., Creese, S., et al. (2011). A conceptual model for privacy policies with consent and revocation requirements. In *Privacy and Identity Management for Life*, pp 258–270. Berlin: Springer.

ODM. (2021). *Operational Data Model*. Accessed February 2021 from www.cdisc.org/odm

openEHR. (2021). Accessed February 2021 from www.openehr.org/downloads/modellingtools

Rahim, F. A., Ismail, Z., & Samy, G. N. (2014). A conceptual model for privacy preferences in healthcare environment. In *8th International Conference on Knowledge Management in Organizations*, pp. 221–228. Dordrecht, Netherlands: Springer.

Samsuri, S., Ahmad, R., & Ismail, Z. (2011). Towards implementing a privacy policy: An observation on existing practices in hospital information system. *The Journal of e-Health Management*, 2011:1–9.

Samy, G. N., Ahmad, R., & Ismail, Z. (2009). Threats to health information security. In: *2009 Fifth International Conference on Information Assurance and Security*, pp 540–543. IEEE. https://ieeexplore.ieee.org/document/5283006

Schwaig, K. S., Segars, A. H., Grover, V., & Fiedler, K. D. (2013). A model of consumers' perceptions of the invasion of information privacy. *Information & Management*, 50(1):1–12.

Tesema, T. (2010). Patient's perception of health information security: The case of selected public and private hospitals in Addis Ababa. In: *2010 Sixth International Conference on Information Assurance and Security (IAS)*, pp. 179–184. Atlanta, GA: IEEE.

Yu, S., Wang, C., Ren, K., & Lou, W. (2010). Achieving secure, scalable, and fine-grained data access control in cloud computing. In *2010 Proceedings IEEE INFOCOM*, pp. 1–9. https://ieeexplore.ieee.org/abstract/document/5462174?casa_token=pRmEJ6afB0gAAAAA:q62r5bst3qCRqNBj4I33Effpy84_P_WY4RbrEQhJsyOezcj8iU0vAQBMyZY51VAFOISF7uCnEw

# 15 Data Science in Health Informatics

*Behzad Soleimani Neysiani, Nasim Soltani, Saeed Doostali, Mohammad Shiralizadeh Dezfoli, Zahra Aminoroaya, and Majid Khoda Karami*

**CONTENTS**



DOI: 10.1201/9781003142751-20

## 15.1 INTRODUCTION

Data science (DS) techniques create extraordinary opportunities in various areas of medicine and healthcare. DS distinguishes intelligent techniques and traditional techniques in healthcare data analysis systems. These algorithms can recognize a pattern in behavior and figure out the logic (Gómez-González et al., 2020). An overview of DS's current application in medicine and healthcare is significant even though DS includes many techniques whose applications are growing every day. DS has been growing into the public healthcare sector and has significantly impacted every healthcare system aspect. Artificial Intelligence (AI) and data mining (DM) are branches of the DS that their related sciences and relations are shown in Figure 15.1, indicating databases generate many problems in this field of study. Visualization can help percept and analyze businesses. The business stakeholders consider their strategies based on statistics and their domain knowledge to extract new knowledge using AI, DM, neurocomputing, and knowledge discovery in databases (KDD) techniques benefiting from machine learning (ML) and pattern recognition techniques. Then DS team shares the new extracted results and presents them to each other to select the valid and best ones. Finally, afterward, their inquisitiveness will solve the problem and find new problems with scientists to address.

Nowadays, the amount of data in almost all sciences, including genetics, earth science, agriculture, education, and medicine, increases dramatically. The amount of data in the electronic and real world is continuously rising. The data generated by the healthcare field has substantial and complex amounts that cannot be analyzed and processed by traditional methods. The healthcare industry has widely generated huge amounts of data, including compliance and regulatory requirements, record keeping, and patient care (Raghupathi & Raghupathi, 2014).

Analyzing these enormous amounts of data to extract novel and usable information or knowledge is a very complicated and time-consuming task. This massive amount of data is called 'big data,' used in many medical and healthcare applications, like disease surveillance, clinical decision-making, and population healthcare management (Raghupathi & Raghupathi, 2014). Extracting valuable knowledge

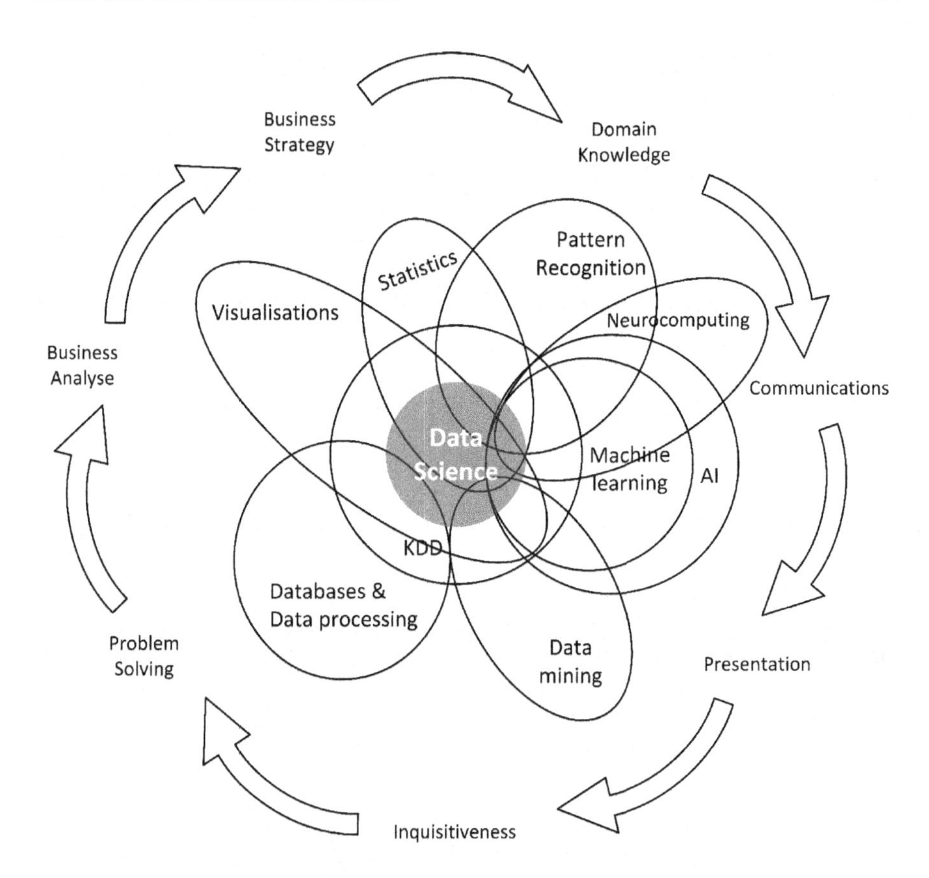


**FIGURE 15.1** Multidisciplinary Data Science.

*Source*: Modified from Kelleher and Tierney (2018) andTierney (2012a, 2012b).

from the whole available data is vital; however, it is time-consuming. Big data refers to data with tremendous value, volume, velocity, variety, and veracity and deals with techniques to process these five V challenges. For example, they analyze massive datasets and statistics to explore patterns that may describe an attack by bioterrorists.

Reports collected from the US healthcare system reached 150 exabytes in 2011. The amount of US healthcare big data is about 1 zettabyte (1021 gigabytes) scale now and will get to a yottabyte (1024 gigabytes) very soon at this rate of growth (Cottle et al., 2013). One of California's healthcare networks, Kaiser Permanente, has about 9 million accounts, and it maybe has about 44 petabytes of data from electronic health records (EHRs), like annotations and images (Cottle et al., 2013). It should be mentioned that almost all reports have many issues like typos, which need to be determined (Soleimani Neysiani & Babamir, 2018) and corrected (Soleimani Neysiani & Babamir, 2019a, 2019b, 2019c).

This study illustrates data science techniques and their application and then classifies healthcare problems. Finally, it introduces a road map for researchers to solve similar problems using current DS tools and techniques.

## 15.2 DATA CHALLENGES IN HEALTHCARE SYSTEMS

Healthcare can be thought of as a multidimensional system created with the unique goals of preventing, diagnosing, and treating impairments in human or health-related issues—the major components of a healthcare system consisting of facilities, professionals, and financing support. Facilities include hospitals, clinics, medicines delivery, treatment technologies, or diagnosis. The physicians and nurses are healthcare professionals and belong to many different sectors like midwifery, medicine, and dentistry, which serve in the healthcare system. The medical records are usually stored in typed reports or handwritten notes, even for a medical examination (AOCNP, 2015), and it has been conventional in Egypt since 1600 BC (Gillum, 2013). Stanley Reiser (1991) believes in stories from all involving peoples in an illness. A healthcare ecosystem is shown in Figure 15.2, which includes individual and public care. Also, it includes software and hardware, treatment and prevention, and core and lateral concerns.

There are four main aims in healthcare systems, including patient experience, reducing costs, population health, and care-team well-being, which should be considered in entire healthcare ecosystems.

There are many challenges and purposes in healthcare systems that have been classified into four categories, as Figure 15.3 (Multiple, 2020) shows, in which data science techniques are applicable in almost all cases (Amisha et al., 2019).

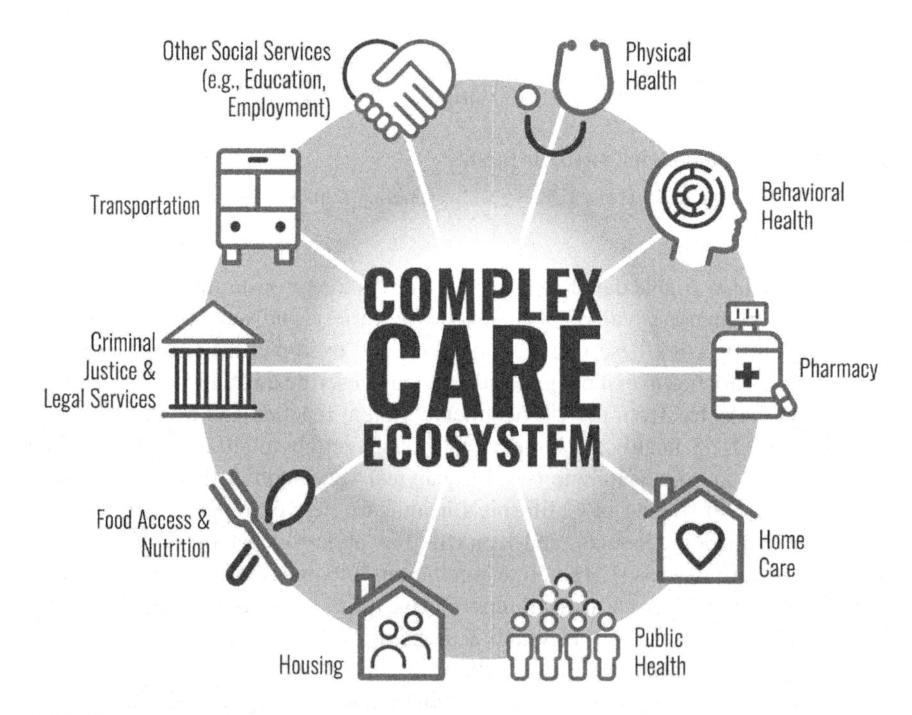


**FIGURE 15.2** Healthcare Complex Ecosystem.

*Source:* Needs (2020).

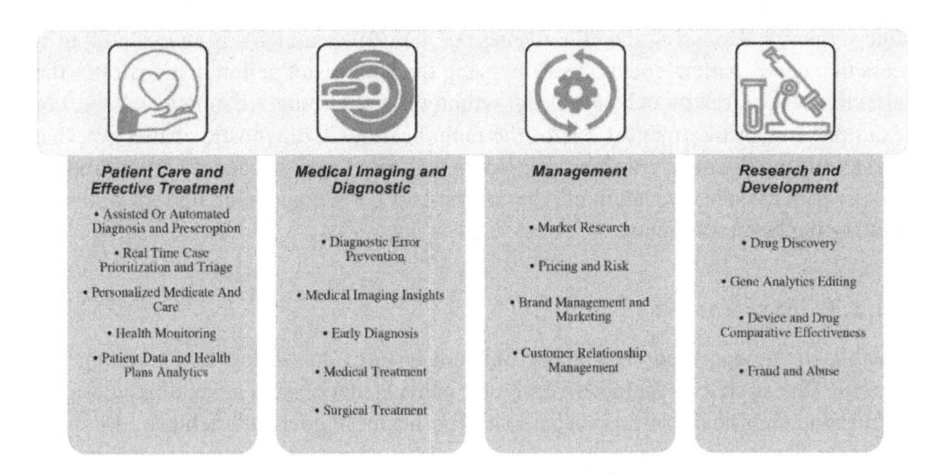


**FIGURE 15.3** Healthcare Challenges.

*Source:* Based on Multiple (2020).

### 15.2.1 Patient Care and Effective Treatment

However, one efficient healthcare application improves diagnostic accuracy, effectively helping healthcare providers diagnose symptoms due to mimic human cognitive functions. The following are two phases of decision-making in the medical world (Chattopadhyay, 2017):

- *Differential Diagnosis (DD):* Doctors perceive all patients' information, check it based on their knowledge for diagnosis, and assign weights to symptoms for those diseases with similar symptoms. Pattern recognition, matching, and selecting constitute most tasks of physicians.
- *Final or Provisional Diagnosis (FD):* Now, the elementary treatments and checkups will begin until determining the final prognosis and treatments.

Automated diagnosis and prescription are two of the main topics that a healthcare system may face due to the enormous number of patients. Real-time prioritizing and triaging of patients at the emergency or registration units can make patients more satisfied. Checking every patient carefully and avoiding a general prescription considering the patient's historical file is very important and critical. There are many plans for health monitoring, analyzing, and planning for both treatment and well-being.

### 15.2.2 Medical Imaging and Diagnostic

It is crucial to prevent diagnostic error; otherwise, a patient will be hurt or die. So, every tool to a better understanding of patient problems like Imaging Insight is vital for physicians. Early diagnosis can prevent costly and dangerous treatments like surgeries. Assessing the effectiveness of medical treatments is an essential process that needs patient checkups. Conveying the analysis of action demonstrates the effectiveness of treatment by checking symptoms, causes, and treatment courses. For example, cancer treatment is one of the main healthcare diagnostic challenges, that is, identifying the most common symptoms that can diagnose cancer earlier, finding cancer patterns, creating an intelligent method for early diagnosing tumors, and suggesting the best treatments.

### 15.2.3 Management

Healthcare management has many challenges like identifying and tracking the chronic illness states of intensive care unit patients, decreasing hospital admissions, and supporting healthcare management. The business goals are achieved by management. The critical interaction management between customers and organizations is customer relationships, which may happen through call centers, ambulatory care settings, and billing departments in healthcare systems.

### 15.2.4 Research and Development (R&D)

R&D helps us find new scientific facts like searching for new drugs, analyzing genomes, and finding their features and applications. It also compares devices and drugs' efficiency on treatment or disease prevention.

## 15.3 HEALTHCARE DATA TYPES FOR ANALYZING

There are various data in healthcare systems that can be classified as Table 15.1 (Jiang et al., 2017). The data source column refers to how the data is achieved, and the example column illustrates some source samples. The data type column refers to

**TABLE 15.1**
**Healthcare Data Types and Sources**

| Data Source | Examples | Data Type | Data Format |
|---|---|---|---|
| *Human-Generated Data* | Physicians' notes, email, and paper documents | Structured & unstructured | ASCII/text |
| *Machine-Generated Data* | Readings from various monitoring devices | Structured & unstructured | Relational tables |
| *Transaction Data* | Billing records and healthcare claims | Structured & unstructured | Relational tables |
| *Biometric Data* | Genomics, genetics, heart rate, blood pressure, X-ray, fingerprints | Structured & unstructured | ASCII/text, images |
| *Social Media Data* | Interaction data from social websites | Unstructured | Text, images, video |
| *Publications* | Clinical research and medical reference material | Unstructured | Text |

*Source:* Fang et al. (2016).

the nature of data that structured data numerically, categorically, or temporally. It can be used efficiently for AI techniques, especially ML algorithms. However, unstructured data, like text or images, cannot be processed quickly and efficiently and even compared to find their similarities or equivalence (Soleimani Neysiani et al., 2020). Finally, the last column indicates the data format in which texts can be saved using ASCII or Unicode standards in computers. The relational tables refer to databases' structures. Multimedia files like images, videos, and audio can be stored, too.

### 15.3.1 Statistical Techniques

Quantitative research is familiar in healthcare studies to check goals and trends and find variables relations or compare groups. Quantitative researchers collect structured numerical data, analyze them to find research problems, and question and test their hypotheses. A statistical hypothesis will test a specific proposition statistically. Statistical analysis conduct a literature review and write one or multiple direct research question(s) (Guetterman, 2019). It will (1) formulate a hypothesis, (2) select and run a statistical test, (3) find an appropriate sample size, (4) preprocess data, (5) calculate descriptive statistics, (6) verify statistical tests assumptions, (7) run the analysis, (8) test fitness of statistical model, (9) write results reports, and (10) determine threats to conclusion validity.

Healthcare studies almost just use descriptive statistics. They show frequencies, averages, and standard deviation (SD), which can miss advanced analytical opportunities, like knowing that patients have favorable treatment attitudes. These attitudes can be expressed using inferential statistical tests to finding significant statistical differences between the two groups or not, which may be more beneficial for physicians.

Statistical analysis aggregates numeric data to inferences variables using (1) descriptive statistics and (2) inferential statistics to generalize situations beyond the actual dataset.

### 15.3.2 Descriptive Statistics

Data aggregation can help with understanding many values easily. Summarizing typical values is known as central tendency measures like the mean, the median, and the mode. Commonly used descriptive statistics are summarized in Table 15.2 (Guetterman, 2019). The values distribution is represented through variability measures, for example, the variance, the SD, and the range. Descriptive statistics have some data distribution indicators or the frequency of the dataset's value, like a histogram plot. Independent and dependent variables are known as outcome and predicted variables, respectively, in research fields such as correlational studies. A dependent variable can be predicted or influenced by some independent variables.

### 15.3.3 Inferential Statistics to Compare Groups with *t* Tests and Analysis of Variance (ANOVA)

Researchers can check a sample and generalize their conclusions to a population using inferential statistics by examining group differences and variables relationships. Some inferential tests are described in Table 15.3.

**TABLE 15.2**
**Descriptive Statistics**

| Goal | Statistical Metric | Description of calculation | Intent |
|---|---|---|---|
| **Measures of Central Tendency** | Mean | Summation of values divided by their numbers. | Describe all responses with the average value |
| | Median | The middle point of ordered values | Crucial when dealing with extreme outliers |
| | Mode | The most frequent value. | The most common value |
| **Measures of Variability** | Variance | An average of the absolute value of the difference data samples vector and its mean | A distribution indicator |
| | Standard deviation | The square root of variance. | Distribution indicator about values' difference from their average |
| | Range | The difference of the maximum from the minimum value | A very general indicator of the distribution |
| | Frequencies | Count occurrences of each value | Distribution of each value occurs |

*Source:* Modified from Guetterman (2019).

**TABLE 15.3**
**Inferential Statistics**

| Statistic | Intent |
|---|---|
| ***t* tests** | It examines a statistically significant means difference between the two groups. |
| **Variance Analysis** | It examines a statistically significant means difference between two or more groups. |
| **Correlation** | It examines a relationship or association between two or more variables. |
| **Regression** | It examines how one or more variables predict another variable. |

*Source:* Modified from Guetterman (2019).

Other statistics then deal with two or more possibilities for determining which group differs from the others (e.g., scheduled comparisons or Bonferroni comparisons). Planned comparisons are established before analyzing the groups' contrast, but other tests are conducted after this, like the Bonferroni comparison.

### 15.3.4 Relationships Examination Using Regression and Correlation

The correlation and regression analysis methods are other general linear models. Correlation reveals systematically together changes in values between two variables. Correlation analysis indicates that two variables are rising and falling together or are not related systematically. A correlation coefficient (r) and its corresponding p-value as confidence interval determine correlation analysis. The p-value is compared with a hypothesized significance condition (e.g., $p < 0.01$) for a statistically significant relationship determination. Otherwise, the coefficients are useless, and the correlation coefficient will be checked.

The 'r' and 'R' are used to indicates a relationship between two and more than two variables, respectively, which the 'R' is called multiple correlations or regression, too (Guetterman, 2019). A correlation coefficient provides the direction and strength of the relationship as two significant pieces of information. The r-statistic and strength ranges are from −1.0 to +1.0 (Guetterman, 2019). A value of 0 indicates no relationship, while an extreme indicates a perfect relationship (Guetterman, 2019). The capacity coefficient (+ or −) indicates the relationship direction (Guetterman, 2019).

When one value increases while the other value tends to decrease, the correlation coefficient becomes negative. Two variables increase and decrease together in the case of a positive coefficient (Guetterman, 2019).

Regression adds a layer beyond correlation to predict one value from another. If an independent variable (X) is predicting a dependent variable (Y), simple linear regression gives an equation ($Y = a \times X + b$) as a line in the space model.

One of the essential components is predicting constants, that is, the interception indicated by b0, the systematic explanation of the change (b1), and the error, which is a residual value and is not counted in the existing equation part of the regression output. Critical pieces of the regression output will be examined to evaluate a regression model (i.e., model fit): (1) determine systematically accounting the dependent variables' variance in the model using F-statistic and its significance, (2) measuring the accounting level of the dependent variables' variance in the model using the r-squared value, (3) the coefficients' importance of each independent variable in the model, and (4) the remainder to investigate the random error in the model, such as outliers, which are potentially essential factors.

Inferential tests depend on basic assumptions: (1) data are typically distributed, (2) observations are independent, and (3) our dependent or outcome variable is continuous. Nonparametric statistics can be used when data do not meet these assumptions (Field, 2013).

## 15.4 AI TECHNIQUES

AI's goal is to build agents to solve every problem. Agents can be made by a simple if–then rule or must learn from their environments and improve their performance like humans. Thus, AI deals with human simulation, and the Turing test expects no one to determine AI from a human. AI generally searches in a problem space model to find the best solution rapidly. The main AI problem is runtime performance; otherwise, it can solve any problem after thousands of decades. The AI search can be guided by heuristics methods or deputed to metaheuristic algorithms (Russell & Norvig, 2002).

A metaheuristic is a partial search algorithm with imperfect information and limited time and computation capacity to generate, find, and select an acceptable solution for a problem with a vast solution space (Bianchi et al., 2009). Metaheuristics samples a subset of total existed solutions; otherwise, it is very time-consuming to be enumerated or explored entirely. Metaheuristics can be used for significant problems if they consider few assumptions of the optimization problem they try to solve (Blum & Roli, 2001). Iterative methods and optimization algorithms can guarantee an optimal global solution for some problems class, but the metaheuristics cannot solve any problems (Blum & Roli, 2001). Almost metaheuristics implement a stochastic optimization, which means the solution is dependent on randomly generated variables set (Bianchi et al., 2009). Metaheuristics usually find reasonable solutions faster than traditional iterative methods, optimization algorithms, or simple heuristics (Blum & Roli, 2001).

The AI cognitive domains and their application subdomains are shown in Figure 15.4 (Bohnhoff, 2019). An AI agent must perceive its environment according to its input, so machines will perceive machine touch, computer audition, and computer

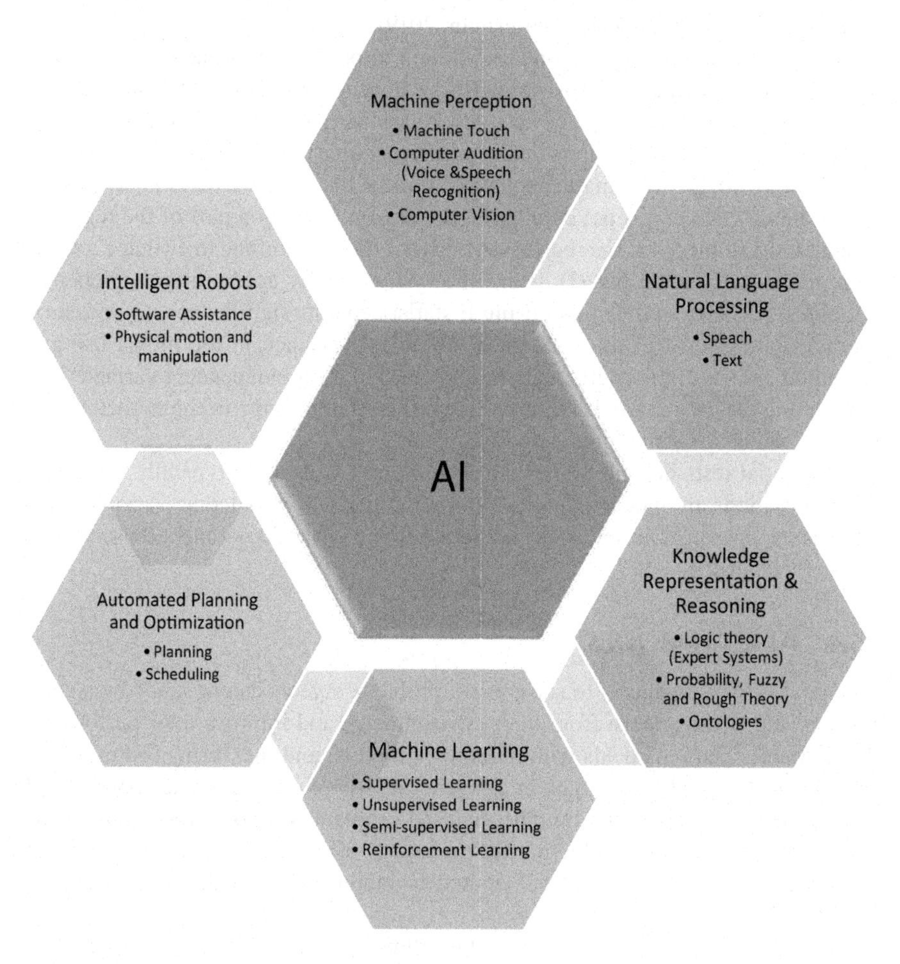


**FIGURE 15.4** AI Domains and Applications.

*Source:* Based on Bohnhoff (2019).

vision to receive environmental data and perceive its environment. The audio data needs speech recognition to be converted into a text format. The Natural Language Processing (NLP) unit must process the voice and text to find the necessary data like objects, events, and their relations. After data gathering and perception, an AI agent needs reasoning and inductive power like a human to combine logical rules and facts and finally decide. Expert systems are product-based systems that can deduct in forward and backward chaining. Forward chaining means to conclude based on facts and rules, but backward chaining considers a goal or conclusion that looks at the rules to find their premise facts and then asks the users to decide whether the decision is valid. Human knowledge is not limited to data, facts, and rules. However, human saves an ontology about objects and their relations.

Another human ability is to extract rules from data using generalization and specialization, classification, or categorization. ML techniques and algorithms do rule extraction tasks and add learning ability to agents. There are four different learning techniques: (1) Supervised learning considers a data label in datasets and teaches the relation of other features to the class label feature till it can find hidden rules in data. (2) Unsupervised learning tries to find data similarity. It divides data into clusters that their inner similarity is high, and inter-similarity is low. Unsupervised learning does not know how many clusters or categories should be made and cannot suggest a label for each cluster. Sometimes there is some relation between objects in various transactions, which can be detected using an unsupervised learning algorithm called frequent pattern mining (Hoseini et al., 2015). Also, recommended systems can use unsupervised algorithms to suggest new methods, tools, and products like drugs or prescriptions (Soleimani Neysiani et al., 2019; Varzaneh et al., 2018). (3) Semi-supervised learning (SSL) merges a small labeled dataset with an unlabeled dataset. So the machine learning technique can cluster data better. The label of each cluster can be deducted based on its labeled data. (4) Reinforcement learning (RL) uses prediction error as a reward or penalty for tuning learning parameters. There is a hybrid version of RL and other learning approaches.

Another human ability is planning and scheduling jobs or tasks. There are usually some constraints, and the AI agent must optimize scheduling to have the best performance. AI search approaches, including heuristic and metaheuristic algorithms, are used for optimization. Finally, an AI agent, like a human, needs to be actuated by the environment, so it needs robotic hardware and software, such as for translating its decision and saying its following action or purpose, moving the robot, and navigating and manipulating its motion.

## 15.5 METAHEURISTIC APPROACHES

Almost all literature on metaheuristics is nature-based experimental research and describes empirical computer results using algorithmic experiments. The following are most of the properties of metaheuristics (Blum & Roli, 2001):

- They are guidance strategies for the search process.
- Their purpose is search-space exploring and finding semi-optimal solutions efficiently.

- They maybe use simple local search procedures or complex learning processes.
- They are usually nondeterministic and approximate.
- They are of general purpose and not problem-specific.

### 15.5.1 Metaheuristic Types

#### 15.5.1.1 Population-Based versus Single-Solution Methods

Single-solution methods try to improve and modifying just one candidate solution (Talbi, 2009), like iterated local search, simulated annealing, guided local search, and variable neighborhood search (Talbi, 2009). Population-based methods save and better multiple-candidate solutions and usually have population properties to narrow the searching direction like evolutionary computation, particle swarm optimization (PSO), and genetic algorithms (GAs; Talbi, 2009). Swarm intelligence is another metaheuristics type with self-organized and decentralized agents in a population or swarms like ant colony optimization (ACO; Dorigo, 1992), PSO, swarm intelligence, and social cognitive optimization.

#### 15.5.1.2 Local Search versus Global Search

Local search algorithms improve a single solution as their search strategy is like a hill-climbing method that does not guarantee a globally optimum solution. Tabu search, simulated annealing, variable neighborhood search, iterated local search, and greedy randomized adaptive search procedure (GRASP) tries to improve local search results (Blum & Roli, 2001), classified as local or global search–based metaheuristics.

The pure global search metaheuristics are population-based, including evolutionary computation, ACO, GA, PSO, rider optimization algorithm, and other newly introduced algorithms (Binu & Kariyappa, 2019).

#### 15.5.1.3 Memetic and Hybridization Algorithms

A hybrid metaheuristic mixes a metaheuristic and other traditional optimization methods, like constraint programming, mathematical programming, and ML. Both hybrid metaheuristic components can be run concurrently and exchange information to narrow the search. Memetic algorithms (Moscato, 2000) use the evolutionary or population synergy by individual learning or local improvement procedures like using a local search algorithm instead of a primary mutation operator in evolutionary algorithms.

#### 15.5.1.4 Metaphor-Based and Nature-Inspired Metaheuristics

The design of nature-inspired metaheuristics is a very active research field. Natural systems are the most inspiration of almost all recently evolutionary computation-based algorithms. Nature is the source of concepts, principles, and mechanisms for artificial computing systems designs and complex computational problems like evolutionary algorithms, simulated annealing, PSO, and ACO. Metaphor-inspired metaheuristics recently have begun to garner criticism from researchers for hiding their lack of novelty behind an elaborate metaphor (Sörensen, 2015).

#### 15.5.1.5 Euler Diagram of Metaheuristic Algorithms

A metaheuristic algorithms classification is shown in Figure 15.5, which includes almost all metaheuristic algorithm types and their samples.

### 15.5.2 Applications in Healthcare Systems

Combinatorial optimization is broadly used in many novel applications such as science, engineering, and health. Hence, several studies have received extensive attention to efficiently provide practical and theoretical knowledge to solve these fields' problems. The optimization of a combinatorial problem tries to optimize the objective function while also satisfying the given constraints. The algorithms proposed in such problems find a proper and feasible disposition, grouping, order, or selection of a finite set of discrete objects (Stützle, 1999).

Grouping problem as a particular type of combinatorial optimization is a well-known approach in numerous real-world applications. Finding a solution has high complexity, so proposing an algorithm to obtain an optimal solution is very difficult. This class of the problem is a nondeterministic polynomial hardness (NP-hard)

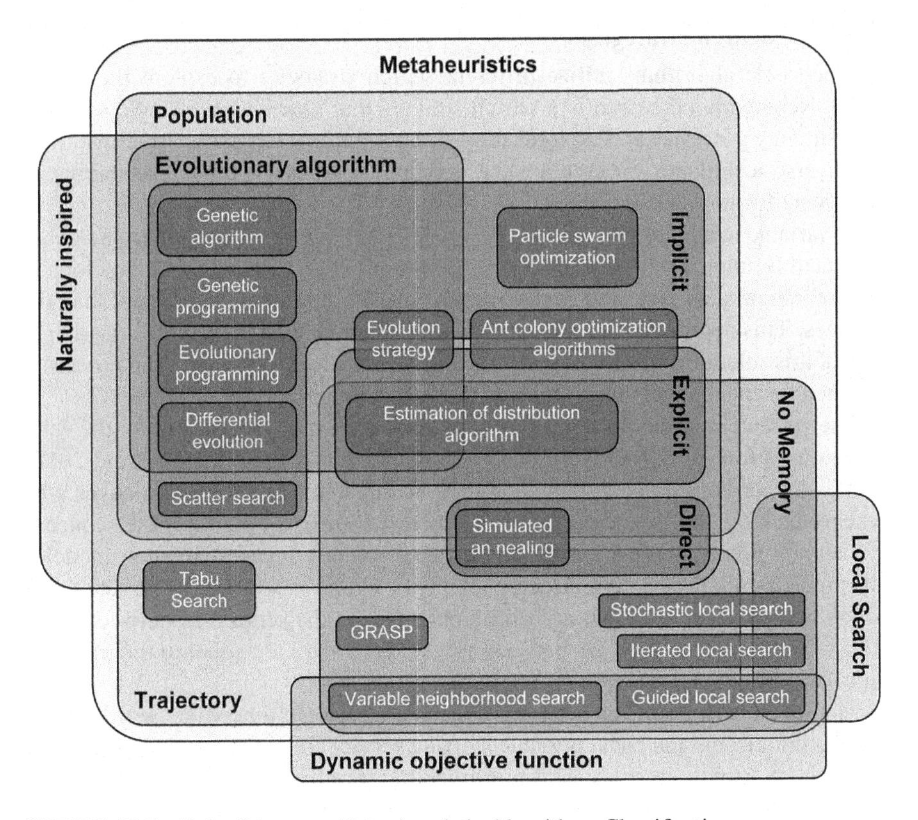


**FIGURE 15.5** Euler Diagram of Metaheuristic Algorithms Classification.

*Source:* Commons (2020) andGame and Vaze (2020).

optimization problem, which implies that finding an optimal solution in a polynomial time is impossible (Johnson & Garey, 1979). In this way, the researchers try to use metaheuristic algorithms to overcome the mentioned drawback.

In a grouping problem, $V$ as a set of items such that $V = \{v_1 \ldots\ldots v_n\}$ is partitioned into the mutually disjoint groups $G_k$, $1 \leq k \leq D$, where $D$ is the number of disjoint groups. In other words, we have $V = \bigcup_{i=1}^{D} G_i$, where for each $i \neq j$, $Gi \cap Gj = \emptyset$. An algorithm needs to find a proper distribution of the items in different subsets to solve the grouping problem, such that each item is precisely in one subset. Moreover, in most grouping problems, we must satisfy a set of constraints, so some groupings are not allowed (Falkenauer, 1994).

Home healthcare is a critical problem in improving medical and paramedical services to patients with a physical challenge or a terminal illness. Let $D$ be the number of caregivers. Then the grouping problem can be defined as distributing $n$ patients in $D$ different groups. Minimizing the aggregate cost incurred with a regular caregiver's cumulative load is the object (Cissé et al., 2017).

As we mentioned, metaheuristic algorithms have been deeply applied in grouping problems since their general nature allows the designers to apply these algorithms to different problems efficiently (Kamalakannan et al., 2017).

#### 15.5.2.1 Search Strategies

Metaheuristic algorithms utilize different search strategies to explore the search space. Neighborhood search is a search strategy that works with a single solution. This strategy generates and explores the current solution's neighbors. Note that there are diverse techniques for creating the neighbor solutions, such as exchanging or modifying two or more variables.

In variable neighborhood search (Pei et al., 2019), the current solution's neighbor is created to improve its quality. The algorithm provides new neighbor solutions in each search process cycle and adopts the new ones if they are more efficient than the old ones. This approach comes from an initial solution; then it generates the neighbors of this solution and applies a local search to get a local optimum. The process is repeated for the first neighbor, which is the best solution.

The variable neighborhood search can provide high-quality solutions for the vehicle routing problem (VRP) within a short computational time (Frifita et al., 2017). VRP is an NP-hard problem that combines routing and scheduling challenges with synchronization and time constraints (Afifi et al., 2013). It optimizes the sequence of visits for home caregivers with starting and finishing times. Afifi et al. modeled this problem by considering a directed graph G = (V, E), in which V is the set of visit points plus the departure and arrival points and E is the set of arcs between these points. The weight of each arc between two points i and j are equal to the traveling time Tij.

Moreover, each point has a service time and a selected time window that specifies the initials and the latest possible starting service times. Note that the primary objective of Afifi et al. study is to minimize the traveling time. Their experiments show that the proposed approach outperforms the existing approaches in half of the instances.

Simulated annealing is another search strategy that emulates the chemical process of metal annealing (Kämpke, 1988). This approach attempts to escape from the local optimum by accepting the lower quality" s neighbors. If the newest solution is better than the current solution, then the algorithm replaces it. At the same time, if a new solution has a lower quality than the current solution, then based on the temperature (i.e., probability of accepting the worse solution), the algorithm decides to replace the new solution or not.

A bi-objective location–allocation–routing model is proposed for the green home healthcare supply chain, including several optimization decisions (Fathollahi-Fard et al., 2019). The main decisions are scheduling and routing decisions of caregivers and assigning patients to the closest pharmacy. Fathollahi-Fard et al. (2019) applied an edited version of simulated annealing as a metaheuristic algorithm to solve their NP-hard problem. In their approach, three new formulas are proposed to update the probability of trusting a new solution that can create a more acceptable balance between exploration and exploitation. Moreover, the Taguchi method and the epsilon constraint method are utilized to calibrate the proposed simulated annealing algorithm and check the algorithm's result in small sizes.

Tabu search uses the memory concept to improve the quality of solutions (Schermer et al., 2019). This approach considers a short-term memory called tabu list to keep recently visited solutions and escape from the local optimum. Tabu search generates the neighbor solutions that do not belong to the list.

The extended version of the periodic vehicle routing problem is proposed for home healthcare logistics (Liu et al., 2014). The extended version includes transportation of particular drugs, blood samples, and medical devices between hospitals and patients' homes. Note that each patient needs a certain number of visits when the daily routing satisfies time-window constraints. Liu et al. (2014) combined Tabu search by various local search plans, containing feasible and infeasible local searches to minimize the maximal routing costs and balance vehicles' workloads. Their experiments demonstrate a local search plan beginning with an infeasible local search with a small probability pursued by a feasible local search with high probability.

#### 15.5.2.2 Evolutionary Algorithms

Evolutionary algorithms as the stochastic search approaches simulate organic evolution mechanisms, such as crossover, mutation, and selection. GA (Khalily-Dermany et al., 2019), the well-known evolutionary algorithm, uses the mentioned mechanisms to address home healthcare problems. This algorithm generates a random initial population of the problem solutions (called chromosomes). It selects the best chromosomes based on the crossover rate for some iteration and combines them to create a set of new solutions. This set is mutated to escape from the local optimum. Note that for selecting the best chromosomes, they are ranked based on their fitness. The GA repeats the mentioned process for a predefined number of iteration or until an acceptable solution is found. The GA has primarily used algorithms for discrete problems that their problem states are discrete, like task scheduling that can be used for shift scheduling in hospitals (Doostali et al., 2020; Soltani et al., 2016).

A fast heuristic approach is introduced to optimize vehicles' workload balance by reducing service time differences (Borchani et al., 2019). They design two different problem-specific algorithms based on GAs and hybrid GAs with variable neighborhood descent search. Borchani et al.'s (2019) experiments show the efficiency and competitiveness of the proposed hybrid method compared to the best-known existing algorithm.

Shi et al. (2017) proposed a hybrid GA and simulated annealing method to optimize the vehicle routing scheduling problem. They considered uncertain demand as a fuzzy variable and designed a fuzzy chance constraint model. Their numerical results for Solomon's and Homberger's benchmark samples demonstrate that the hybrid method performs efficiently. Following the work of Shi et al. (2017), the authors applied different metaheuristics, including GA, simulated annealing, bat algorithm, and firefly algorithm (FA), to solve a home healthcare routing and scheduling challenge stochastic travel and service times (Shi et al., 2018).

An expansion of the GA called Grouping GA is introduced that incorporates a group-based representation scheme (Falkenauer, 1993). This meta-heuristic considers the groups as the unit instead of each group's elements and adopts a GA (i.e., crossover, mutation, and selection) to work efficiently in group-based solutions encoding.

Two metaheuristic grouping methods, grouping GA and grouping PSO (GPSO), are proposed to address the home healthcare problem's scheduling problem (Mutingi & Mbohwa, 2014). These two methods use unique grouping operators' strengths to solve the problem within a reasonable computation time by considering time and capacity constraints.

#### 15.5.2.3 Swarm Intelligence Algorithms

All swarm intelligence algorithms such as ACO (Evtimov & Fidanova, 2018), PSO (Sanchez et al., 2018), and FA (Ebadifard et al., 2018) simulate the collective self-organized behavior of natural systems. In fact, in these algorithms, a population's elements work together to find the best solution (called computational intelligence). The swarm intelligence algorithms usually have better performance for continuous-state problems than discrete-state problems (Soltani Soulegan et al., 2021).

ACO is a metaheuristic that develops the ability to find the shortest path of real ants. This algorithm considers a fully connected graph for encoding a given optimization problem. Each node of this graph shows a component of solutions, and each edge denotes the connection between the components. Each component has a value called pheromone, which can be modified and read by ants. The pheromone is a memory for the algorithm. Each ant creates a solution for each iteration problem and, according to a stochastic mechanism, updates pheromone to generate a more desirable solution for the next iteration.

A modified version of ACO is proposed to provide proper scheduling and routing plans for home healthcare problems in Chinese communities (Zhang et al., 2018). The mixed-integer programming model considers patients' intensive distributions and their vertical locations in Chinese communities different from Europe's exact problem. Moreover, Zhang et al. (2018) consider time windows, uncertain service time, and match qualities as the primary factors in their modelling. The numerical results portray that their approach is operative, and decision-makers can gain helpful advice from it.

Note that ACO is designed to optimize discrete search spaces. In contrast, PSO is developed to solve ongoing problems. This algorithm simulates the swarming and flocking behaviors of animals. PSO generates a population of particles, which are the candidate solutions for the problem. All particles have a random position and velocity.

Moreover, three vectors are defined in d-dimensional space: velocity, position, and memory. These vectors help the particles to remember their fittest known position funded so far. In each iteration, based on the velocities, all particles move to a new position. Each particle updates the velocity based on its best position (local best) and the best particle position (global best). Then the fitness of the particles is evaluated, and the best global position is updated.

A PSO-based method is proposed to schedule home care workers in an actual situation in the UK (Akjiratikarl et al., 2007). The main objectives in Akjiratikarl et al. (2007) approach are exploiting a systematic method to improve the existing schedule of home care workers and providing a suitable approach to enable PSO to be effectively applied to this kind of challenge. To this end, they defined a particle as a multidimensional point in space that priority's assignment corresponding care activities are represented. They also used Taguchi design of experiments to find the best value of PSO parameters and compared their results through the existing manual solution and those obtained by the AiMES Centre at the University of Liverpool using ILOG.

GPSO is a new swarm intelligence technique that combines the grouping representation scheme and a grouping variation operator with the traditional PSO operators (Mutingi & Mbohwa, 2014).

FA is another swarm intelligence method that emulates the social communication of fireflies. This algorithm encodes a solution to the problem by firefly and evaluates its quality with its brightness. Each firefly attracts other fireflies based on its brightness. FA generates a random population of fireflies, which are distributed in the search space. Then, in each iteration, fireflies update their positions to find the best solution. Each firefly compares its brightness with other fireflies and then updates its position, considering brighter fireflies.

An approach based on FA is introduced to solve home care routine by Dekhici et al. (2019). They provided a suitable discretization and correction of the solution to adopt FA in this combinatorial problem. They also consider two different constraints: hard and soft. The precedence between care and patient care synchronization is considered hard constraints, while time-window preferences are soft. Hence, their primary purpose is to minimize the total routing duration while the maximum constraints satisfaction is achieved. The numerical results demonstrate that the discrete version of FA gives a better total travel time and satisfies the constraints better.

## 15.6 DM TECHNIQUES

KDD and DM techniques provide the framework and methods to convert the data into useful information for data-driven decision purposes. DM supports diverse techniques to extract invaluable information or knowledge from data. Different fields of science can apply these types of methods. Multiple research analyses are published about DM applications such as for defense, banking, insurance, education, telecommunications, and medicine. DM applications in the healthcare systems can benefit all parties involved.

DM also utilizes various techniques (such as classification, clustering, regression, and association rules) and algorithms (such as decision trees [DTs], GA, nearest-neighbor method) to analyze the massive amount of raw or multidimensional data. In other words, DM uses for extracting hidden knowledge from massive clinical or medical databases to enhance decision support, prevention, diagnosis, and treatment in the healthcare systems. This knowledge provides invaluable information. Moreover, DM can identify patient information and various disease symptoms by finding relationships between various features (Lee et al., 2013). Some manual analysis and interpretation are executed in the traditional model for transforming data to knowledge. For example, doctors or specialists generally analyze current trends, disease, and healthcare data in medical centers, then make a report, and use it for decision-making or planning for medical diagnosis and treatments. This manual data analysis is expensive, slow, time-consuming, and highly subjective. However, various data processing steps are applied for KDD, including the following:

- *Selection:* It should select a target based on the data mining task.
- *Preprocessing:* Data that are incorrect, noisy, inconsistent, without quality should be removed.
- *Transformation:* Smoothing, aggregation, normalization, and attribute/feature selection are considered in this step.
- *Data mining:* Extracting meaningful patterns and assigning data mining approaches are applied in the data mining step.
- *Interpretation/evaluation:* This step includes statistical validation and qualitative analysis.

Data mining has two primary tasks, as shown in Figure 15.6 (Kodeeshwari & Ilakkiya, 2017; Winkler, 2015):

- *Predictive tasks:* The prediction of unknown variables and future knowledge are extracted from varied techniques or algorithms. These tasks include classification and association rules.
- *Descriptive tasks:* These tasks elaborate the data or discover human-understandable patterns and portray complete results in comprehensible tables and diagrams.

The remainder of this section provides a short review of standard data mining tasks. Classification is called supervised learning. It takes some of the data (named as the training set), which have a records collection wherein each record contains a set of attributes and defines one attribute named as a class. The classification's primary goal is to produce a model to accurately predict a class attribute's value in previously untouched records. A test set can apply to predict the accuracy of the proposed

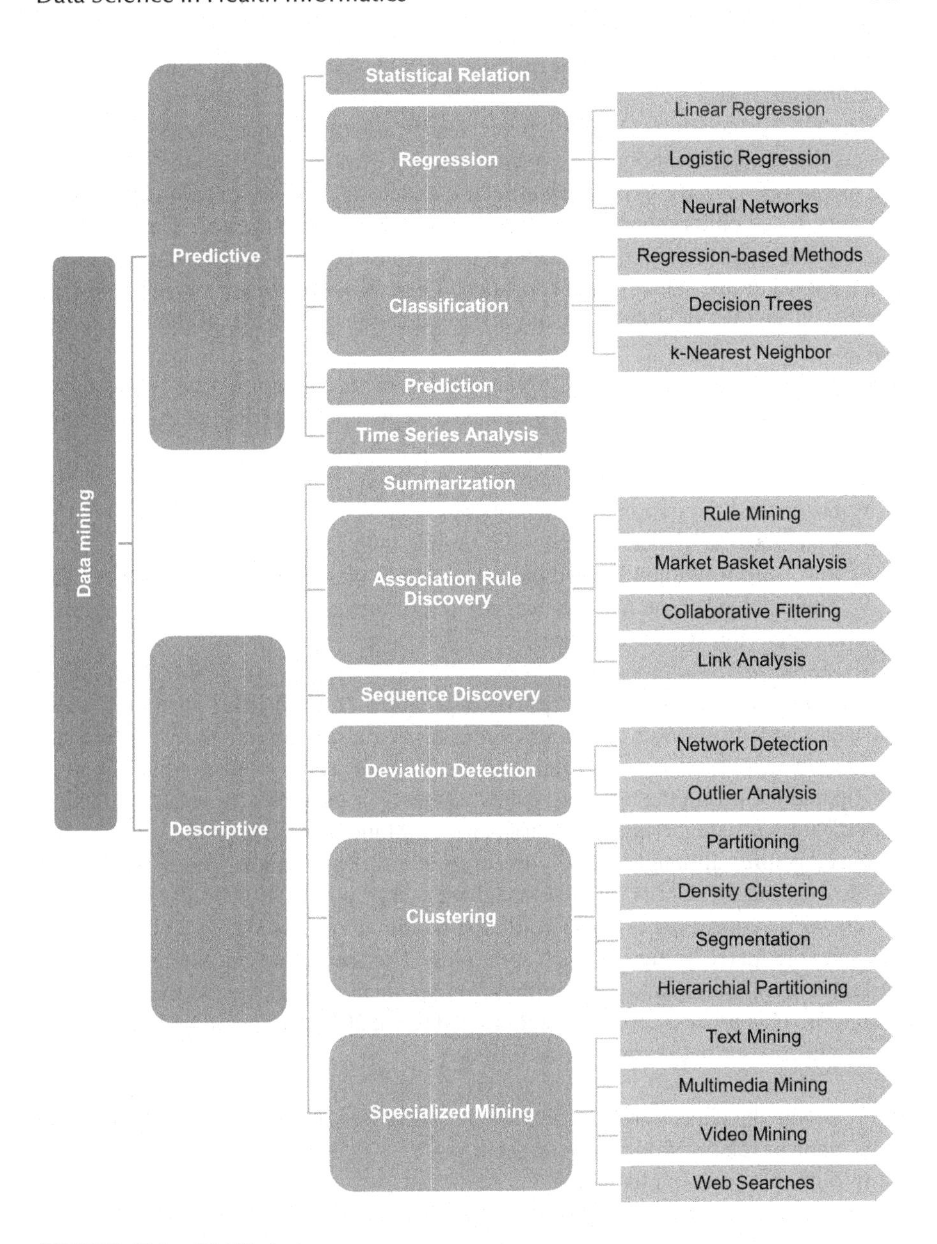


**FIGURE 15.6** DM Techniques.

*Source:* Based on Kodeeshwari and Ilakkiya (2017) and Winkler (2015).

model (Kumar & Verma, 2012). Classifying tumor cells and analyzing the effectiveness of treatment are some of the classification applications in healthcare systems.

Several classification techniques and algorithms are proposed, such as DT induction (ID3 & C4.5, and Hunt's Algorithm), memory-based methods (such as k-nearest neighbor [KNN]), rule-based methods, Bayesian classification genetic programming, naïve Bayes (NB), support vector machines (SVMs), artificial neural networks, and ensemble methods.

The KNN algorithm is a well-known classification technique. The KNN method selects the nearest neighbor of a group consisting of K records from the set of training records close to the experimental record and decides on the experimental record category based on its superiority or the label associated with them. In simpler terms, this method selects the categories with the highest number of records assigned to that category in the selected neighborhood. Therefore, the category with the highest number of nearest neighbors among all k categories is considered the new record category.

The essential opinion of the KNN method is that if a creature quacks like a duck and walks sort of a duck, it must be a duck. The general procedure of this algorithm is as follows:

- Determine the parameter k (number of nearest neighbors).
- Calculate the distance of the input sample with all training samples.
- Sort training samples based on distance and select k-nearest neighbors.
- The class selection has the majority in the immediate neighborhood as an estimate for the class sample.

The same as classification, regression attempts to predict an attribute's value (variable) based on other attributes' value(s). The main contrast between classification and regression is associated with the kind of target attribute that has to be anticipated regarding the other attributes' values. For example, the target variable in classification is categorical, is numeric, or is continuous in regression.

Furthermore, although classes are created in classification, no classes are in regression, and all data is divided into diverse split points. Notice that the number of "errors" is equivalent to a square of distinction among the number of actual values and anticipated values for each split point. The split point error across different variables is compared, and the minimum split point error is chosen as the split point/root node. This process recursively continued (Loh, 2007). To put it another way, the main objectives of regression are as follow:

- Divide the dataset into two continuous variables; then describe the associations or relationships between them.
- Find the value of attributes.
- Predict the value of one attribute regarding the value of other attributes.
- Control the accuracy of prediction.

As the primary well-known regression technique, the neural network (NN) is a simplified model that works based on the human brain's function. This network's basis is the simulation of a large number of small processing units that are interconnected. Each of these units is called a neuron. Neurons are layered, and there are usually three layers in a NN: an input layer, one or more hidden layers (or middle layers), and an output layer. Units (neurons) are connected with different weights. The input data are delivered to the input layer and propagated through the network between the layers. Finally, the results will be received from an output neuron in the output layer. NNs can learn. This learning is done by experimenting with specific records and adjusting the weight of communication between neurons (see Figure 15.7). The most common model of NNs used is the multilayer perceptron model or networks with error propagation. A nonlinear function is usually applied to the inputs weighted sum, and the model output is constructed. The problem input variables are entered in the initial or input layer, and the response variables are placed in the final or output layer. In these layers, a node is considered for each variable. The layers between the two are called hidden layers. There is no definite method to determine the number of nodes required in these layers, and the appropriate number of these nodes is often determined by trial and error. Each node in each layer is connected to all nodes in its next layer is a fully interconnected network. Figure 15.7 shows that the number of neurons in each layer is independent of the number of neurons in the other layers.

A set of data points listed in time is a time series (Liu et al., 2018). It is a sequence of successive equally spaced points. Time-series evaluation contains several techniques to analyze time-series data to extract meaningful statistics data and other data characteristics. This analysis can be valuable to watch how given medical variables change over time.

Descriptive tasks include summarization, association rule discovery, sequence discovery, deviation detection, clustering, and specialized mining.

Summarization is the operation of shortening a set of data (Wikipedia, 2021) to create a subset of data representing the original content's essential information. For example, it can summarize the document of the patients infected with the coronavirus. In addition to text, this method can summarize images and videos. Text summarization finds the most informative sentences, while image and video summarization finds the most representative images and most important frames, respectively.

One of the crucial techniques of data mining is association rule mining (ARM). It attempts to extract frequent patterns (Hoseini et al., 2015) and meaningful relationships among different items (Zhao & Bhowmick, 2003). Many association rule

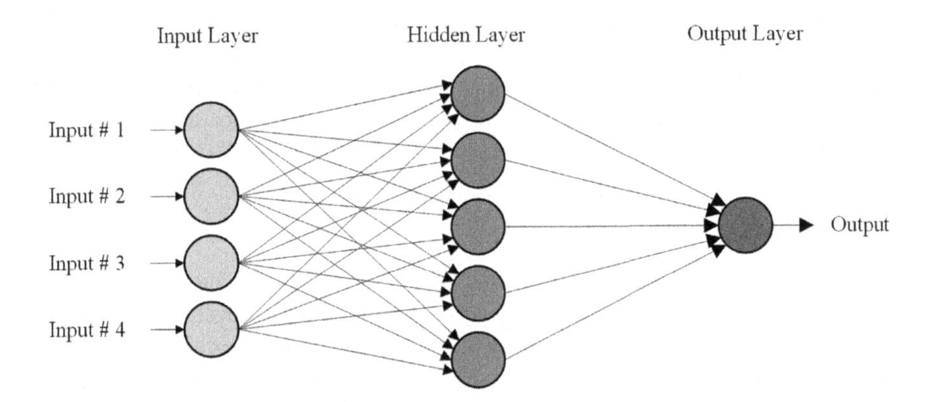


**FIGURE 15.7** An Example of an NN with Four Inputs, One Output, and One Hidden Layer.

*Source:* Hassan et al. (2015).

mining techniques are proposed, such as (1) the Apriori algorithm and its extension, which is called AprioriTID, DIC, STEM, ICAP (Incremental Constrained Apriori); (2) CARMA (Continuous Association Rule Mining Algorithm); (3) RARM (Rapid Association Rule Mining); (4) FP-Tree algorithm, (5) Goethals FP-Growth; (6) Broglet's FP-Growth; (7) Eclat and SaM algorithm; 8) COFI-Tree and CT-PRO; and (9) Recursive Elimination.

Sequential pattern mining is a structured data mining method associated with linked patterns among data examples that the values are delivered in a sequence. This method evaluates specific factors, similar occurrence frequency, duration, or values in a set of sequences for finding meaningful hidden patterns. The most critical application of sequence discovery is to find gene sequences in new diseases.

Building a novel model that explains the most critical changes in the data from previously measured values is the goal of deviation detection. In this method, finding the values that deviate significantly from the norm has received extensive attention for improving the quality of data. Discovering faulty sensors and filtering spurious reports from different sensors are essential parts of setting up the body sensor network. Even if we are confident of the quality of data, the identification of outliers is still a valuable process.

Clustering as unsupervised learning divides the data into groups (called clusters) regarding their identical attributes. Science and statistics, pattern recognition, image processing, web applications, DNA analysis, and GIS extensively support clustering due to existing various clustering algorithms such as hierarchical methods, partitioning methods, grid-based methods, and constraint-based clustering.

Finally, specialized mining such as text mining, multimedia mining, video mining, and web searches can help medical professionals to discover valuable data about new patients.

## 15.7 DM TOOLS

These days, many commercial and educational software for data mining in various fields of data has been introduced to the world of science and technology. Each of them is focused on specific algorithms according to the primary type of data they are exploring. Accurate and scientific comparison of these tools can be made in many different aspects, such as (1) various input data types and formats, (2) possible volume for data processing, (3) implemented algorithms, (4) result evaluation methods, (5) illustration methods, (6) preprocessing methods for data processing, (7) interfaces, and (8) the price and availability of software.

### 15.7.1 Orange DM

Orange software is based on components written in the Python programming language for ML and DM software fields. Orange software was developed at the bioinformatics laboratory at the computer and information science faculty, Ljubljana University, Slovenia (JavatPoint, 2018b). Moreover, it supports visualization:

- The units of Orange are labeled "widgets." Preprocessing and visual data image range the widgets to the assessment of algorithms and predictive modelling. They deliver primary functionalities, include demonstration data table, features selection, reading data, training learning algorithms, and visualization (JavatPoint, 2018b). Besides, Orange provides a more enjoyable and interactive atmosphere to keen analytical tools (JavatPoint, 2018b). It is pretty exciting to operate.

### 15.7.2 SAS DM

This DM tool is a product of the SAS Institute (MIT, 2017) that stands for statistical analysis and data management. SAS can extract, change, and manage various resources' information accompanied by a graphical user interface (GUI) for non-technical users (JavatPoint, 2018b).

Users can analyze massive amounts of data and provide accurate insight for on-time decision-making goals by SAS data miner. Not only is it suitable for data mining, optimization, but it is also practical for text mining purposes.

### 15.7.3 DataMelt DM

DataMelt—also well known as DMelt—is a visualization and computational tool which provides an interactive framework for data analysis. It is mainly designed for engineers, scientists, and students. This multiplatform and portable tool, which is written in JAVA, includes scientific and mathematical libraries. Most important, DataMelt can support big data, DM, and statistical analysis.

- **Scientific libraries:** These are used for drawing the 2D/3D plots.
- **Mathematical libraries:** These include random number generation, algorithms usage, and curve fitting.

### 15.7.4 Rattle

Ratte is a GUI-based DM tool that uses the R stats programming language. Offering remarkable data mining features supports, Rattle has a well-developed and comprehensive user interface. Also, duplicate code for any GUI operation is produced by integrated log code in Ratte (JavatPoint, 2018b).

The produced dataset by this tool can be viewed and edited. Rattle provides other services to review the code, so researchers can use it for several objectives and extend its code without limitation.

### 15.7.5 Rapid Miner

Rapid Miner is a popular predictive analysis system as a software brand and a company name that created it (JavatPoint, 2018b). It is developed in the JAVA programming language. The tool proposes an integrated environment for diverse purposes such as text mining, deep learning, machine learning, and predictive analysis. Hence, it can be used for a wide range of applications, including commercial applications, company applications, training, education, research, application development, machine learning (JavatPoint, 2018b).

Data mining and machine learning users can use this tool to perform all the necessary steps, from preparing the initial information to visualizing the results, evaluating, validating, and optimizing the output in an integrated environment. This software platform's core is open source and free, based on which many free and commercial products have been written. Also, it uses various algorithms to prepare and model information which the vast range of options led many academic projects with a user-friendly graphical environment. The strengths of this tool include the following:

- Provides reports and copies of the algorithm implementation steps
- Well groomed and decorated
- Good graphical view
- Ability to adapt to output files of some software such as Excel
- Ability to correct and debug very quickly
- Existence of relevant video tutorials prepared for this software which are on the web
- Documentation includes a guide for many operators in the software
- The possibility of simultaneously implementing different learning algorithms in the software and comparing them with each other is considered
- Special features in this tool like running processes as background job
- Works on various operating systems such as Windows, Linux, and Macintosh systems, because of the implementation and development of this software using the Java language
- Provides text mining features
- Rapid Miner will add all model learning algorithms of WEKA data mining software after each update

It should be mentioned that this software was known as the most widely used DM software in 2014.

## 15.8 HEALTHCARE DM

DM is effectively used for many sectors. For example, it enables visualizing the customer response and predicting their profitability for banking systems. It serves other similar sectors such as telecom, manufacturing, automotive industry, education, and healthcare.

By considering the growth in electronic health records, data mining has remarkable potential for medical services (Javatpoint, 2018a). Digital approaches bring about data that are easily assessable and minimize human efforts than writing information in a paper. The computer saves a massive amount of patient data accurately, improving the data management system's quality. However, healthcare has not consistently been fast enough to incorporate the latest research into everyday practice (Javatpoint, 2018a). In the following, some applications of DM approaches in healthcare are introduced.

Several algorithms, comprising KNNs, DTs, NB, logistic regression, multilayer perceptron, discriminant analysis, and SVM diagnosis, are applied for breast cancers

(Senturk & Kara, 2014). The experimental results declared that SVM's classification accuracy was higher than others by applying these algorithms. Every ML algorithm has some parameters that should be justified, like k in the KNN algorithm (Bahadorpour et al., 2017).

As another application, NN classification techniques and model-based DM techniques are compared for accuracy in detecting breast cancers (Ghassem Pour et al., 2012). The experimental results expressed that the ensemble-oriented method can be added to enhance the outcomes of both techniques. The NN approach with the ensemble-oriented method had the highest classification accuracy than DM techniques.

Another study applied 14 features of 303 people who underwent coronary angiography. For a more accurate diagnosis of coronary artery disease, the results of three classification methods, including NN, simple Bayesian, and KNN, are combined using Schaffer–Demster's evidence combination theory. The results showed that the mean accuracy, sensitivity, and specificity in the proposed method were 90.1%, 89.09%, and 91.3%, respectively, which were higher than each of the classifiers participating in the combination and compared to similar research and had better accuracy in the diagnosis of people with coronary artery diseases (Ghassem Pour et al., 2012).

Other studies have used databases from four different datasets to diagnose heart disease. The data were collected from four sources (the Cleveland Clinic Foundation; the Hungarian Institute of Cardiology; Long Beach Medical Center, California; and University Hospital Zurich, Switzerland). In these databases, data were measured with 76 different characteristics or variables directly or indirectly related to heart disease. In data preprocessing, 14 important features were selected to minimize the number of variables and processing time for implementing DM bioinformatics models (Ghassem Pour et al., 2012). The target variable in this study was the presence or absence of heart disease. In each of the subjects, one of these two conditions was recorded: (1) the value of the target variable is equal to 1 to indicate the presence of heart disease, and (2) it is equal to zero to indicate the absence or presence of disease. This study showed that DM models could be used as a tool based on ML with extraordinary accuracy in distinguishing the groups with heart diseases from others.

In the unsupervised clustering algorithm, relief has the best performance in both k-means and medoids algorithms with 76%. The best performance with 80% in NN algorithms is related to the perceptron algorithm for the backup vector machine dataset. The best performance in the core Bayes algorithm for the dataset is 85% uncertainty. The algorithm's best performance is to call the evolutionary SVM equal to 100% in the study dataset and the information rate. The DT's best performance is the backup vector machine dataset with 84% (Ghassem Pour et al., 2012). This study showed that thallium scan results are the best features to diagnose these two groups for the first time. Moreover, it introduces a machine-based learning system to distinguish 100% of sick people from healthy ones accurately (Ghassem Pour et al., 2012). So the system can aid in cardiac medical diagnostic clinics (Ghasemi et al., 2020).

The DM tools are hugely used for the COVID-19 epidemic in 2020 in at least 1,305 articles (Albahri et al., 2020; Nguyen, 2020). Several machine learning algorithms can predict epidemic behavior and patient status using KNN, DT, NB, SVM, logistic regression, and other tools and algorithms (Albahri et al., 2020).

## 15.9 TRIAD HEALTHCARE SYSTEM APPROACH

The best procedure to use data mining in practice is a triad system approach (Javatpoint, 2018a), shown in Figure 15.8.

- The analytics system: It collects, comprehends, and standardizes healthcare data.
- The content system: It standardizes working knowledge for care delivery like evidence-based best practices. This system makes it possible for organizations to assign the latest medical confirmation to practice quickly.
- The deployment system: Change management in a hierarchical structure is the deployment system goal.

## 15.10 ANALYSIS OF VARIOUS DISEASE DETECTION RESULTS

Classification and prediction are mainly used data mining techniques for disease detection in healthcare systems. Binary classification is used to determine the presence or absence as a binary target or label attribute for each example record in a dataset. It is possible to have a multiple label and a multivalue label, which their validation performance is usually lower than their binary classification (Soleimani Neysiani & Shiralizadeh Dezfoli, 2020). Even though their validation performance is usually more than information retrieval techniques as another AI domain technique, which

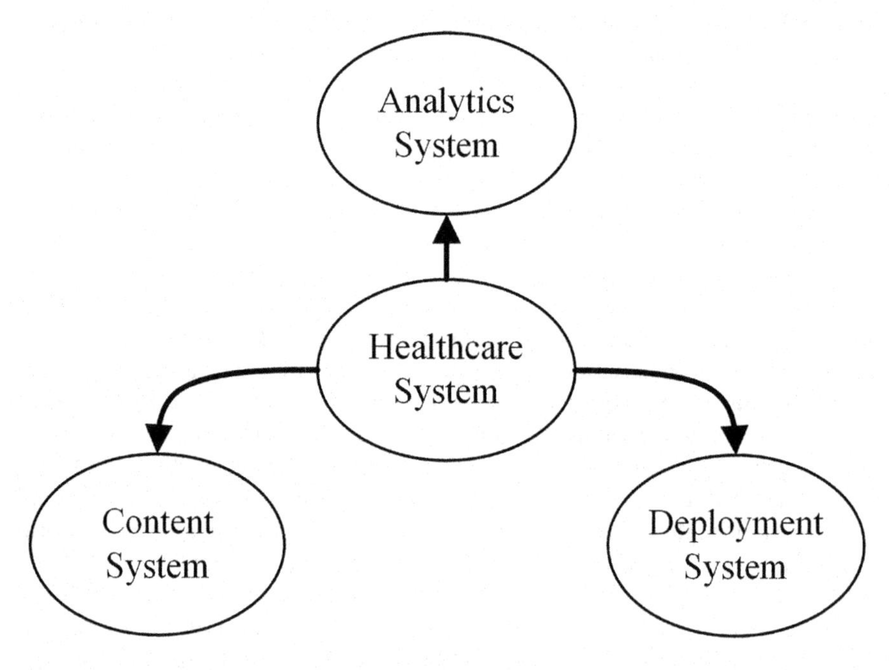


**FIGURE 15.8** Three-System Approach in Healthcare.

*Source:* Based on Javatpoint (2018a).

usually uses heuristic methods (Soleimani Neysiani & Babamir, 2020), every multi-label and multivalue label problem can be evaluated using similar metrics for binary classification. Every classifier predicts the target or label attribute, and its prediction will be compared with the actual value for training or validation dataset to evaluate the classifier validation performance. There are four states between the actual value and predicted value as they can be the same or not for binary values. The total correct predictions for both binary values per total record is called accuracy.

The accuracy metric for rare disease detection like cancers is useless because if the classifier considers all examples as healthy, its accuracy will be high. After all, there are few examples of that rare disease. For example, if there are four patients between 1,000 examples and the classifier use a default prediction as healthy for all example, its accuracy will be (1,000 −4)/1000 =0.996 =99.6%; that seems to be perfect, but it is not good at all because physicians need to classifiers find those four patients good. If the classifier considered 100 examples as cancer cases and three were correct, it performed better than the prior 99.6% accuracy. In these cases, precision and recall metrics are used. The precision is the correct, true prediction ratio per total true predictions, which is 3/100 =0.03 =3% in the mentioned sample and seems to be very low in this case. The recall is the correct, true prediction ratio per all true actual examples, 3/4 =0.75 =75% in this sample, which seems to be good. The F1-score is a harmonic average of precision and recall that can be calculated using the 2 * precision * recall/(precision + recall) equation and will be 0.05=5% in this case, and it is low because of low precision. So the F1 score is more critical for rare disease detection problems. This example shows the validation performance metric selection importance in healthcare systems.

Table 15.4 portrays the main disease detection problems and their results (Durairaj & Ranjani, 2013). The diseases are common problems in humans. The conventional

**TABLE 15.4**
**Comparative Analysis of DM in Healthcare System**

| # | Types of Disease | Data Mining Tool | Technique | Algorithm | Traditional Method | Accuracy Level % for DM Application |
|---|---|---|---|---|---|---|
| 1. | Tuberculosis | WEKA | Naïve Bayes classifier | KNN | Probability statistics | 78 |
| 2. | Heart disease | ODND, NCC2 | Classification | Naïve | Probability | 60 |
| 3. | Kidney dialysis | RST | Classification | Decision-making | Statistics | 76 |
| 4. | Diabetes mellitus | ANN | Classification | C4.5 Algorithm | Neural network | 82 |
| 5. | Blood bank sector | WEKA | Classification | J48 | – | 90 |
| 6. | Dengue | SPSS Modeler | – | C5.0 | Statistics | 80 |
| 7. | Hepatitis C | SNP | Information | Gain | Decision rule | 74 |

*Source:* Durairaj and Ranjani (2013) andJavatpoint (2018a).

statistical application methods are also given and compared to analyze the data mining applications' influence to identify the disease. The results show that DM can be more than 74% accurate (except for heart disease). This accuracy is close to 90% in the blood bank sector, which can be very promising.

The figure shows the bar graph of the data from Table 15.4 to determine healthcare problems' accuracy. The bar chart shows the predicted accuracy level of diverse DM applications, which indicate an excellent accuracy that seems they can be used as a helpful tool for physicians, but their F1 score should be considered, too.

## 15.11 DM BOONS AND CHALLENGES IN HEALTHCARE

The use of medical DM techniques has grown exponentially. DM has led to the optimal use of the large amount of data stored by clinical systems and the discovery of knowledge contained within these data. Many methods of diagnosing diseases have side effects and high costs. Therefore, researchers are looking for cheap and more accurate methods to diagnose diseases. Existing studies have used features collected from patients and various data mining algorithms to increase diagnostic accuracy.

The field of medicine and health is one of the crucial sectors in industrial societies. There are much data in this area, often organized, and DM facilitates the methodical analysis of such cases. Knowledge extracted from the massive amount of data related to illness records and medical records of individuals using the DM process can identify the laws governing the creation, growth, and spread of disease. It also provides valuable information to identify the causes of disease occurrence, diagnosis, prognosis, and treatment. Diseases should be provided to health professionals according to the prevailing environmental factors. The result is an increase in life expectancy and peace for the community. Some of the reasons for using DM in medicine and health are as follows:

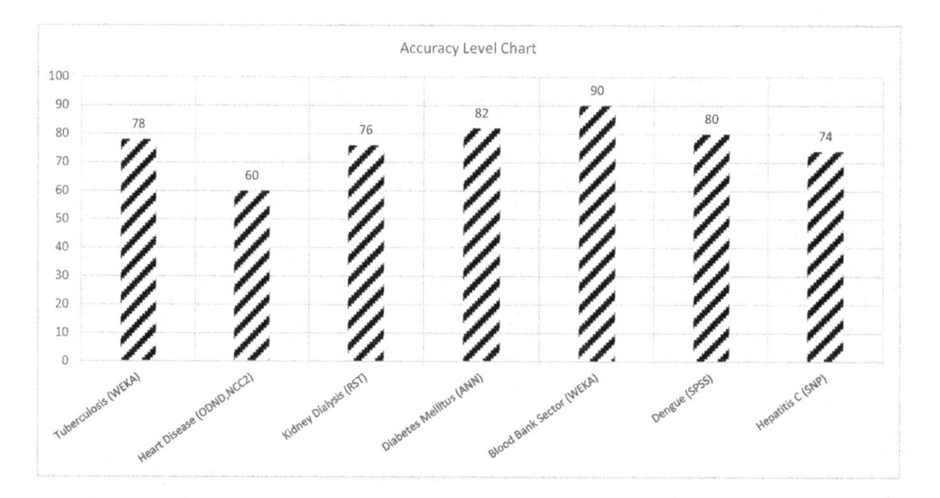


**FIGURE 15.9** Accuracy-Level Chart of DM in Healthcare System.

*Source:* Based on Durairaj and Ranjani (2013) andJavatpoint (2018a).

**Data overflow:** Having a large volume of data makes it difficult to use and sift through and extract knowledge. Some experts believe that medical science has slowed down due to the high volume and complexity of medical information. For this purpose, we need to use computers and DM tools.

**Health policymaking:** Unusual events can be identified and resolved before the situation worsens using integrated databases and DM.

**More cost savings:** The early diagnosis of diseases through DM can prevent the use of high-cost hospital tests to diagnose diseases and significantly reduce the cost of diagnosis and treatment.

**Identification and noninvasive decision-making:** Some laboratory methods for identifying invasive diseases are costly and painful for the patient.

**Side effects of drugs:** Some chemical drugs that are found to be harmless in the short term have harmful effects after a while.

Because DM is used in different fields, it has made data mining interdisciplinary. Sampling, estimation, and hypothetical tests of statistics; search algorithms; modelling techniques; and AI and ML theories are some of the areas used in DM techniques. Other tools may be used, such as optimization, evolutionary computing, information theory, signal processing, visualization, and information retrieval, depending on the areas in which DM techniques are used.

There is no consensus on the type of domains involved in DM and the extent of these interactions in the formation of DM, and depending on the type of application and the defined problem area, different domains may be involved in the data mining process.

The link between DM and other domains is shown in Figure 15.1. From this perspective, DM knowledge meets new needs, such as predicting and discovering hidden patterns in data, by using various sciences such as statistics, mathematics, and AI. It also serves as a bridge between statistics, AI, modelling, ML, and visual representation.

Healthcare systems are critical because they deal with human life. Formal verification of any tools, algorithm, and computer systems suggestion needs analytical tools and methods (Javani et al., 2015; Soleimani Neysiani & Babamir, 2015). Also, after formal verification, their results should be validated by physicians.

DM can be used for many tasks (Javatpoint, 2018a):

- Automating the healthcare institutions' workflows
- Reducing decision-making efforts
- Finding new valuable knowledge
- Information and knowledge support using predictive models
- Improving diagnosis and treatment planning processes
- Finding the connection among different subsystems using biomedical signal processing

The DM challenges can be summarized as follow (Javatpoint, 2018a):

- Vast and heterogeneous raw data
- Different data sources such as laboratory results, patients' conversations, and doctors' reviews
- Incorrect, inconsistent, and missing data such as diverse formats
- Error rates for diagnoses and treatments

- Evaluating knowledge integrity
- Evaluating the impact of specific data modifications on the statistical significance of individual patterns and build up universal measures for all DM algorithms

## 15.12 ML TECHNIQUES CATEGORIZATION

Modern AI techniques contain machine learning methods for structured data, such as the classical SVM and NN, and modern DL, as well as NLP for unstructured data.

A categorization of ML techniques is shown in Table 15.5 (Gavrilova, 2020), which includes unsupervised learning, supervised learning, SSL, and RL techniques. Each technique is used for multiple applications and has many algorithms, which are described in Table 15.5.

**TABLE 15.5**
**ML Techniques**

| | Sample Algorithms | | Applications | |
|---|---|---|---|---|
| **Unsupervised learning** | Clustering<br>• K-means<br>• X-means | Dimensionally reduction<br>• PCA<br>• Feature selection<br>• Instance-based learning<br>• Active learning | Clustering<br>• Customer segmentation<br>• Targeted marketing<br>• Recommended systems | Dimensionally Reduction<br>• Big data visualization<br>• Meaningful compression<br>• Feature elicitation<br>• Structure discovery<br>• Elicitation<br>• Market basket analysis<br>• Disease common symptoms analysis |
| **Supervised learning** | Classification<br>*Categorical Target Variable*<br>• Naïve Bayes<br>• K-nearest neighbor (KNN)<br>• Decision tree<br>• Random forest<br>• Neural metworks<br>• (MLP)<br>• (SOM)<br>• (DL) | Regression<br>*Continuous Target Variable*<br>• Linear regression<br>• Logistic regression<br>• Support vector machine (SVM) | Classification<br>*Categorical Binary/Multiclass Target Variable*<br>• Diagnostics and Recognition<br>• Customer retention<br>• Image classification<br>• Identity fraud detection<br>• Anomaly (unusual occurrence) detection | Regression<br>*Continuous Target Variable Predict Values*<br>• Prediction like advertising popularity, population growth<br>• Forecasting like market<br>• Estimating life expectancy<br>• Process optimization<br>• New insights |
| **Semi-Supervised Learning** | Categorical Target Variable<br>Classification and Clustering<br>*With Target Variable Just for Some Samples*<br>• Generative models<br>• Low-density separation<br>• Graph-based methods<br>• Heuristic approaches | | Categorical Target Variable<br>Classification and Clustering<br>*With Target Variable Just for Some Samples*<br>• Text or document classification<br>• Localizing objects<br>• Lane finding on GPS data<br>• Dialogue systems and chat-bots | |

| | Sample Algorithms | | Applications | |
|---|---|---|---|---|
| | Categorical Target Variable | | Categorical Target Variable | |
| | Classification | Control | Classification | Control |
| | *Has Target Variable* | *Without Target Variable* | *Has Target Variable* | *Without Target Variable* |
| **Reinforcement Learning** | • Criterion of optimality<br>• Policy<br>• State-value function<br>• Brute force<br>• Value function<br>• Monte Carlo methods<br>• Temporal difference methods<br>• Direct policy search | • Associative reinforcement learning<br>• Deep reinforcement learning<br>• Inverse reinforcement learning<br>• Safe Reinforcement Learning | • Precision medicine<br>• Learning tasks like medical imaging, diagnostic systems, treatment, and prediction of diseases<br>• Skill acquisition<br>• Real-time decisions | • Smart records<br>• Discovery and growth<br>• Reward function concern<br>• Self-Care systems like insulin pump and dynamic treatment regimen<br>• Robot navigation for driverless Surgery<br>• Dialogue systems and chat-bots<br>• Rehabilitation<br>• Health management systems |

*Source:* Based on Gavrilova (2020).

### 15.12.1 Unsupervised Learning Techniques

Unsupervised learning is like kindergarten instructions in which the teacher gives some objects to children and asks them to categorize similar objects in the same groups (Wikipedia, 2020d). Children may put a toy and a flower in the same group because they have the same color. The teacher expects them to distinguish their shapes, not their colors. The unsupervised learning base concept is similar to similarity detection, which is used by some similarity measurements. The distance metrics show dissimilarity and can be reversed or negated to show similarity. So, simple subtraction or Euclidian distance can be used for similarity detection. Unsupervised learning can help split data into some groups. The number of groups can be determined or assigned to the algorithms to find the optimum value. Grouping data is known as clustering, too. Clustering helps the analysis find differences and similarities of objects in a better view. Interestingly, many machine learning techniques use other machine learning techniques to improve their performance. Dimension reduction is an ML application sample for ML techniques that lead ML to find redundant data and eliminate them for better runtime and/or validation performance. When we have hundreds of apples and just one orange, it is not easy to find the orange, but dimension reduction helps us. The detail of algorithms is not essential, but some are listed in Table 15.5 for future searches. Clustering can be used for customer, patient, disease, drug, and genomes segmentation. It can also determine similar customer or users and suggest them new objects based on their similar user's interests.

### 15.12.2 Supervised Learning Techniques

Supervised learning is like teaching in which the teacher calls out something, like shapes or colors, and introduces it to the students (Wikipedia, 2020c). After many repetitions and retraining, the teacher asked students to predict the name of a previous and/or new object(s). If the students remember the object correctly, it will be an accurate prediction. Otherwise, it is a false prediction. If the target feature of objects is nominal, the prediction is called classification, and for numeral values, it is called regression. The predictive algorithms are appropriate to detect any behavior or event based on some objects to learn or train the machine learning technique.

### 15.12.3 SSL Techniques

SSL uses a small labeled dataset to cluster or predict a large dataset (Wikipedia, 2020b). For example, if we have many genomes and do not know how to categorize them, we can use an SSL algorithm and label some genomes; then the SSL algorithm tries to categorized other genomes and labeling them (Zhang et al., 2015). It can be used to do the same behavior for the future based on prior behaviors (Wongchaisuwat et al., 2016).

### 15.12.4 RL Techniques

RL uses a punishment and encouragement mechanism for learning based on external environment feedback. This technique monitors the effect of its decision and tries to improve its efficiency based on some specific metrics (Coronato et al., 2020). This technique is beneficial for vast and challenging tasks, even for humans (Team, 2020), which use different algorithms (Wikipedia, 2020a).

## 15.13 ANALYSIS OF ML AND DM TECHNIQUES

This section analyzes studies and research in the last 20 years extracted using publish or perish tool. Figures 15.10 and 15.11 show the number of articles, including search terms in their title and title plus the abstract. The results show that until 2015, DM in papers' titles has been increasing, but in 2017 and 2018, it decreased. In contrast to the title, DM in the content of papers in healthcare has continuously been increasing (see Figure 15.11). The results of Google searches for data mining in the last 20 years are also shown in Figure 15.12. It shows that the desire to search in this field has decreased. Similar research has been done for ML. Figures 15.13 and 15.14 show the number of articles, including search terms in their title and title plus the abstract. The results state that in contrast to data mining, machine learning in both contexts has been increasing every year. Finally, the absorbing result (shown in Figure 15.15) is the significant search for ML in healthcare, especially since 2017.

A summary of ML algorithms in healthcare examples with its dataset examples, advantages, and disadvantages are shown in Table 15.6. It shows various algorithms in three categories: classification, regression, and clustering (Table 15.6).

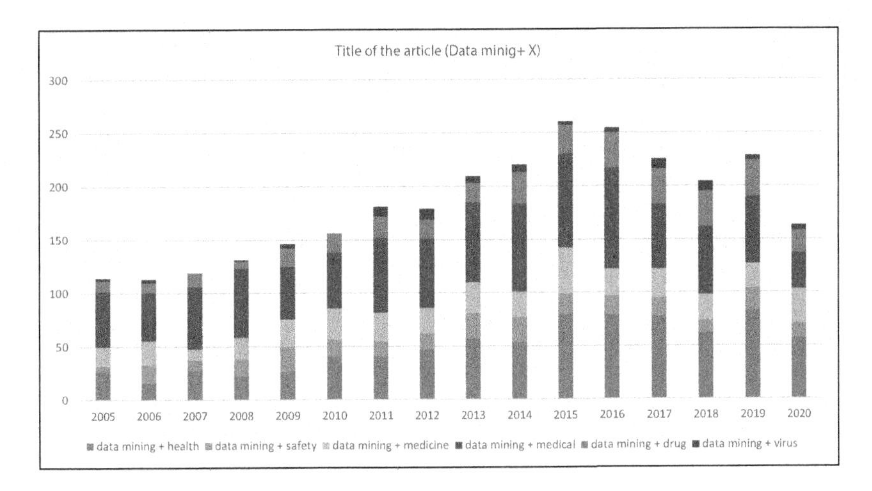


**FIGURE 15.10** DM in Titles

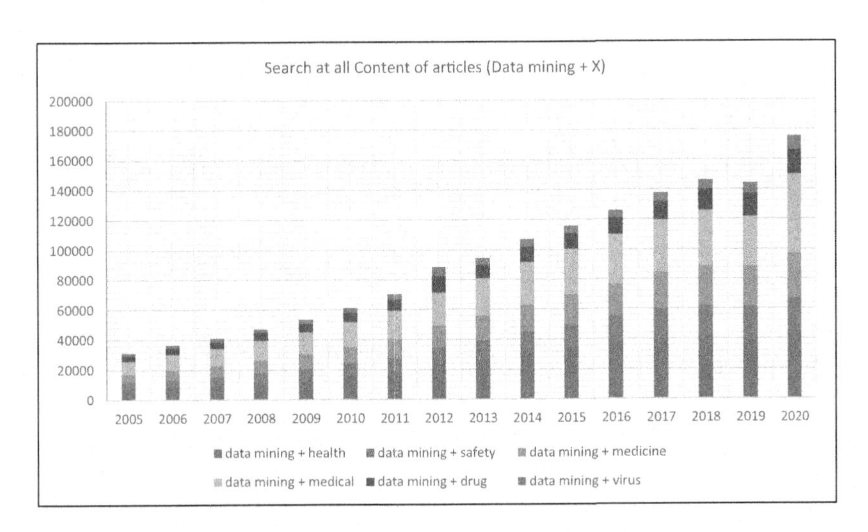


**FIGURE 15.11** DM in Content.

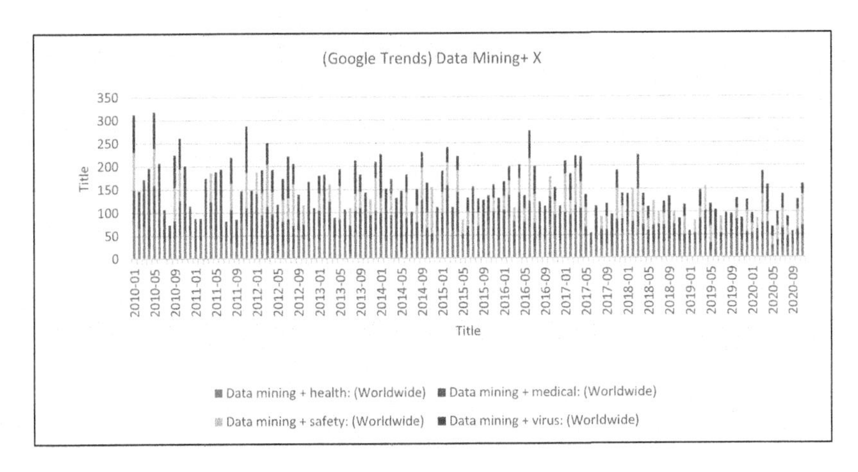


**FIGURE 15.12** DM in Google Trends.

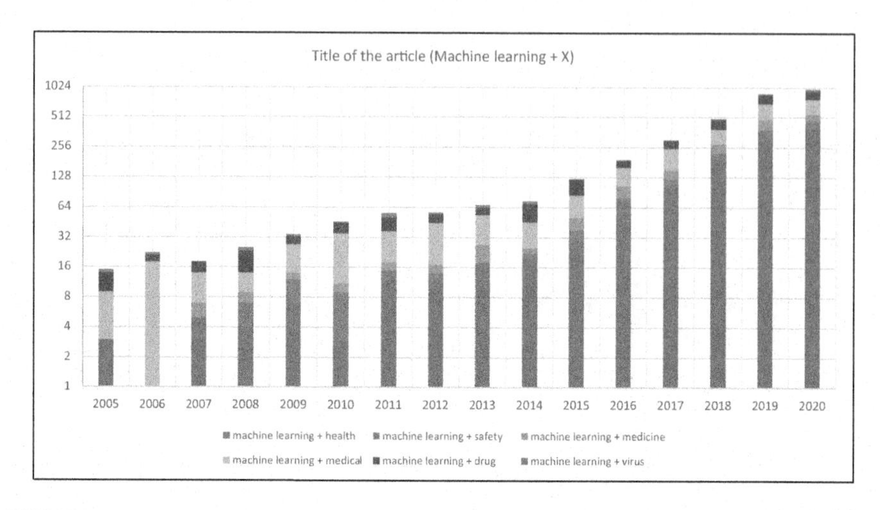


**FIGURE 15.13** ML in Titles.

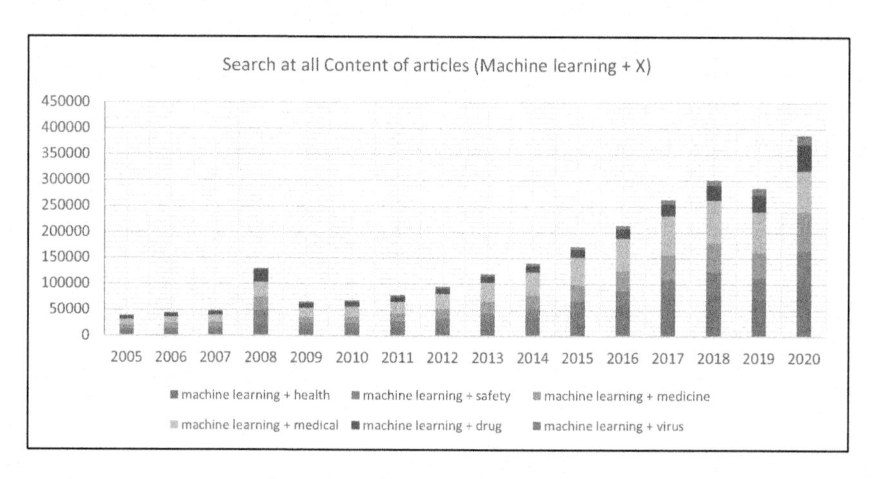


**FIGURE 15.14** ML in Contents.

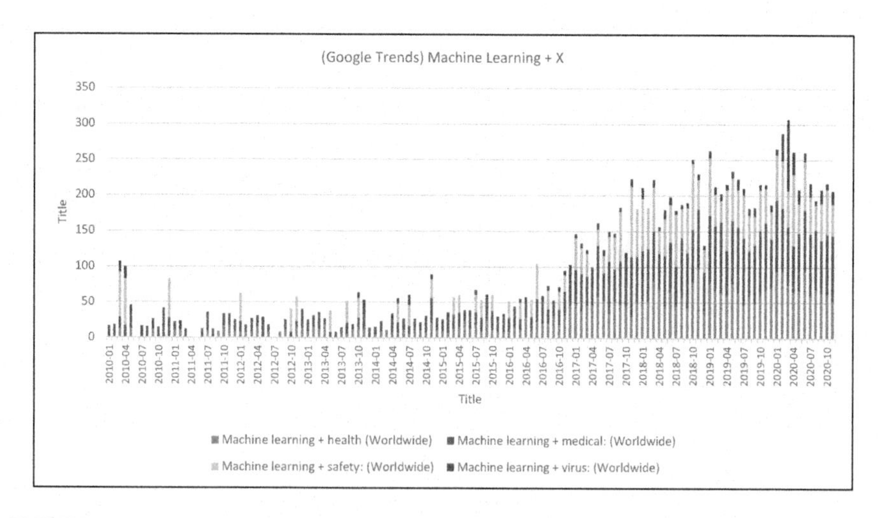


**FIGURE 15.15** Machine Learning in Google Trends

**TABLE 15.6**
**Summary of ML Algorithms in Healthcare Informatics**

| ML Category | Algorithm | Healthcare Examples | Dataset Examples | Advantage | Disadvantage |
|---|---|---|---|---|---|
| **Classification** | Decision Tree | Tree brain MRI classification, medical prediction | ADNI, hemodialysis | Simple, easy to implement | Space limitation, overfitting |
| | SVM | Image-based MR classification, children's health | NCCHD | High accuracy | Slow training, computationally expensive |
| | Neural Network | Cancer, blood glucose level prediction, heart rate variability recognition | Cleveland, acute nephritis diagnosis | Handle noisy data, detect nonlinearly relationship | Slow, computationally expensive, black-box models, low accuracy |
| | Sparse | EHR count data, heartbeats classification, tumor classification, gene expression | Colon cancer, MIT-BIH ECG, DE-SynPUF | Efficiency, handles imbalanced data, fast, compression | Computationally expensive |
| | Deep Learning | Registration of MR brain images, healthcare decision-making, Alzheimer diagnosis | ADNI, Huntington disease | Handles large dataset, deals with deep architecture, generalization, unsupervised feature learning, supports multitask learning and semi-supervised learning | Difficult to interpret, computationally expensive |
| | Ensemble | Microarray data classification, drug treatment response prediction, morality rate prediction, Alzheimer's classification | ADNI | Overcomes overfitting, generalization, predictive, high performance | Hard to analyze, computationally expensive |
| | Other Classifiers | Hospital infection analysis, anomalies detection, health monitoring, drug reaction signal generation, health risk assessment | ECG data, ADRs | Depends on method | Depends on method |
| **Regression** | - | Brain imaging analysis, battery health diagnosis | CKD, Lition batteries | Depends on method | Depends on method |
| **Clustering** | Partitioning | Risk of readmission prediction, depression clustering | NIS and MHS, ADNI | Handles large datasets, fast, simple | High sensitivity to initialization, noise, and outliers |
| | Hierarchical | Microarray data clustering, patients grouping based on length of stay in hospital | Microarray datasets, HES | Visualization capability | Poor visualization for extensive data, slow, use a tremendous amount of memory, low accuracy |
| | Density-based | Biomedical image clustering, finding bicliques in a network | Skin lesion images, BMC biomedical images | Detect outliers and arbitrary shapes, handle non-static and complex data | Not good for large datasets, slow, tricky parameter selection |

*Source:* Modified from Fang et al. (2016).

## 15.14 CONCLUSION

DS is a multidisciplinary combination of AI, statistics, DM, big data, and many other fields that aim to solve human problems. The healthcare system has invaluable effects on human life since it could raise its quality and decrease patients' treatment period. AI, DM, and ML algorithms are widely used to solve many healthcare problems and improve system performance metrics.

The experimental results of DS and AI applications in healthcare studies indicate a hopeful future. It is expected that other researchers can use DS techniques in their studies and implement operational and applicative systems to improve human life levels. Also, computer engineering and data scientists have been trying to improve the DS tools and techniques. Operational systems need extensive educational efforts for end users to trust and learn how to use tools efficiently. Expert users like physicians should consider the validation and verification of tools and algorithms.

DS in the healthcare domain can be fascinating for both computer and physician-scientists. Computer scientists can try to improve the efficiency of their techniques and algorithms. Also, they can implement their novel approaches by cooperating with physicians and improve the life quality for humans.

Figure 15.16 shows a word cloud obtained from the books (DuBois, 2019; Matheny et al., 2019; Panesar, 2019; Topol, 2019) in this field containing related concepts and keywords. Word clouds highlight the related words in a domain based on their frequency as the word importance, the most prominent word, and the most important word. The built word cloud shows concepts, tools, algorithms, metrics, and clinical objects. The words in this figure can be directive keywords for future work.

**FIGURE 15.16** Healthcare and DS Word Cloud.

## REFERENCES

Afifi, S., Dang, D.-C., & Moukrim, A. (2013). A simulated annealing algorithm for the vehicle routing problem with time windows and synchronization constraints. In G. Nicosia & P. Pardalos (Eds.), *Learning and Intelligent Optimization* [LION 2013]. *International Conference on Learning and Intelligent Optimization, Berlin, Heidelberg*, Berlin, Heidelberg: Springer.

Akjiratikarl, C., Yenradee, P., & Drake, P. R. (2007). PSO-based algorithm for home care worker scheduling in the UK. *Computers & Industrial Engineering, 53*(4), 559–583.

Albahri, A., Hamid, R. A., Alwan, J. K., Al-Qays, Z., Zaidan, A., Zaidan, B., Albahri, A., AlAmoodi, A., Khlaf, J. M., & Almahdi, E. (2020). Role of biological data mining and machine learning techniques in detecting and diagnosing the novel coronavirus (COVID-19): A systematic review. *Journal of Medical Systems, 44*, 1–11.

Amisha, P. M., Pathania, M., & Rathaur, V. K. (2019). Overview of artificial intelligence in medicine. *Journal of Family Medicine and Primary Care, 8*(7), 2328.

Aocnp, D. (2015). The evolution of the electronic health record. *Clinical Journal of Oncology Nursing, 19*(2), 153.

Bahadorpour, M., Soleimani Neysiani, B., & Shahraki, M. N. (2017). Determining optimal number of neighbors in item-based kNN collaborative filtering algorithm for learning preferences of new users. *Journal of Telecommunication, Electronic and Computer Engineering (JTEC), 9*(3), 163–167.

Bianchi, L., Dorigo, M., Gambardella, L. M., & Gutjahr, W. J. (2009). A survey on metaheuristics for stochastic combinatorial optimization. *Natural Computing, 8*(2), 239–287. https://doi.org/10.1007/s11047-008-9098-4

Binu, D., & Kariyappa, B. S. (2019). RideNN: A new rider optimization algorithm-based neural network for fault diagnosis in analog circuits. *IEEE Transactions on Instrumentation and Measurement, 68*(1), 2–26. https://doi.org/10.1109/TIM.2018.2836058

Blum, C., & Roli, A. (2001). Metaheuristics in combinatorial optimization: Overview and conceptual comparison. *ACM Computing Surveys, 35*, 268–308. https://doi.org/10.1145/937503.937505

Bohnhoff, T. (2019). *A Five-Minute Guide to Artificial Intelligence*. Retrieved 12/13/2020 from https://medium.com/appanion/a-five-minute-guide-to-artificial-intelligence-c4262be85fd3

Borchani, R., Masmoudi, M., & Jarboui, B. (2019). Hybrid Genetic Algorithm for Home Healthcare routing and scheduling problem. In *2019 6th International Conference on Control, Decision and Information Technologies (CoDIT)*. Retrieved from https://ieeexplore.ieee.org/abstract/document/8820532.

Chattopadhyay, S. (2017). A neuro-fuzzy approach for the diagnosis of depression. *Applied Computing and Informatics, 13*(1), 10–18.

Cissé, M., Yalçındağ, S., Kergosien, Y., Şahin, E., Lenté, C., & Matta, A. (2017). OR problems related to Home Healthcare: A review of relevant routing and scheduling problems. *Operations Research for Healthcare, 13*, 1–22.

Commons, W. (2020). *File:Metaheuristics Classification Fr.svg – Wikimedia Commons, the Free Media Repository*. Retrieved 12/12/2020 from https://commons.wikimedia.org/w/index.php?title=File:Metaheuristics_classification_fr.svg&oldid=451307143

Coronato, A., Naeem, M., De Pietro, G., & Paragliola, G. (2020). Reinforcement learning for intelligent healthcare applications: A survey. *Artificial Intelligence in Medicine, 109*, 101964. https://doi.org/10.1016/j.artmed.2020.101964

Cottle, M., Hoover, W., Kanwal, S., Kohn, M., Strome, T., & Treister, N. (2013). Transforming Healthcare Through Big Data Strategies for leveraging big data in the healthcare industry. *Institute for Health Technology Transformation*. Retrieved from http://ihealthtran.com/big-data-in-healthcare

Dekhici, L., Redjem, R., Belkadi, K., & El Mhamedi, A. (2019). Discretization of the firefly algorithm for home care. *Canadian Journal of Electrical and Computer Engineering*, *42*(1), 20–26.

Doostali, S., Babamir, S. M., Shiralizadeh Dezfoli, M., & Soleimani Neysiani, B. (2020, December 22–23). IoT-based model in smart urban traffic control: Graph theory and genetic algorithm. In *2020 11th International Conference on Information and Knowledge Technology (IKT)*. Tehran, Iran. Retrieved from https://ieeexplore.ieee.org/abstract/document/9345623

Dorigo, M. (1992). *Optimization, Learning and Natural Algorithms*. PhD Thesis, Politecnico di Milano.

DuBois, K. N. (2019). Deep medicine: How artificial intelligence can make healthcare human again. *Perspectives on Science and Christian Faith*, *71*(3), 199–201.

Durairaj, M., & Ranjani, V. (2013). Data mining applications in healthcare sector: A study. *International Journal of Scientific & Technology Research*, *2*(10), 29–35. Retrieved from www.ijstr.org/research-paper-publishing.php?month=oct2013

Ebadifard, F., Doostali, S., & Babamir, S. M. (2018). A firefly-based task scheduling algorithm for the cloud computing environment: Formal verification and simulation analyses. In *2018 9th International Symposium on Telecommunications (IST)*. Tehran, Iran. Retrieved from https://ieeexplore.ieee.org/abstract/document/8661088

Evtimov, G., & Fidanova, S. (2018). Ant colony optimization algorithm for 1D cutting stock problem. In K. Georgiev, M. Todorov, & I. Georgiev (Eds.), *Advanced Computing in Industrial Mathematics, Studies in Computational Intelligence* (pp. 25–31). Berlin: Springer International Publishing. https://doi.org/10.1007/978-3-319-65530-7_3

Falkenauer, E. (1993). The grouping genetic algorithms: Widening the scope of the GA's. *JORBEL-Belgian Journal of Operations Research, Statistics, and Computer Science*, *33*(1–2), 79–102.

Falkenauer, E. (1994). A new representation and operators for genetic algorithms applied to grouping problems. *Evolutionary Computation*, *2*(2), 123–144.

Fang, R., Pouyanfar, S., Yang, Y., Chen, S.-C., & Iyengar, S. (2016). Computational health informatics in the big data age: A survey. *ACM Computing Surveys (CSUR)*, *49*(1), 1–36. https://doi.org/10.1145/2932707

Fathollahi-Fard, A. M., Govindan, K., Hajiaghaei-Keshteli, M., & Ahmadi, A. (2019). A green home healthcare supply chain: New modified simulated annealing algorithms. *Journal of Cleaner Production*, *240*, 118200.

Field, A. (2013). *Discovering Statistics Using IBM SPSS Statistics* (M. Carmichael, Ed. 4 ed.). London: Sage.

Frifita, S., Masmoudi, M., & Euchi, J. (2017). General variable neighborhood search for home healthcare routing and scheduling problem with time windows and synchronized visits. *Electronic Notes in Discrete Mathematics*, *58*, 63–70.

Game, P. S., & Vaze, D. (2020). Bio-inspired optimization: Metaheuristic algorithms for optimization. *arXiv preprint arXiv:2003.11637.*

Garey, M. R., & Johnson, D. (1979). *Computers and Intractability: A Guide to the Theory of NP-Completeness*. New York, NY: WH Freeman and Company. https://ci.nii.ac.jp/naid/10006421402/en/

Gavrilova, Y. (2020). *How to Choose a Machine Learning Technique*. Retrieved 12/13/2020 from https://serokell.io/blog/how-to-choose-ml-technique

Ghasemi, F., Soleimani Neysiani, B., & Nematbakhsh, N. (2020). Feature selection in pre-diagnosis heart coronary artery disease detection: A heuristic approach for feature selection based on Information Gain Ratio and Gini Index. *In 2020 6th International Conference on Web Research (ICWR).* Tehran, Iran. Retrieved from https://ieeexplore.ieee.org/abstract/document/9122285

Ghassem Pour, S., McLeod, P., Verma, B., & Maeder, A. (2012, 2016–10–25 T01:16:03Z). Comparing data mining with ensemble classification of breast cancer masses in digital mammograms. *Second Australian Workshop on Artificial Intelligence in Health.* Aachen, Germany.

Gillum, R. F. (2013). From papyrus to the electronic tablet: A brief history of the clinical medical record with lessons for the digital age. *The American Journal of Medicine, 126*(10), 853–857.

Gómez-González, E., Gomez, E., Márquez-Rivas, J., Guerrero-Claro, M., Fernández-Lizaranzu, I., Relimpio-López, M. I., Dorado, M. E., Mayorga-Buiza, M. J., Izquierdo-Ayuso, G., & Capitán-Morales, L. (2020). Artificial intelligence in medicine and healthcare: A review and classification of current and near-future applications and their ethical and social Impact. *arXiv preprint arXiv:2001.09778.*

Guetterman, T. C. (2019). Basics of statistics for primary care research. *Family Medicine and Community Health, 7*(2), e000067. https://doi.org/10.1136/fmch-2018-000067

Hassan, H., Negm, A., Zahran, M., & Saavedra, O. (2015). Assessment of Artificial Neural Network for bathymetry estimation using High Resolution Satellite imagery in Shallow Lakes: Case study El Burullus lake. *International Water Technology Journal, 5.*

Hoseini, M. S., Shahraki, M. N., & Soleimani Neysiani, B. (2015). A new algorithm for mining frequent patterns in can tree. *In 2015 2nd International Conference on Knowledge-Based Engineering and Innovation (KBEI).* Tehran, Iran. Retrieved from https://ieeexplore.ieee.org/abstract/document/7436153

Javani, M., Soleimani Neysiani, B., & Babamir, S. M. (2015). Formal verification of UML statecharts using the LOTOS formal language. *In 2015 2nd International Conference on Knowledge-Based Engineering and Innovation (KBEI).* Tehran, Iran. Retrieved from https://ieeexplore.ieee.org/abstract/document/7436139

JavatPoint. (2018a). *Data Mining in Healthcare.* Retrieved 3/1/2021 from www.javatpoint.com/data-mining-in-healthcare

JavatPoint. (2018b). *Data Mining tools.* Retrieved 3/1/2021 from www.javatpoint.com/data-mining-tools

Jiang, F., Jiang, Y., Zhi, H., Dong, Y., Li, H., Ma, S., Wang, Y., Dong, Q., Shen, H., & Wang, Y. (2017). Artificial intelligence in healthcare: Past, present and future. *Stroke and Vascular Neurology, 2*(4), 230–243.

Kamalakannan, R., Sudhakara Pandian, R., Sornakumar, T., & Mahapatra, S. (2017). An ant colony optimization algorithm for cellular manufacturing system. *Applied Mechanics and Materials, 854,* 133–141.

Kämpke, T. (1988). Simulated annealing: Use of a new tool in bin packing. *Annals of Operations Research, 16*(1), 327–332.

Kelleher, J. D., & Tierney, B. (2018). *Data Science.* Cambridge, MA: MIT Press.

Khalily-Dermany, M., Nadjafi-Arani, M. J., & Doostali, S. (2019). Combining topology control and network coding to optimize lifetime in wireless-sensor networks. *Computer Networks, 162,* 106859.

Kodeeshwari, R., & Ilakkiya, K. T. (2017). Different types of data mining techniques used in agriculture-a survey. *International Journal of Advanced Engineering Research and Science, 4*(6), 17–23. https://doi.org/10.22161/ijaers.4.6.3

Kumar, R., & Verma, R. (2012). Classification algorithms for data mining: A survey. *International Journal of Innovations in Engineering and Technology (IJIET), 1*(2), 7–14.

Lee, S. I., Ghasemzadeh, H., Mortazavi, B., Lan, M., Alshurafa, N., Ong, M., & Sarrafzadeh, M. (2013). Remote patient monitoring: What impact can data analytics have on cost? In *WH '13 Proceedings of the 4th Conference on Wireless Health*. Baltimore, MD. https://dl.acm.org/doi/abs/10.1145/2534088.2534108

Liu, R., Xie, X., & Garaix, T. (2014). Hybridization of tabu search with feasible and infeasible local searches for periodic home healthcare logistics. *Omega, 47*, 17–32.

Liu, Y., Chen, J., Wu, S., Liu, Z., & Chao, H. (2018). Incremental fuzzy C medoids clustering of time series data using dynamic time warping distance. *PLoS ONE, 13*(5), e0197499.

Loh, W.-Y. (2007). Classification and regression tree methods. In *Encyclopedia of Statistics in Quality and Reliability*. Madison, WI: Wiley.

Matheny, M., Israni, S. T., Ahmed, M., & Whicher, D. (2019). *Artificial Intelligence in Healthcare: The Hope, the Hype, the Promise, the Peril.* NAM Special Publication. Washington, DC: National Academy of Medicine, 154.

MIT. (2017). *SAS at MIT.* Retrieved 3/1/2021 from http://web.mit.edu/sas/www/

Moscato, P. (2000). *On Evolution, Search, Optimization, Genetic Algorithms and Martial Arts – Towards Memetic Algorithms.* Caltech Concurrent Computation Program.

Multiple, A. (2020). *15 AI Applications/Usecases/Examples in Healthcare in 2020.* Retrieved 12/13/2020 from https://research.aimultiple.com/healthcare-ai/

Mutingi, M., & Mbohwa, C. (2014). Home Healthcare staff scheduling: Effective grouping approaches. In *IAENG Transactions on Engineering Sciences-Special Issue of the International Multi-Conference of Engineers and Computer Scientists, IMECS and World Congress on Engineering.* London: CRC Press, Taylor & Francis Group.

Needs, T. N. C. f. C. H. a. S. (2020). *Regional Complex Care Convenings.* Retrieved 12/13/2020 from www.nationalcomplex.care/regional-convenings/

Nguyen, T. T. (2020). Artificial intelligence in the battle against coronavirus (COVID-19): A survey and future research directions. *Preprint.* https://doi.org/10.13140/RG.2.2.36491.23846/1

Panesar, A. (2019). *Machine Learning and AI for Healthcare* (pp. 1–73). Berkeley, CA: Springer, Apress.

Pei, Z., Wang, Z., & Yang, Y. (2019). Research of order batching variable neighborhood search algorithm based on saving mileage. In *3rd International Conference on Mechatronics Engineering and Information Technology (ICMEIT 2019)* (pp. 196–203). Dalian, China: Atlantis Press.

Raghupathi, W., & Raghupathi, V. (2014). Big data analytics in healthcare: Promise and potential. *Health Information Science and Systems, 2*(1), 3. https://doi.org/10.1186/2047-2501-2-3

Reiser, S. J. (1991). The clinical record in medicine part 1: Learning from cases. *Annals of Internal Medicine, 114*(10), 902–907.

Russell, S., & Norvig, P. (2002). *Artificial Intelligence: A Modern Approach.* Hoboken, NJ: Pearson.

Sanchez, I. A. L., Vargas, J. M., Santos, C. A., Mendoza, M. G., & Moctezuma, C. J. M. (2018). Solving binary cutting stock with matheuristics using particle swarm optimization and simulated annealing. *Soft Computing, 22*(18), 6111–6119.

Schermer, D., Moeini, M., & Wendt, O. (2019). A hybrid VNS/Tabu search algorithm for solving the vehicle routing problem with drones and en route operations. *Computers & Operations Research, 109*, 134–158.

Senturk, Z. K., & Kara, R. (2014). Breast cancer diagnosis via data mining: Performance analysis of seven different algorithms. *Computer Science & Engineering, 4*(1), 35.

Shi, Y., Boudouh, T., & Grunder, O. (2017). A hybrid genetic algorithm for a home healthcare routing problem with time window and fuzzy demand. *Expert Systems with Applications, 72*, 160–176.

Shi, Y., Boudouh, T., Grunder, O., & Wang, D. (2018). Modelling and solving simultaneous delivery and pick-up problem with stochastic travel and service times in home healthcare. *Expert Systems with Applications, 102*, 218–233.

Soleimani Neysiani, B., & Babamir, S. M. (2015). Automatic verification of UML state chart by BOGOR model checking tool: Automatic formal verification of network and distributed systems. In *2015 2nd International Conference on Knowledge-Based Engineering and Innovation (KBEI)*. IEEE, Tehran, Iran.

Soleimani Neysiani, B., & Babamir, S. M. (2018). Automatic typos detection in bug reports. In *IEEE 12th International Conference Application of Information and Communication Technologies*. IEEE, Almaty, Kazakhstan.

Soleimani Neysiani, B., & Babamir, S. M. (2019a). Automatic interconnected lexical typo correction in bug reports of software triage systems. In *International Conference on Contemporary Issues in Data Science*. Zanjan, Iran. Retrieved from https://www.researchgate.net/publication/331315272

Soleimani Neysiani, B., & Babamir, S. M. (2019b, 2020/1/2). Effect of typos correction on the validation performance of duplicate bug reports detection. In *10th International Conference on Information and Knowledge Technology (IKT)*. IEEE, Tehran, Iran.

Soleimani Neysiani, B., & Babamir, S. M. (2019c). Fast language-independent correction of interconnected typos to finding longest terms: Using trie for typo detection and correction. In *24th International Conference on Information Technology (IVUS)*. Lithuania. Retrieved from http://ceur-ws.org/Vol-2470/

Soleimani Neysiani, B., & Babamir, S. M. (2020). Automatic duplicate bug report detection using information retrieval-based versus machine learning-based approaches. In *IEEE 6th International Conference on Web Research (ICWR)*. IEEE, Tehran, Iran. Retrieved from https://ieeexplore.ieee.org/abstract/document/9122288

Soleimani Neysiani, B., Babamir, S. M., & Aritsugi, M. (2020). Efficient feature extraction model for validation performance improvement of duplicate bug report detection in software bug triage systems. *Information and Software Technology, 126*, 106344–106363. https://doi.org/10.1016/j.infsof.2020.106344

Soleimani Neysiani, B., & Shiralizadeh Dezfoli, M. (2020). Identifying affecting factors on prediction of students' educational statuses: A case study of educational data mining in Ashrafi Esfahani University of Isfahan of Iran .In *10th International Conference on Information and Knowledge Technology (IKT)*. Tehran, Iran. Retrieved from https://www.researchgate.net/publication/344289949

Soleimani Neysiani, B., Soltani, N., Mofidi, R., & Nadimi-Shahraki, M. H. (2019). Improve performance of association rule-based collaborative filtering recommendation systems using genetic algorithm. *International Journal of Information Technology and Computer Science, 11*(2), 48–55.

Soltani Soulegan, N., Barekatain, B., & Soleimani Neysiani, B. (2016). *Job scheduling based on single and multi objective meta-heuristic algorithms in cloud computing: A survey*. In *International Conference on Information Technology, Communications and Telecommunications (IRICT)*. Tehran, Iran. Retrieved from https://www.sid.ir/en/Seminar/ViewPaper.aspx?ID=7678

Soltani Soulegan, N., Barekatain, B., & Soleimani Neysiani, B. (2021). MTC: Minimizing time and cost of cloud task scheduling based on customers and providers needs using genetic algorithm. *International Journal of Intelligent Systems & Applications, 13*(2).

Sörensen, K. (2015). Metaheuristics – the metaphor exposed. *International Transactions in Operational Research*, *22*(1), 3–18.

Stützle, T. (1999). *Local Search Algorithms for Combinatorial Problems: Analysis, Improvements, and New Applications*. Darmstadt, Germany: IOS Press.

Talbi, E.-G. (2009). *Metaheuristics: From Design to Implementation* (Vol. 74). New York, NY: John Wiley & Sons.

Team, G. L. (2020, 2/17/2020). *Use of Reinforcement Learning in Healthcare*. Retrieved 3/1/2021 from www.mygreatlearning.com/blog/reinforcement-learning-in-healthcare/

Tierney, B. (2012a). *Data Science Is Multidisciplinary*. Retrieved 12/15/2020 from https://oralytics.com/2012/06/13/data-science-is-multidisciplinary/

Tierney, B. (2012b). *Data Science Is Multidisciplinary*. Retrieved 4/1/2020 from https://brendantierneydatamining.blogspot.com/2016/10/data-science-is-multidisciplinary.html

Topol, E. (2019). *Deep Medicine: How Artificial Intelligence Can Make Healthcare Human Again*. New York, NY: Hachette UK.

Varzaneh, H. H., Soleimani Neysiani, B., Ziafat, H., & Soltani, N. (2018). Recommendation systems based on association rule mining for a target object by evolutionary algorithms. *Emerging Science Journal*, *2*(2), 100–107.

Wikipedia. (2020a, 3/1/2021). *Reinforcement Learning*. Retrieved 3/2/2021 from https://en.wikipedia.org/wiki/Reinforcement_learning

Wikipedia. (2020b, 2/16/2021). *Semi-supervised Learning*. Retrieved 3/2/2021 from https://en.wikipedia.org/wiki/Semi-supervised_learning

Wikipedia. (2020c, 3/1/2021). *Supervised Learning*. Retrieved 3/2/2021 from https://en.wikipedia.org/wiki/Supervised_learning

Wikipedia. (2020d, 2/16/2021). *Unsupervised Learning*. Retrieved 3/2/2021 from https://en.wikipedia.org/wiki/Unsupervised_learning

Wikipedia. (2021, 3/10/2021). *Automatic Summarization*. Retrieved 3/9/2021 from https://en.wikipedia.org/wiki/Automatic_summarization

Winkler, R. (2015). An evolving computational platform for biological mass spectrometry: workflows, statistics and data mining with MASSyPup64. *PeerJ*, *3*, 34. https://doi.org/10.7717/peerj.1401

Wongchaisuwat, P., Klabjan, D., & Jonnalagadda, S. R. (2016). A semi-supervised learning approach to enhance healthcare community – based question answering: A case study in alcoholism. *JMIR Medical Informatics*, *4*(3), e24.

Zhang, G., Ou, S., Huang, Y., & Wang, C. (2015). Semi-supervised learning methods for large scale healthcare data analysis. *International Journal of Computers in Healthcare*, *2*, 98. https://doi.org/10.1504/IJCIH.2015.069788

Zhang, T., Yang, X., Chen, Q., Bai, L., & Chen, W. (2018). Modified ACO for home health-care scheduling and routing problem in Chinese communities. In *2018 IEEE 15th International Conference on Networking, Sensing and Control (ICNSC)*. IEEE, Zhuhai, China. https://ieeexplore.ieee.org/abstract/document/8361373/

Zhao, Q., & Bhowmick, S. S. (2003). Association rule mining: A survey. *Nanyang Technological University, Singapore*, *135*.

# Index

Note: page numbers in *italics* indicate figures; **bold** page numbers indicate tables

## K



## L



## M



## N



## O



## P



## R

## S



## T



## U



## V



## W



## X

Printed in the United States
by Baker & Taylor Publisher Services